Physical Ultrasonics of Composites

Physical Ultrasonics of Composites

Stanislav I. Rokhlin
Dale E. Chimenti
Peter B. Nagy

OXFORD
UNIVERSITY PRESS
2011

OXFORD
UNIVERSITY PRESS

Oxford University Press, Inc., publishes works that further
Oxford University's objective of excellence
in research, scholarship, and education.

Oxford New York
Auckland Cape Town Dar es Salaam Hong Kong Karachi
Kuala Lumpur Madrid Melbourne Mexico City Nairobi
New Delhi Shanghai Taipei Toronto

With offices in
Argentina Austria Brazil Chile Czech Republic France Greece
Guatemala Hungary Italy Japan Poland Portugal Singapore
South Korea Switzerland Thailand Turkey Ukraine Vietnam

Published by Oxford University Press, Inc.
198 Madison Avenue, New York, NY 10016

www.oup.com

Library of Congress Cataloging-in-Publication Data
Rokhlin, S. I. (Stanislav I.)
Physical ultrasonics of composites / Stanislav I. Rokhlin, Dale E. Chimenti, Peter B. Nagy.
p. cm.
ISBN 978-0-19-507960-9 (hardcover : alk. paper) 1. Composite materials–Testing.
2. Ultrasonic testing. I. Chimenti, Dale E. II. Nagy, P. B. (Peter B.), 1952- III. Title.
TA418.9.C6R65 2010
620.1′1874—dc22 2010009171

To our teachers:

L. G. Merkulov, B. W. Maxfield, and P. Greguss

Contents

Preface

The motivation to write this book sprang from a realization that composite materials, by their unique nature, pose special problems, challenges, and opportunities for those seeking to characterize their internal constitution and mechanical behavior using ultrasonics. Although there are many excellent texts and monographs on composite mechanics and on ultrasonics, no book, recent or vintage, has adequately covered the topics we feel are essential for a comprehensive understanding of the use of ultrasound in the inspection of composites. Each of us has taught graduate courses on subjects related to material in this book, and the experience of these lectures and our own research has informed every chapter. We have striven for a thorough, yet compartmentalized, treatment in which not all topics would necessarily depend on a reading of each preceding chapter. The material, however, is all here for the reader interested in a complete grasp of the subject.

This book consists roughly of two major themes: mechanics and elasticity of composites are treated first, followed by their ultrasonic behavior and applications. We first consider the basics of physical properties of composites and their elastic behavior as homogeneous, anisotropic materials, followed by a discussion of wave mechanics in anisotropic media. This is not a book of theory, however, so we have included only as much theoretical discussion as necessary to motivate our subsequent analysis of ultrasonic waves in composite laminates. In the second part, we examine ultrasonic waves in bulk composites with emphasis on solutions of the elastic wave equation in anisotropic media, ways of visualizing those solutions, and the relationships among the field variables of wave motion. Reflection and transmission at interfaces are treated, followed by the characterization of composites by volumetric ultrasonic waves. Guided waves are dealt with next, in both plates and rods, an important topic for the characterization of thin laminates. We discuss, too, the effects of fluid loading on guided waves in composites and the occurrence of *leaky waves*. Bulk waves in layered multiaxial composites are treated next, where we offer a way to avoid the numerical instabilities inherent in the classical transfer matrix formalism. Following these preceding topics, we address waves in periodically layered media, a subject of clear interest in laminated composites. We develop a model and formalism appropriate for the calculation and measurement of reflection and transmission voltage signals performed with real, diffracting transducers. Finally, the last chapter in the book is devoted to a discussion of the emerging field of air-coupled ultrasonics.

We have attempted to review some of the numerous theoretical and experimental results and applications that are now available in the literature on this subject. The field is far too broad, however, for us to claim complete coverage. We apologize in advance to any researchers whose work we may not have included; our decisions regarding references should not be construed as a judgment on anyone's research. There are many excellent comprehensive reviews of this field, but that is not the character of the present volume.

It is our sincere wish that this book will be of help to students and to researchers interested in learning ultrasonics as applied to the characterization of composites. We hope that it may also serve as a reference for workers in the field. In that connection we have attempted to provide many derived results and procedures.

Aesop's Fables teach us that "In union there is strength." As investigators, we know that most successes we may have had in research can be counted in terms of our collaborations. This book has benefited from the work of many people whose assistance and contributions we gratefully acknowledge. Collectively, the authors of the book you hold have published nearly 250 research articles on the topics covered in this volume. We have not, of course, done this work alone. Our collaborators include undergraduates, graduate students, postdoctoral fellows, visiting faculty, and professional colleagues. Happily, we also collaborate with each other.

Each of us in turn would like to take this opportunity to thank our partners in research whose work, in many cases, we are reporting here. SIR thanks Wei Huang, Andrey Degtyar, Anton Lavrentyev, Weicheng (David) Wang, Ya-Cherng Chu, Ningyu Wang, and especially Lugen Wang and Ken Bolland, among his graduate students, postdocs, and collaborators. DEC wishes to express his gratitude to graduate students Han Zhang, Dong Fei, and Junho Song, to postdoctoral researchers Che-Hua Yang, Ali Safaeinili, Oleg Lobkis, and Steve Holland, and to visiting senior investigators Adnan Nayfeh and Bert Auld. PBN gratefully acknowledges the work of graduate student Waled Hassan and joins DEC in thanking Adnan Nayfeh for many excellent collaborations.

Our research efforts have been pursued not only in our home institutions, but also abroad, where friendly colleagues have offered to host us for extended research stays. Principal among these opportunities have been the Imperial College ultrasonics group in London headed by Peter Cawley, Gerd Busse's Institute for Plastics Testing at Universität Stuttgart, and the incomparable hospitality *á la Bordelaise* of Bernard Hosten at his Laboratoire de Mécanique Physique at the Université de Bordeaux.

Our understanding and appreciation of ultrasonics has been influenced by fruitful interactions with many professional colleagues. Among these generous people (and in no particular order) we wish to name Jan Achenbach, Adnan Nayfeh, Bruce Thompson, Wolfgang Sachse, Bert Auld, Robert E. Green, Moshe Rosen, Paul Höller, Peter Cawley, Mike Lowe, Bernard Hosten, Leo Felsen, Smaine Zeroug, Subhendu Datta, Ron Roberts, Larry Jacobs, Marc Deschamps, Ajit Mal, David K. Hsu, and Vikram Kinra. One person, however, has a special place in all our careers. He achieved this status by communicating to us his enthusiasm for ultrasonics and by making for each of us a home in his home. Whether as visiting faculty, adjunct professor, or scientific collaborator, Laszlo Adler invited us, mentored us, and inspired us. It is with profound gratitude and a deep sense of affection that we acknowledge his contributions to this book and to our careers. Finally, we wish to thank our wives, Rita, Linda, and Lucy, whose patient indulgence of our passion for ultrasonics has never been forgotten, nor taken for granted.

The editors at Oxford University Press have our gratitude for their supportive and critical role during the manuscript preparation and revisions, in particular Jeremy Lewis and Hallie Stebbins. We also wish to thank our scientific colleagues for generously allowing us to reprint figures from our joint research articles.

Columbus	Stanislav I. Rokhlin
Ames	Dale E. Chimenti
Cincinnati	Peter B. Nagy

January 2010

Introduction

The search for ever lighter, stiffer, stronger, and tougher substances from which to construct aerospace vehicles has led inexorably toward more highly engineered, multi-phase, anisotropic, low-density materials—toward composites. The novelty and expense of composite materials has meant, historically, that they are thoroughly inspected prior to use, typically using ultrasonics. But, the strong, almost pathological, elastic anisotropy of a uniaxial laminate has significant implications for its elastic behavior and for the propagation of ultrasound. For this reason, the use of ultrasonic waves to characterize the mechanical condition of composites has presented engineers inspecting these advanced materials with special challenges. Our purpose in this book is to explain, describe, and demonstrate the behavior of ultrasound in composite materials and its use in their nondestructive characterization.

In the early years of aviation, airplanes were built of the best materials then available, wood and cloth. Although the application of different woods and doped cloth reached a high level of sophistication [1–3], these materials simply could not accommodate the stresses imposed by the higher structural and aerodynamic loads of larger, faster airplanes. Eventually, wood and cloth were superseded by metals. The recognition that finely distributed precipitates impeded atomic slip in aluminum [4] and thereby increased both strength and stiffness dramatically meant that such alloys would be suitable for stressed-skin monocoque aircraft structures. The subsequent development of copper-bearing, age-hardened aluminum alloys [5] demonstrated greatly increased stiffness and tensile strength over pure aluminum, while maintaining the required high fracture toughness. About this time, Dix at Alcoa discovered that a thin coating of pure aluminum would substantially increase the corrosion resistance of age-hardened aluminum alloys [6].

The concept of composite materials, which are composed of a hardening or setting substance that acts as a load-transfer medium combined with a nonmetallic fibrous reinforcement, was first patented in England by DeHavilland Aircraft Company in the 1930s [7]. With the war effort in full swing, however, it was only in the mid-1940s that the first glass–fiber laminate aircraft fuselage was built and flight tested at Wright Field, Dayton, OH [8]. Progress in the development of fiber-reinforced composites was relatively slow at first while better resins and fibers were being sought. The concept of engineered materials got a dramatic boost from the 1952 discovery, by Conyers Herring at Bell Laboratories in New Jersey, that tin whiskers, 1 μm in diameter, possessed elastic and plastic behavior close to the theoretical limit for a perfect crystal. Ultimate tensile strength in whiskers can be almost 1000 times greater than ordinary polycrystalline bulk tin [9]. Despite the many attempts made to reduce this advance to a practical system, including the fabrication of ceramic whiskers, glass fibers, and metallic fibers, none of these attempts was entirely successful in achieving a continuous fiber with low weight, high modulus, and high strength.

Pursuing the same objective of near-perfect material performance in micron-sized fibers by the pyrolitic transformation of continuous polymer filaments proved much more efficient in the case of pure carbon [10]. This approach led to two types of graphite fibers of extraordinary stiffness-to-weight ratio, one based on rayon from Union Carbide in the United States, and the other based on polyacrylonitrile from Morganite and Courtaulds in England. Together with boron fibers, also initially produced in a complex and costly process [11], graphite fibers were the first continuous filamentary reinforcement employed in engineered materials that have since become known as *advanced composites*. By the mid-1960s, high-modulus continuous-fiber composite materials were under active study and development for eventual aircraft (and spacecraft) applications. Although composites had many clear advantages over conventional built-up metal structures, the new materials made their appearance slowly, first in limited applications in high-performance military aircraft and spacecraft, later in commercial and general aviation, and eventually in consumer products as diverse as snowboards, tennis rackets, and bicycles. The most common method of fabrication in essentially all the early aircraft applications was the hand layup of pre-impregnated fiber rovings to form stiffness-tailored laminates after curing or setting of the resin matrix.

Introduction of the new technology proceeded cautiously owing to two major factors: lack of field experience with composites and their high production and fabrication costs. Early applications included the boron-epoxy rudder on the McDonnell-Douglas F-4 Phantom, boron-epoxy vertical stabilizers on the Grumman F-14, boron-epoxy skins on the McDonnell-Douglas F-15 Eagle empennage, and the carbon-epoxy speedbrakes on the F-15. The General Dynamics F-16 Falcon also had carbon-epoxy empennage skins. In the late 1970s, the McDonnell-Douglas AV-8B Harrier made extensive use of carbon-epoxy in the wing skins, torque boxes, and forward fuselage [12]. Graphite-epoxy composites played a major role in the McDonnell-Douglas F/A-18 with applications to the speedbrake, stabilators, vertical tails, and rudders [13]. By the mid-1980s, DARPA, NASA, and the US Air Force relied on the advantages of high-performance composites in charging Grumman Aerospace to build the X-29A, whose radical forward-swept wing design could function only if the wing airfoil surfaces were fabricated from stiffness-tailored advanced composites [14]. Finally, at the time this book went to press, Boeing Commercial Airplanes had just launched the 787 Dreamliner passenger aircraft, 50% of whose primary structure (by weight) is built from composite materials [15]. If calculated by volume, the percentage of advanced composites in the 787 is 80%.

Other factors, in addition to cost, led early in their application to the currently accepted engineering practice of subjecting safety-critical composite structures to full nondestructive inspection. These factors are the composite materials' peculiar sensitivity to types of damage and modes of failure not witnessed in metallic structures. Composite fabrication defects include lack of proper resin cure, incomplete resin-fiber wetting, evolution of gas porosity during cure, included foreign matter, ply delaminations, and others. During service, composites are sensitive to moisture uptake, especially in hot/wet environments, delamination growth, and impact damage. This last item is especially pernicious because it can occur with little visible superficial indication of subsurface damage. It has been found [16] that impact-induced nonlinear stress waves tend to cause ply delamination in a zone that widens with depth in the laminate, where the plies at the far surface may be completely destroyed. A composite structure with this type of damage is in local failure and has a greatly reduced compressive strength. The recent introduction of several advanced commercial transports and the appearance earlier in the current decade of composite low-observable military aircraft serve to illustrate cogently the extensive use of ultrasonic inspection methods to qualify these air vehicles for safe flight. From the beginning, ultrasonics has been the tool of choice to inspect composites, because the likely critical defects are mechanical in nature. These defects and the composite's material properties are most easily, and inexpensively, detected and elucidated in ultrasonic nondestructive inspection.

The history of ultrasonics research and applications is a long and glorious one, reaching well back into the 19th century. Some earlier texts contain excellent recapitulations of the historical development of ultrasonics [17–19], and the reader is referred to these for further details of work leading to applications of ultrasound to materials inspection and nondestructive characterization.

With the upsurge of rearmament for World War II, the demand for quality assurance in machines on which lives (and perhaps battles) depended grew substantially. All existing forms of product quality assessment were taxed to their limit, and new ones needed to be invented [20]. In this charged industrial environment, Firestone in the US [21–25], Sokolov in Russia [26–29], and Kruse [30], Czerlinsky [31], and Trost [32] in Germany began to exploit ultrasound for various types of nondestructive materials testing and characterization. These developments continued with even greater activity after 1945, as war production gave way to consumer products. Nonetheless, most of the safety-critical applications that justified the expense of ultrasonic inspection were concentrated in the defense industry. A thorough review of this entire era and a very complete bibliography can be found in the monograph by Krautkrämer and Krautkrämer [33] and in the historical review article by Graff [19].

By the 1950s, many materials research laboratories had begun investigating the structure–property relationships that were to dominate advances in materials science and technology for decades. Among the many laboratory tools brought to bear on these questions, ultrasonics is especially useful for several reasons. Mechanical waves can be designed to sample the entire bulk of a specimen, or only its surface. By adjusting the frequency, and therefore the wavelength of the sound, different materials characteristics, corresponding to different length scales, can be examined. Moreover, essentially all the information gathered in an ultrasonic measurement yields valuable insights on the material under study. Wavespeeds provide access to elastic stiffnesses [34–36] and, indirectly, to interatomic potentials, lattice specific heats, and other solid-state parameters. Attenuation and scattering can be related to grain structure in metals [37–41], or to the more subtle structural details in polymers [42]. Ultrasonic energy appearing at harmonics of the fundamental permits examination of the elusive nonlinear behavior of materials [43–45]. Variations in ultrasonic velocity with applied stress can be used to infer the stress state of a metal or the third-order elastic constants [46–49]. The complicated case of sound wave propagation in elastic–plastic media has also been analyzed [50, 51].

In the same time frame, the study of ultrasonic excitations in finite media—surfaces, plates, and bars—was proceeding owing to both the scientific and technological importance of these waves. By the mid-1940s, Firestone and Ling [52] had explored the use of guided ultrasonic plate waves at the request of Sperry Products, Inc. for the nondestructive inspection of sheet metal. These workers also initiated development of ultrasonic inspection instruments, and Sperry was for many years in the forefront of the development of ultrasound equipment and methods. Firestone was the first investigator to denote repeatedly the elastic guided wave modes in a plate as "Lamb" waves, referring to the first comprehensive theoretical expression for the velocity dispersion of guided plate waves derived by Horace Lamb [53] decades earlier. This name is now the accepted terminology for guided waves in a traction-free, isotropic plate that are polarized in a plane normal to the plate surface. Shortly thereafter, some of the experimental phenomenology and analytical behavior of fluid-coupled guided plate waves was established by Osborne and Hart [54], Schoch [55], Worlton [56], and Frederick and Worlton [57]. Most experiments up to 1961 were performed specifically to test the validity of Lamb's original theory.

Perhaps the most common structural form of aerospace composite materials during the past 30 or 40 years has been the laminate, essentially a plate when considered on a local scale. Plate waves, therefore, are one natural choice for quantitative characterization of composites, but so are bulk ultrasonic waves propagating into and through composite laminates. The nature of a laminate made from fibers and matrix is that its design imparts extra stiffness in just the directions needed to achieve its purpose. In advanced composites, stiffness arises from continuous, high-modulus fibers. To achieve a balance between stiffness and reliability, composites are almost always fabricated with fibers in more than one direction. This situation implies a multilayered structure, and often the layering is periodic, leading to characteristic ultrasonic phenomena. As ultrasonic probes are seldom larger than a dozen or so acoustic wavelengths in diameter, understanding how the interaction of sound with generation and detection devices influences the received signals is a critical element in making accurate measurements. Finally, the low mass density and low out-of-plane stiffness of composites make them ideal candidates for ultrasonic coupling in air, where measurements can be done simply and without the bother of water immersion. In this book, we have attempted to address the fundamentals of elastic wave propagation in anisotropic inhomogeneous media and the special ultrasonic problems that arise in the elastic property characterization of composites.

Bibliography

1. A. Flatters, "A new or improved process for rendering fabrics gas and waterproof," British Patent 129,455 (1918).
2. N. A. T. N. Feary, "Improvements in flexible material for aeroplanes," British Patent 149,745 (1919).
3. T. F. Tesse, "Coated aeroplane cloth and process of making same," US Patent 1,521,055 (1925).
4. Z. Jeffries and R. S. Archer, "The slip interference theory of the hardening of metals," Chem. Met. Eng. **24**, 1057–1067 (1921).

5. R. S. Archer and Z. Jeffries, "Aluminum-base alloy and method of treating it," US Patent 1,472,738 and "Aluminum-base alloy," US Patent 1,472,739 (1923).
6. E. H. Dix, Jr, "A new corrosion-resistant aluminum product," NACA Technical Note No. 259 (1927); " Corrosion-resistant aluminum alloy articles and method of making the same," Canadian Patent 280,337 (1928).
7. N. A. De Bruyne, "Improvements relating to the manufacture of material and articles from resinous substances," British Patent 470,331 (1937).
8. G. B. Rheinfrank, Jr and W. A. Norman, "Application of glass laminates to aircraft," Mod. Plastics **21**, 94–99 (1944).
9. C. Herring and J. K. Galt, "Elastic and plastic properties of very small metal specimens," Phys. Rev. **85**, 1060–1061 (1952).
10. F. S. Galasso, *High Modulus Fibers and Composites* (Gordon and Breach, New York, 1969).
11. H. M. Otte and H. A. Lipsitt, "On the interpretation of electron diffraction patterns from 'amorphous' boron," Phys. Stat. Sol. **13**, 429–438 (1966).
12. J. Kelly, J. Dorr, A. Spero, and K. Corona, "Development of organic matrix composites for transport aircraft, marine, and civil structures," in *Composite Properties and Applications*, ed. A. Miravete, (Woodhead Publishing Ltd, Cambridge), pp. 283–290 (1993).
13. A. A. Baker, S. Dutton, and D. W. Kelly, *Composite Materials for Aircraft Structures*, 2nd ed (AIAA, Reston, 2004), Chap. 1.
14. B. N. Pamadi, *Performance, Stability, Dynamics, and Control of Airplanes*, 2nd ed (AIAA, Reston, 2004), Sect. 1.11.
15. "Boeing 787: A Matter of Materials—Special Report: Anatomy of a Supply Chain," IndustryWeek.com, December 1, 2007.
16. M. V. Hosur, U. K. Vaidya, J. W. Gillespie, Jr., and S. Jeelani, "Ultrasonic evaluation of ballistic impact damage in thick-section, twill-weave S2-glass/vinyl-ester laminates," in *Review of Progress in Quantitative Nondestructive Evaluation*, eds. D. O. Thompson and D. E. Chimenti, vol. **19** (American Institute of Physics, New York, 2000), pp. 1231–1238.
17. K. F. Graff, *Elastic Waves in Solids* (Dover Press, New York, 1992).
18. J. D. Achenbach, *Wave Propagation in Elastic Solids* (North-Holland, Amsterdam, 1973).
19. K. F. Graff, "A history of ultrasonics," in *Physical Acoustics*, vol. XV, eds. W. P. Mason and R. N. Thurston (Academic Press, New York, 1981), pp. 2–90.
20. D. M. Forney, and D. E. Chimenti, "Nondestructive evaluation: Coming of age," in *Encyclopedia Britannica 1986 Yearbook on Science and the Future* (Britannica, Chicago, 1985), pp. 86–105.
21. F. A. Firestone, "Flaw detecting device and measuring instrument," US Patent 2,280,226 (1940).
22. F. A. Firestone, "Resonance inspection method," US Patent 2,439,131 (1943).
23. F. A. Firestone, "Surface and shear wave method and apparatus," US Patent 2,439,139 (1943).
24. F. A. Firestone, "Supersonic reflectoscope, an instrument for inspecting the interior of solid parts by means of sound waves," J. Acoust. Soc. Am. **17**, 287–299 (1945).
25. F. A. Firestone and J. R. Frederick, "Refinements in supersonic reflectoscopy. Polarized sound," J. Acoust. Soc. Am. **18**, 200–211 (1946).
26. S. Ya. Sokolov, "Ultrasonic waves and their application," (in Russian), Zhur. Tekh. Fiz. **2**, 522–544 (1935).

27. S. Ya. Sokolov, "Ultrasonic methods for determining internal flaws in metal objects," (in Russian), Zavodskaya Laboratoriya **4**, 1468–1473 (1935).
28. S. Ya. Sokolov, "Means for indicating flaws in materials," US Patent 2,164,125 (1937).
29. S. Ya. Sokolov, "Ultrasonic methods for investigating the properties of heat-treated steel and for determining internal flaws in metal objects," (in Russian), Zhur. Tekh. Fiz. **11**, 160–169 (1941).
30. F. Kruse, "On materials testing through the use of ultrasound," (in German), Akust. Z. **6**, 137–149 (1941).
31. E. Czerlinsky, "Nondestructive testing with ultrasound," (in German), Deutsche Luftfahrtforschung: Untersuchungen und Mitteilungen, Nr. 1069 (1943).
32. A. Trost, "Detection of material separation in sheet metal using ultrasound," (in German), Z. Ver. dtsch. Ing. **87**, 352–354 (1943).
33. J. Krautkrämer and H. Krautkrämer, *Ultrasonic Testing of Materials*, 2nd ed (Springer-Verlag, Berlin, 1977).
34. W. P. Mason, *Physical Acoustics and the Properties of Solids* (Van Nostrand, New York, 1958).
35. W. Koltonski and I. Malescki, "Ultrasonic method for the exploration of the properties and structure of mineral layers," Acustica **8**, 307–314 (1958).
36. R. F. S. Hearmon, *An Introduction to Applied Anisotropic Elasticity* (Oxford University Press, London, 1961).
37. W. P. Mason and H. J. McSkimin, "Attenuation and scattering of high frequency sound waves in metals and glasses," J. Acoust. Soc. Am. **19**, 464–473 (1947).
38. W. P. Mason and H. J. McSkimin, "Energy losses of sound waves in metals due to scattering and diffusion," J. Appl. Phys. **19**, 940–946 (1948).
39. W. Roth, "Scattering of ultrasonic radiation in polycrystalline metals," J. Appl. Phys. **19**, 901–910 (1948).
40. R. L. Roderick and R. Truell, "The measurement of ultrasonic attenuation in solids by the pulse technique and some results in steel," J. Appl. Phys. **23**, 267–279 (1952).
41. P. F. Sullivan and E. P. Papadakis, "Ultrasonic double refraction in worked metals," J. Acoust. Soc. Am. **33**, 1622–1624 (1961).
42. H. Kolsky, *Stress Waves in Solids* (Dover, New York, 1963).
43. R. F. S. Hearmon, "Third-order elastic coefficients," Acta Crystallogr. **6**, 331–340 (1953).
44. A. Seeger and O. Buck, "The experimental measurement of higher order elastic constants," (in German), Z. Naturforsch. **A15**, 1056–1067 (1960).
45. M. A. Breazeale and J. Ford, "Ultrasonic studies of the nonlinear behavior of solids," J. Appl. Phys. **36**, 3486–3490 (1965).
46. D. S. Hughes and J. L. Kelly, "Second-order elastic deformations of solids," Phys. Rev. **92**, 1145–1149 (1953).
47. R. N. Thurston and K. Brugger, "Third-order elastic constants and the velocity of small-amplitude elastic waves in homogeneous stressed media," Phys. Rev. **133**, A1604–1610 (1964).
48. R. A. Toupin and B. Bernstein, "Sound waves in deformed perfectly elastic materials. Acoustoelastic effect," J. Acoust. Soc. Am. **33**, 216–255 (1961).
49. M. R. James and O. Buck, "Quantitative nondestructive measurements of residual stress," in *CRC Critical Reviews in Solid State and Material Sciences*, vol. **10** (CRC Press, Akron, 1980), pp. 61–105.

50. A. E. Green and P. M. Naghdi, "A general theory of an elastic-plastic continuum," Arch. Rat. Mech. Anal. **18**, 2551–2581 (1965).
51. E. H. Lee and D. T. Liu, "Finite-strain elastic-plastic theory with applications to plane-wave analysis," J. Appl. Phys. **38**, 19–27 (1967).
52. F. A. Firestone and D. S. Ling, *Report on the Propagation of Waves in Plates*, Tech. Rept. No. 50-6001, Sperry Products, Inc, Danbury CT (1945).
53. H. Lamb, "On waves in an elastic plate," Proc. Roy. Soc. **A93**, 114–128 (1917).
54. M. F. M. Osborne and S. D. Hart, "Transmission, reflection, and guiding of an exponential pulse by a steel plate in water. I. Theory," J. Acoust. Soc. Am. **17**, 1–18 (1945).
55. A. Schoch, "Sound transmission through plates," (in German), Acustica **2**, 1–18 (1952).
56. D. C. Worlton, "Experimental confirmation of Lamb waves at megacycle frequencies," J. Appl. Phys. **32**, 967–971 (1961).
57. C. L. Frederick and D. C. Worlton, "Ultrasonic thickness measurements with Lamb waves," J. Nondestruct. Test. **20**, 51–55 (1962).

Physical Ultrasonics of Composites

1

Fundamentals of Composite Elastic Properties

1.1 Introduction

In this chapter, we review the mechanical behavior of composites considered from a macroscopic perspective, *i.e.*, the microscopic heterogeneity of the material is ignored in this treatment. Our objective is to provide an overview of the basic composite constitutive behavior and to set the notation for the subsequent chapters. To establish this framework, we draw on concepts from continuum mechanics and elasticity, both of which are also covered by specialized books on these topics. The results in this chapter are important for us because they provide the theoretical framework for all the elastic wave phenomena we describe in detail in the subsequent chapters.

1.2 Mechanical Relationships

1.2.1 Stress

Stress in a solid body is measured in force per unit area; there are normal stresses, acting along a normal to the infinitesimal element of the area, and shear (tangential) stresses, acting in the plane of the element. Let us assume that an infinitesimal traction force $d\mathbf{T}$ acts on an infinitesimal surface element $d\mathbf{A} = \mathbf{n}dA$, where $\mathbf{n}$ denotes the normal unit vector of the surface element. In index notation, the stress tensor is then defined through

$$dT_i = \sigma_{ij} dA_j. \tag{1.1}$$

The sign convention for the stress tensor in a Cartesian coordinate system is shown in Fig. 1.1. The choice of coordinate system is arbitrary, but for the sake of simplicity and concreteness, let us develop the relationships

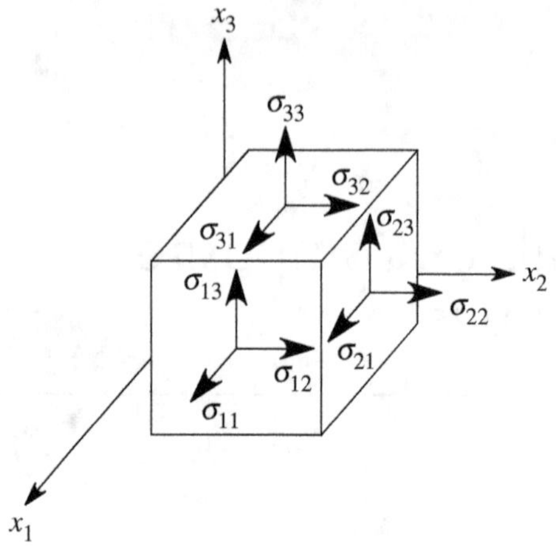

Figure 1.1. Convention for stress tensor notation in a Cartesian coordinate system.

in a Cartesian system. They can all be generalized at a later time. Only the stress components acting on the surface elements with positive normal vectors are shown for clarity. On the surface elements with negative normal vectors, the stress directions are opposite. Conventionally, the first index indicates the normal of the surface the stress component is acting upon and the second index indicates the direction of the resulting traction force (however, we will show shortly that equilibrium conditions require that the stress tensor be symmetric, therefore the order of the indices is only of academic importance). For example, σ_{11} is the normal stress acting on the x_2, x_3 plane in the x_1 direction, σ_{12} is the shear stress acting on the same plane in the x_2 direction, and so forth.

To make all possible connections among spatial coordinates, there are three normal and six shear stress elements, although only three independent shear stress components are needed to completely characterize the state of stress at any point in the material. When the cubic volume element in Fig. 1.1 is in equilibrium, the requirement of zero net moment in any direction leads to the important conclusion that the stress tensor is symmetric, *i.e.*,

$$\sigma_{ij} = \sigma_{ji}, \quad i, j = 1, 2, 3. \tag{1.2}$$

The sign of the stress will depend on the direction of the applied traction force on the surface and on the orientation of the surface itself; the sign is positive if the surface normal and the force direction are both positive or negative. Otherwise, the stress is taken to be negative. Figure 1.2 further illustrates this sign convention for normal and shear stresses. Normal stresses are conventionally positive in tension and negative in compression. It is important to realize that the sign of the normal stress acting on a given surface will not change for any rotation of the reference coordinate system, but that of the shear stress might, *i.e.*, the difference between positive and

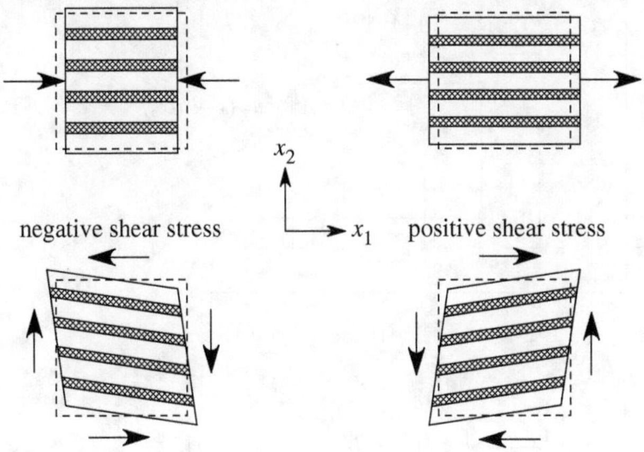

Figure 1.2. Sign convention for normal and shear stresses in a Cartesian coordinate system (in this example, the x_1 direction coincides with the fiber direction of the composite specimen).

negative shear stresses does not have the same physical significance as the difference between tension and compression.

1.2.2 Strain

Under load, the elastic body deforms, leading to a change of its volume and shape. As a result, every particle (mass point) of the body (defined by its radius vector **r**) is displaced to a new position with a new radius vector (**r**+**dr**) as shown in Fig. 1.3 in the coordinate projections. The vector **u**, characterizing a field of displacement of body particles, is called a displacement vector. Using Fig. 1.3(a), the normal strains in the x_1 and x_2 directions are defined as the normalized change in length and are given by

$$\epsilon_{11} = \frac{u_1(x_1 + dx_1) - u_1(x_1)}{dx_1} = \frac{\partial u_1}{\partial x_1} \tag{1.3}$$

and

$$\epsilon_{22} = \frac{u_2(x_2 + dx_2) - u_2(x_2)}{dx_2} = \frac{\partial u_2}{\partial x_2}, \tag{1.4}$$

and likewise for the x_3 coordinate. Under shear loading, a body will be deformed as shown in Fig. 1.3(b). The engineering shear strain $\gamma_{12} = \alpha_1 + \alpha_2$ is defined as the decrease in the angle between two lines that were parallel to the x_1 and x_2 axes in the undeformed state. For small deformations, the

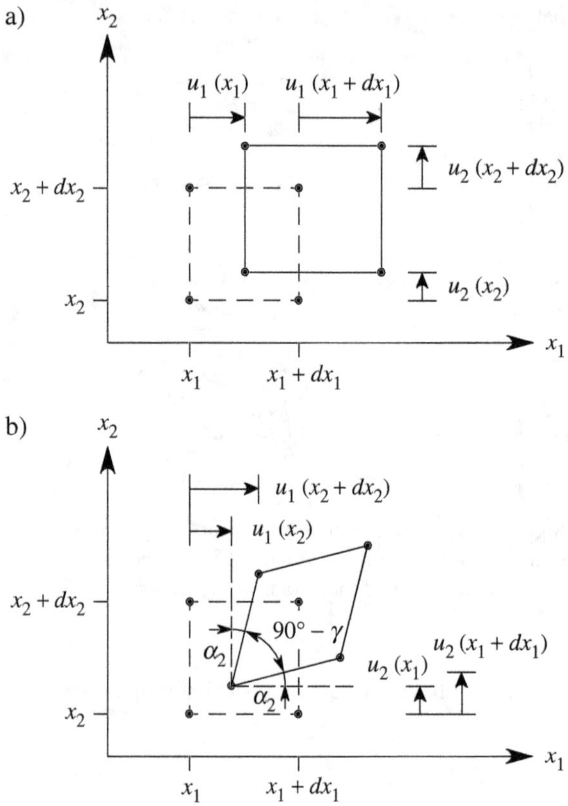

Figure 1.3. Normal (a) and shear (b) strains.

expression for shear strain can be linearized to yield

$$\epsilon_{12} = \frac{1}{2}\gamma_{12} = \frac{1}{2}\left(\frac{\partial u_1}{\partial x_2} + \frac{\partial u_2}{\partial x_1}\right), \tag{1.5}$$

where we exploited that $\tan\alpha \approx \alpha$. Similar relationships apply for the remaining directions. For a more detailed exposition of stress and strain in solids, the reader may refer to reference works on elasticity, such as the classic treatment by Timoshenko and Goodier [1], or to books on composite mechanics, such as the ones by Hahn and Tsai [2], Jones [3], or Daniel and Ishai [4]. Continuum mechanics is covered in many volumes, such as those by Chung [5] or Frederick and Chang [6].

1.2.3 Constitutive equation

The general relationships between stresses and strains are called constitutive relationships. These relationships are often very complicated as not only anisotropic, but nonlinear, plastic, and other effects also must be included accurately to describe the behavior of the deformed material. These effects can be best investigated in the simplest state of loading namely, in uniaxial tension or compression.

Uniaxial tensile loading

Figure 1.4 shows a schematic diagram of the uniaxial tension experiment most often used for metals to measure the stiffness of a material in the form of a thin bar. Figure 1.5 illustrates the main trends of the stress–strain relation typically observed for metals under uniaxial tensile loading. Generally, there is an essentially linear initial part where the behavior of the material is elastic, followed by a plastic region where the stress–strain relationship becomes exceedingly nonlinear. Note that polymer matrix composites usually fail without exhibiting a strongly nonlinear plastic region [3, 4]. In most engineering solids, including composites, for small strains the relationship connecting stress and strain can be approximated by a linear equation. This simple fact is illustrated in Fig. 1.5. In ultrasonic wave propagation, the induced stress and strain levels are many orders of magnitude below the elastic limit and nonlinear effects are negligible in most cases. (In conventional linear ultrasonic measurements, the strains do not exceed 10^{-7}). The initial linear slope of the stress–strain curve is expressed in Hooke's law, which is an experimentally derived relationship

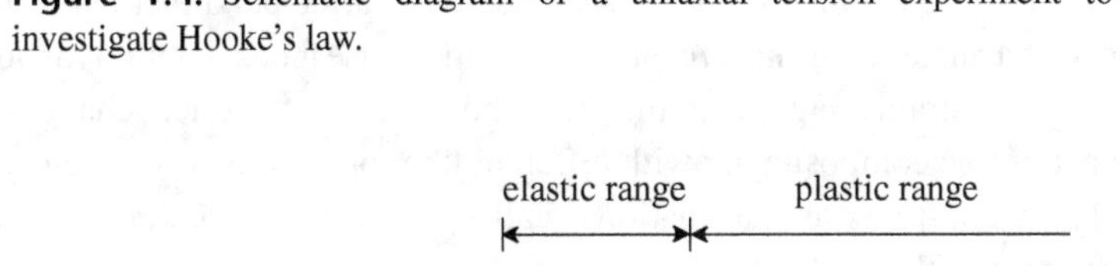

Figure 1.4. Schematic diagram of a uniaxial tension experiment to investigate Hooke's law.

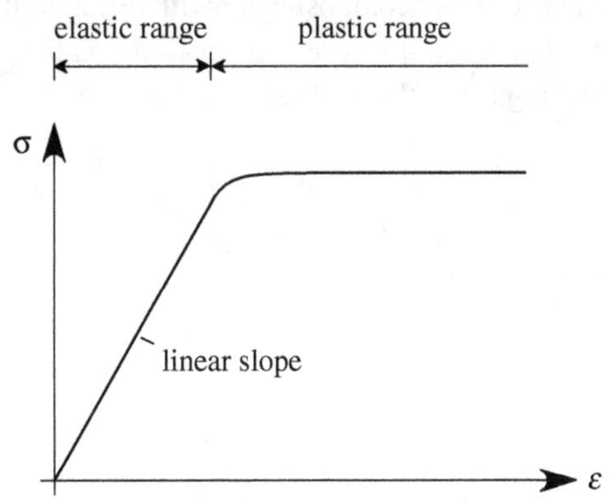

Figure 1.5. A schematic diagram of the stress–strain relation for uniaxial tensile loading.

that states the proportionality between stress σ and strain ϵ. In this section (and throughout the book), we make a clear distinction between vector and tensor notation common in composite mechanics and the notation more commonly used in elasticity. In later sections, we address this important question directly. For now, we are somewhat informal in our choice of notation, but the meaning should be clear nonetheless.

In its simplest form, Hooke's law says that in the case of uniaxial tension and compression

$$\sigma = E\epsilon \tag{1.6}$$

or

$$\epsilon = S\sigma, \tag{1.7}$$

where E is Young's modulus and S is the compliance. The stress can be calculated from

$$\sigma = \frac{P}{A} \tag{1.8}$$

and the strain is

$$\epsilon = \frac{\Delta\ell}{\ell}, \tag{1.9}$$

where P is the applied load, A is the cross-sectional area, ℓ is the original length of the bar without deformation, and $\Delta\ell$ is the elongation of the bar in the direction of the applied load.

As was indicated in Fig. 1.4, because of the well-known Poisson effect, deformation will also occur in the direction normal to the applied load. To distinguish between axial and lateral strains as well as between loads applied in different directions with respect to the principal directions of the anisotropic material, from here on we are going to use index notation for both stresses and strains. Figure 1.6 illustrates how the Poisson effect changes in an anisotropic composite bar with different fiber orientations. The negative ratio of lateral-to-axial deformation is called the *Poisson ratio*. For uniaxial loading along the x_1 direction, it can be expressed as

$$\nu_{12} = -\frac{\epsilon_{22}}{\epsilon_{11}} \quad \text{or} \quad \nu_x = -\frac{\epsilon_y}{\epsilon_x}, \tag{1.10}$$

and from Eq. (1.7) the axial strain is

$$\epsilon_{11} = \frac{\sigma_{11}}{E_1} \quad \text{or} \quad \epsilon_x = \frac{\sigma_x}{E_x}. \tag{1.11}$$

Here and in the following paragraphs, for clarity, we provide the corresponding formulas not only in tensor notation, but also in simpler engineering

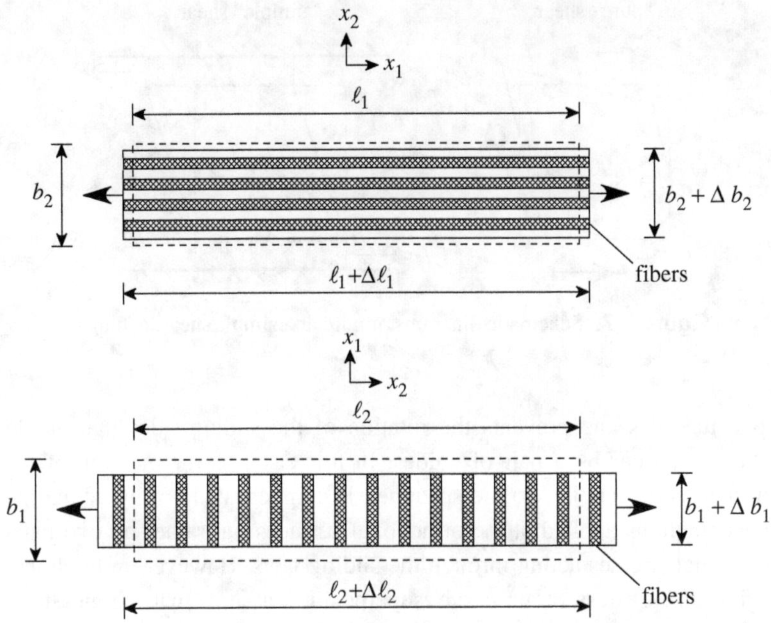

Figure 1.6. Poisson effect in an anisotropic composite bar with different fiber orientations.

notation using x, y Cartesian coordinate indices. The lateral strain ϵ_{22} in the x_2 direction will be

$$\epsilon_{22} = -\nu_{12}\frac{\sigma_{11}}{E_1} \quad \text{or} \quad \epsilon_y = -\nu_x \frac{\sigma_x}{E_x}. \tag{1.12}$$

By a similar approach for a composite loaded along the x_2 direction

$$\epsilon_{22} = \frac{\sigma_{22}}{E_2} \quad \text{or} \quad \epsilon_y = \frac{\sigma_y}{E_y} \tag{1.13}$$

and

$$\epsilon_{11} = -\nu_{21}\frac{\sigma_{22}}{E_2} \quad \text{or} \quad \epsilon_x = -\nu_y \frac{\sigma_y}{E_y}. \tag{1.14}$$

Planar shear loading

The schematic diagrams of two types of planar shear loading are shown in Fig. 1.7. "Pure" shear occurs as shown on the left-hand diagram of Fig. 1.7 without length change of the sides of the initial square element. This condition would be rather difficult to achieve on an externally loaded large test piece; therefore in engineering measurements, we often use an arrangement also shown in Fig. 1.7 that loads the specimen only on two

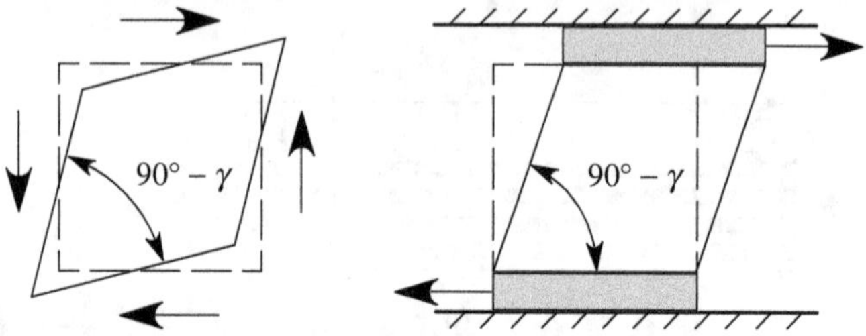

Figure 1.7. Schematic diagrams of pure and simple shear loadings.

opposite sides and prevents the rotation of the sliding sides that would otherwise occur by a pair of guides. In this case, the momentum of the shear tractions applied to the specimen is compensated not by additional shear tractions applied at the orthogonal surfaces but rather by two pairs of normal stresses acting through the sliding jaws. However, in the linear region of the stress–strain curve, superposition assures that the measured engineering shear strain is caused only by the shear stress component. The relation between shear stress σ_{12} and shear strain ϵ_{12} is given by

$$2\epsilon_{12} = \frac{\sigma_{12}}{G} \quad \text{or} \quad \gamma_{xy} = \frac{\tau_{xy}}{G}, \tag{1.15}$$

where G is the shear modulus and, by convention, $2\epsilon_{12} = \gamma_{xy}$.

In the general case of plane stress loading, *i.e.*, when both normal and shear stresses are applied simultaneously in the x_1, x_2 plane

$$\epsilon_{11} = \frac{\sigma_{11}}{E_1} - \nu_{21}\frac{\sigma_{22}}{E_2} \quad \text{or} \quad \epsilon_x = \frac{\sigma_x}{E_x} - \nu_y\frac{\sigma_y}{E_y}, \tag{1.16}$$

$$\epsilon_{22} = \frac{\sigma_{22}}{E_2} - \nu_{12}\frac{\sigma_{11}}{E_1} \quad \text{or} \quad \epsilon_y = \frac{\sigma_y}{E_y} - \nu_x\frac{\sigma_x}{E_x}, \tag{1.17}$$

and

$$2\epsilon_{12} = \frac{\sigma_{12}}{G} \quad \text{or} \quad \gamma_{xy} = \frac{\tau_{xy}}{G}, \tag{1.18}$$

where the principle of superposition has been applied.

1.3 Generalized Hooke's Law

In rigorous tensor notation for a generally anisotropic linear medium, Hooke's law can be written as

$$\epsilon_{ij} = s_{ijk\ell}\sigma_{k\ell} \tag{1.19}$$

in terms of the elastic compliance tensor $s_{ijk\ell}$ or

$$\sigma_{ij} = c_{ijk\ell}\epsilon_{k\ell} \tag{1.20}$$

in terms of the elastic stiffness tensor $c_{ijk\ell}$ and $i, j, k, \ell = 1, 2, 3$. In general, a fourth-rank tensor is needed to connect two second-rank tensors, such as the stress and strain tensors. Later, we shall see that because of symmetry, transformation properties, and conditions on the strain energy density, a lower dimensional mathematical entity will suffice to describe the elastic behavior of solids. Using the symmetry of the stress tensor σ_{ij}, as expressed in Eq. (1.2), we can write

$$\sigma_{ij} = \begin{bmatrix} \sigma_{11} & \sigma_{12} & \sigma_{13} \\ \sigma_{12} & \sigma_{22} & \sigma_{23} \\ \sigma_{13} & \sigma_{23} & \sigma_{33} \end{bmatrix}. \tag{1.21}$$

The linearized strain tensor is given by [1]

$$\epsilon_{ij} = \frac{1}{2}\left(\frac{\partial u_i}{\partial x_j} + \frac{\partial u_j}{\partial x_i}\right) \tag{1.22}$$

and from the expression above, it is also clearly symmetric, *i.e.*,

$$\epsilon_{ij} = \epsilon_{ji}.$$

Therefore, exploiting these symmetry properties we have the following relations for the compliance and stiffness tensors

$$s_{ijk\ell} = s_{jik\ell} = s_{ij\ell k} = s_{ji\ell k} \tag{1.23}$$

and

$$c_{ijk\ell} = c_{jik\ell} = c_{ij\ell k} = c_{ji\ell k}. \tag{1.24}$$

Using the symmetry properties shown above, we may adopt a simpler notational convention, used extensively in elasticity, to write the stress and strain tensors σ_{ij} and $\epsilon_{k\ell}$ and, by implication, the stiffness and compliance tensors $c_{ijk\ell}$ and $s_{ijk\ell}$. The reasoning here is simple: both stress and strain are symmetric, implying that each has, at most, only six independent elements. To connect two mathematical sets each with six independent elements requires an entity with at most 36 elements. Such an entity can be conveniently constructed in the form of a 6×6 square array. This new representation has the added benefit of being easily expressible on a single page in matrix format. To make the correspondence between the two systems, namely tensor and matrix, we follow the established convention and require $11 \to 1$, $22 \to 2$, $33 \to 3$, $12 \to 6$, $13 \to 5$, $23 \to 4$. In each case, quantities with permuted subscripts have the same value; for example, $23 \to 4$ and $32 \to 4$. To complete the simplified matrix picture, we must apply this dimensional reduction to the stress and strain tensors also, converting them into row or column vectors with six elements each.

Traditionally, these six-element vectors are expressed informally by a quantity often also labeled by the Greek letter σ_i. A single subscript indicates that the abbreviated vectorial notation is used instead of the tensor σ_{ij}. To better distinguish vectors from second-rank tensor quantities, let us use $\boldsymbol{\sigma}_i$ to denote the vector stress and $\boldsymbol{\varepsilon}_i$ to denote the vector strain,

$$\begin{array}{llll} \sigma_{11} = \boldsymbol{\sigma}_1 & \sigma_{23} = \tau_{23} = \boldsymbol{\sigma}_4 & \epsilon_{11} = \boldsymbol{\varepsilon}_1 & 2\epsilon_{23} = \gamma_{23} = \boldsymbol{\varepsilon}_4 \\ \sigma_{22} = \boldsymbol{\sigma}_2 & \sigma_{13} = \tau_{13} = \boldsymbol{\sigma}_5 & \epsilon_{22} = \boldsymbol{\varepsilon}_2 & 2\epsilon_{13} = \gamma_{13} = \boldsymbol{\varepsilon}_5 \\ \sigma_{33} = \boldsymbol{\sigma}_3 & \sigma_{12} = \tau_{12} = \boldsymbol{\sigma}_6 & \epsilon_{33} = \boldsymbol{\varepsilon}_3 & 2\epsilon_{12} = \gamma_{12} = \boldsymbol{\varepsilon}_6. \end{array} \tag{1.25}$$

Applying the same notation convention to the compliance tensor $s_{ijk\ell}$, we have for the compliance matrix S_{mn} [8],

$$\begin{aligned} S_{mn} &= s_{ijk\ell} && \text{for } m \text{ and } n = 1, 2, 3, \\ S_{mn} &= 2s_{ijk\ell} && \text{for } m \text{ or } n = 4, 5, 6, \\ S_{mn} &= 4s_{ijk\ell} && \text{for } m \text{ and } n = 4, 5, 6. \end{aligned} \tag{1.26}$$

It should be mentioned that the relationships between the stiffness constants in full tensor and abbreviated matrix notation are much simpler than the corresponding relationships for the compliance constants given by Eq. (1.26), namely

$$C_{mn} = c_{ijk\ell}, \tag{1.27}$$

where the previously described correspondence between the tensor (i, j, k, ℓ) and matrix (m, n) indices must be used. In the abbreviated matrix notation, the constitutive equations Eqs. (1.19) and (1.20) can be rewritten as

$$\boldsymbol{\varepsilon}_i = S_{ij}\boldsymbol{\sigma}_j \tag{1.28}$$

and

$$\boldsymbol{\sigma}_i = C_{ij}\boldsymbol{\varepsilon}_j, \tag{1.29}$$

respectively, with $i, j = 1, 2 \ldots 6$. For example, the explicit relation of Eq. (1.28) for $\boldsymbol{\varepsilon}_i$ becomes

$$\begin{bmatrix} \boldsymbol{\varepsilon}_1 \\ \boldsymbol{\varepsilon}_2 \\ \boldsymbol{\varepsilon}_3 \\ \boldsymbol{\varepsilon}_4 \\ \boldsymbol{\varepsilon}_5 \\ \boldsymbol{\varepsilon}_6 \end{bmatrix} = \begin{bmatrix} S_{11} & S_{12} & S_{13} & S_{14} & S_{15} & S_{16} \\ S_{21} & S_{22} & S_{23} & S_{24} & S_{25} & S_{26} \\ S_{31} & S_{32} & S_{33} & S_{34} & S_{35} & S_{36} \\ S_{41} & S_{42} & S_{43} & S_{44} & S_{45} & S_{46} \\ S_{51} & S_{52} & S_{53} & S_{54} & S_{55} & S_{56} \\ S_{61} & S_{62} & S_{63} & S_{64} & S_{65} & S_{66} \end{bmatrix} \begin{bmatrix} \boldsymbol{\sigma}_1 \\ \boldsymbol{\sigma}_2 \\ \boldsymbol{\sigma}_3 \\ \boldsymbol{\sigma}_4 \\ \boldsymbol{\sigma}_5 \\ \boldsymbol{\sigma}_6 \end{bmatrix}. \tag{1.30}$$

Under the condition that the strain energy density $\mathcal{U}$ associated with the elastic deformation has a quadratic form, both the elastic stiffness and compliance matrices will be symmetric

$$\mathcal{U} = \frac{1}{2}\varepsilon_i\sigma_i \quad = S_{ij}\sigma_j\sigma_i$$
$$= \frac{1}{2}\varepsilon_j\sigma_j \quad = S_{ji}\sigma_i\sigma_j, \tag{1.31}$$

and the same argument can be applied to C_{ij}. The result is that

$$S_{ij} = S_{ji} \quad C_{ij} = C_{ji},$$

which reduces the total number of independent elastic constants (compliances or stiffnesses) to 21. Here is how this works: the 36-element elastic matrix is itself symmetric, as noted above, leaving 15 independent off-diagonal elements plus 6 elements on the diagonal, for a total of 21.

1.4 Stress–Strain Relations for Media of Higher Symmetries

Most often, engineering materials such as composites will require only a much lower degree of anisotropy for their mechanical modeling than the completely general case we have just considered in Section 1.3. Depending on the microscopic and macroscopic composite geometry, layup sequence, reinforcement, and structural configuration, a higher order of material symmetry than one described by 21 independent elastic parameters will likely be entirely sufficient for both stress analysis and descriptions of elastic wave propagation in these materials. In view of this fact, it is appropriate now to consider certain cases of higher material symmetry and the simplifications they allow.

1.4.1 Monoclinic symmetry

If a single plane of mirror symmetry exists in the material, say, for example, one that is coincident with the x_1x_2 plane, then a transformation of the form $x_3 \rightarrow -x_3$ must leave the stiffness tensor unchanged, according to the nature of the postulated symmetry. If we implement such a transformation (outlined later in Sections 1.7.1 and 1.7.2) upon a stiffness matrix C_{ij}, similar in form to the compliance matrix given in Eq. (1.30), we find an odd result. Some elements are not transformed into themselves, as required by the symmetry and the transformation operation. Instead, some elements C_{ab} transform into their *negative* values $-C_{ab}$. For these elements, the existence of a plane of mirror symmetry in x_1x_2 implies that they must vanish. In this particular instance, the vanishing matrix elements are

$$C_{14},\quad C_{15},\quad C_{24},\quad C_{25},\quad C_{34},\quad C_{35},\quad C_{46},\quad \text{and}\quad C_{56}. \tag{1.32}$$

The remaining matrix elements (on and above the diagonal) are nonzero and independent, and the matrix is still symmetric, of course. The result is

$$\begin{bmatrix} \sigma_1 \\ \sigma_2 \\ \sigma_3 \\ \sigma_4 \\ \sigma_5 \\ \sigma_6 \end{bmatrix} = \begin{bmatrix} C_{11} & C_{12} & C_{13} & 0 & 0 & C_{16} \\ C_{12} & C_{22} & C_{23} & 0 & 0 & C_{26} \\ C_{13} & C_{23} & C_{33} & 0 & 0 & C_{36} \\ 0 & 0 & 0 & C_{44} & C_{45} & 0 \\ 0 & 0 & 0 & C_{45} & C_{55} & 0 \\ C_{16} & C_{26} & C_{36} & 0 & 0 & C_{66} \end{bmatrix} \begin{bmatrix} \varepsilon_1 \\ \varepsilon_2 \\ \varepsilon_3 \\ \varepsilon_4 \\ \varepsilon_5 \\ \varepsilon_6 \end{bmatrix}. \tag{1.33}$$

Although there are few composite materials that exhibit this type of symmetry natively, monoclinic material symmetry is important because it is often the resulting *effective* symmetry after a rotational coordinate transformation of materials of much higher symmetry.

1.4.2 Orthotropic symmetry

Adding a second plane of mirror symmetry implies the existence of a third plane of symmetry as well, and therefore an orthotropic material possesses three orthogonal planes of mirror symmetry. Figure 1.8 shows the schematic diagram of a unidirectional composite material of orthotropic symmetry with the x_1x_2, x_2x_3, and x_1x_3 coordinate planes being the three planes of symmetry. In comparison to materials of monoclinic symmetry, the additional symmetry planes in orthogonal materials require that further matrix elements vanish leaving nine independent elements. The stress–strain relation for orthotropic media can be written as

$$\begin{bmatrix} \sigma_1 \\ \sigma_2 \\ \sigma_3 \\ \sigma_4 \\ \sigma_5 \\ \sigma_6 \end{bmatrix} = \begin{bmatrix} C_{11} & C_{12} & C_{13} & 0 & 0 & 0 \\ C_{12} & C_{22} & C_{23} & 0 & 0 & 0 \\ C_{13} & C_{23} & C_{33} & 0 & 0 & 0 \\ 0 & 0 & 0 & C_{44} & 0 & 0 \\ 0 & 0 & 0 & 0 & C_{55} & 0 \\ 0 & 0 & 0 & 0 & 0 & C_{66} \end{bmatrix} \begin{bmatrix} \varepsilon_1 \\ \varepsilon_2 \\ \varepsilon_3 \\ \varepsilon_4 \\ \varepsilon_5 \\ \varepsilon_6 \end{bmatrix}. \tag{1.34}$$

All the elastic stiffnesses in the C_{ij} matrix above are independent. The orthotropic symmetry class includes finely laminated cross-ply composites and some fiber-placed composites, such as carbon–carbon.

1.4.3 Transversely isotropic symmetry

Figure 1.9 illustrates the symmetry characteristics of a transversely isotropic material. This symmetry class consists of an axis of symmetry normal to a

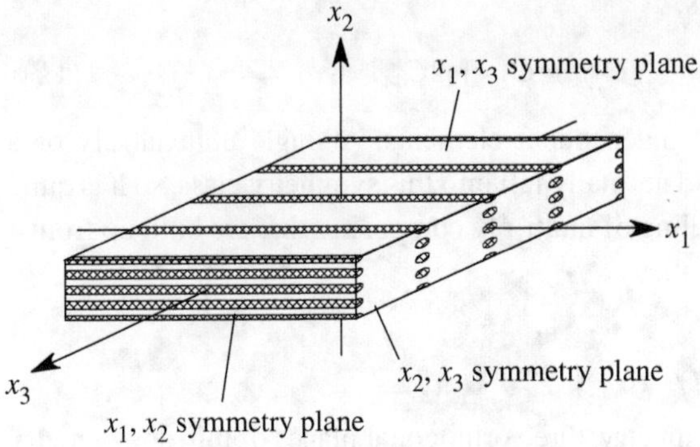

Figure 1.8. Unidirectional composite material of orthotropic symmetry.

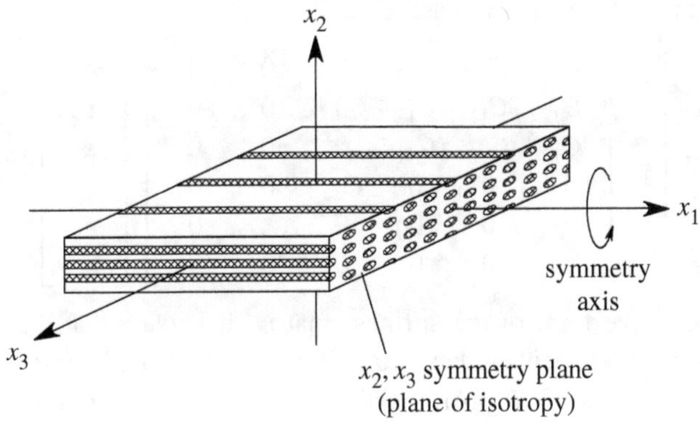

Figure 1.9. Transversely isotropic unidirectional composite material.

plane of isotropy. The equivalent crystal symmetry class is *hexagonal*. In this case, we chose the plane of isotropy to be the x_2x_3 plane and the x_1 axis is the axis of symmetry. Then, after the application of appropriate symmetry operations as outlined above, the remaining nonzero matrix elements in the stress–strain relation are given as

$$\begin{bmatrix} \sigma_1 \\ \sigma_2 \\ \sigma_3 \\ \sigma_4 \\ \sigma_5 \\ \sigma_6 \end{bmatrix} = \begin{bmatrix} C_{11} & C_{12} & C_{12} & 0 & 0 & 0 \\ C_{12} & C_{22}^{\dagger} & C_{23} & 0 & 0 & 0 \\ C_{12} & C_{23} & C_{22}^{\dagger} & 0 & 0 & 0 \\ 0 & 0 & 0 & C_{44} & 0 & 0 \\ 0 & 0 & 0 & 0 & C_{55} & 0 \\ 0 & 0 & 0 & 0 & 0 & C_{55} \end{bmatrix} \begin{bmatrix} \varepsilon_1 \\ \varepsilon_2 \\ \varepsilon_3 \\ \varepsilon_4 \\ \varepsilon_5 \\ \varepsilon_6 \end{bmatrix}, \quad (1.35)$$

where C_{22} is marked with a † because although it is not zero, it is also not independent. This element is given by the following linear combination of the remaining elements,

$$C_{22} = C_{23} + 2C_{44}, \tag{1.36}$$

leaving a total of five independent elements. A single uniaxial ply or a uniaxial laminate would normally fall into this symmetry class, so it is quite important for the modeling of multi-ply composites that are built up from a combination of single plies.

1.4.4 Cubic symmetry

In the case of cubic symmetry, three orthogonal planes of mirror symmetry and three orthogonal four-fold symmetry axes exist simultaneously. In this symmetry class, there are very few independent matrix elements, and the stress–strain relation is given as

$$\begin{bmatrix} \sigma_1 \\ \sigma_2 \\ \sigma_3 \\ \sigma_4 \\ \sigma_5 \\ \sigma_6 \end{bmatrix} = \begin{bmatrix} C_{11} & C_{12} & C_{12} & 0 & 0 & 0 \\ C_{12} & C_{11} & C_{12} & 0 & 0 & 0 \\ C_{12} & C_{12} & C_{11} & 0 & 0 & 0 \\ 0 & 0 & 0 & C_{44} & 0 & 0 \\ 0 & 0 & 0 & 0 & C_{44} & 0 \\ 0 & 0 & 0 & 0 & 0 & C_{44} \end{bmatrix} \begin{bmatrix} \varepsilon_1 \\ \varepsilon_2 \\ \varepsilon_3 \\ \varepsilon_4 \\ \varepsilon_5 \\ \varepsilon_6 \end{bmatrix}. \tag{1.37}$$

From the above form of the stiffness matrix, it is clear that the number of independent elements reduces to only three. Most single-crystal metals (*e.g.*, aluminum, steel, copper, nickel) fall into the cubic symmetry class, and so does diamond. This type of symmetry, however, is not often met with in the analysis of composite materials, though some limited types of composites, such as fiber-placed composites, can be best described by this material symmetry.

1.4.5 Isotropic material

When every direction in three-dimensional 3-D space is taken to be equivalent, *i.e.*, when any axis in space is a symmetry axis and any plane is a symmetry plane, a further simplification occurs. This is the case of isotropic symmetry, and for this class of materials the stress–strain relation is as simple as possible,

$$\begin{bmatrix} \sigma_1 \\ \sigma_2 \\ \sigma_3 \\ \sigma_4 \\ \sigma_5 \\ \sigma_6 \end{bmatrix} = \begin{bmatrix} C_{11}^{\dagger} & C_{12} & C_{12} & 0 & 0 & 0 \\ C_{12} & C_{11}^{\dagger} & C_{12} & 0 & 0 & 0 \\ C_{12} & C_{12} & C_{11}^{\dagger} & 0 & 0 & 0 \\ 0 & 0 & 0 & C_{44} & 0 & 0 \\ 0 & 0 & 0 & 0 & C_{44} & 0 \\ 0 & 0 & 0 & 0 & 0 & C_{44} \end{bmatrix} \begin{bmatrix} \varepsilon_1 \\ \varepsilon_2 \\ \varepsilon_3 \\ \varepsilon_4 \\ \varepsilon_5 \\ \varepsilon_6 \end{bmatrix}, \tag{1.38}$$

where the symbol † again signals a linear dependence on remaining matrix elements. In this case, $C_{11} = C_{12} + 2C_{44}$. This fact has led to a more common terminology when dealing with isotropic materials. Let C_{12} and C_{44} be denoted by λ and μ, respectively. Then, $C_{11} = \lambda + 2\mu$, where λ and μ are called the Lamé constants.

G is used alternatively with μ to denote the isotropic shear modulus and $\lambda/(2\lambda + 2\mu)$ is the isotropic Poisson ratio ν. Accordingly, $\mu(3\lambda + 2\mu)/(\lambda + \mu)$ is the isotropic Young's modulus E, and the isotropic bulk modulus K is given by $\lambda + 2\mu/3$. There is essentially no continuous fiber composite in the isotropic symmetry class, but random chopped-fiber composites and certain particulate composites can demonstrate this highest degree of symmetry. Most fine-grained polycrystalline metals and alloys, depending on their fabrication, show an elastic symmetry that is close to isotropic.

1.5 Relation between Stiffness and Compliance

From Eqs. (1.28) and (1.29) there exists a relation between C_{ij} and S_{ij}, which can be written as follows

$$S_{ij} = C_{ij}^{-1} \qquad C_{ij} = S_{ij}^{-1}, \tag{1.39}$$

where C_{ij}^{-1} and S_{ij}^{-1} are the inverse matrices of C_{ij} and S_{ij}, respectively. For orthotropic media, the elements of C_{ij} can be found from Eq. (1.34),

$$\begin{aligned}
C_{11} &= (S_{22}S_{33} - S_{23}^2)/S \\
C_{22} &= (S_{33}S_{11} - S_{13}^2)/S \\
C_{33} &= (S_{11}S_{22} - S_{12}^2)/S \\
C_{12} &= (S_{13}S_{23} - S_{12}S_{33})/S \\
C_{13} &= (S_{12}S_{23} - S_{13}S_{22})/S \\
C_{23} &= (S_{12}S_{13} - S_{23}S_{11})/S \\
C_{44} &= 1/S_{44}, \quad C_{55} = 1/S_{55}, \quad C_{66} = 1/S_{66},
\end{aligned} \tag{1.40}$$

where S is the subdeterminant of the first three rows and columns of S_{ij}

$$S = S_{11}S_{22}S_{33} - S_{11}S_{23}^2 - S_{22}S_{13}^2 - S_{33}S_{12}^2 + 2S_{12}S_{23}S_{13}. \tag{1.41}$$

Experimentally, the stiffness C_{ij} and compliance S_{ij} elements can be assessed by either acoustic velocity measurements in the material or from engineering elastic constants obtained from static (tensile, shear, bending, or torsion) measurements. These engineering elastic constants are defined in the next section.

1.5.1 Engineering constants for orthotropic materials

In the engineering literature, alternative means are often used to express elastic properties. In Section 1.2.3, we have already introduced the directional Young's moduli, Poisson's ratios, and shear moduli for the simpler case of plane stress. For orthotropic materials, the S_{ij} compliance elements are given in terms of this alternative set of material properties by

$$S_{ij} = \begin{bmatrix} E_1^{-1} & -\nu_{21}E_2^{-1} & -\nu_{31}E_3^{-1} & 0 & 0 & 0 \\ -\nu_{12}E_1^{-1} & E_2^{-1} & -\nu_{32}E_3^{-1} & 0 & 0 & 0 \\ -\nu_{13}E_1^{-1} & -\nu_{23}E_2^{-1} & E_3^{-1} & 0 & 0 & 0 \\ 0 & 0 & 0 & G_{23}^{-1} & 0 & 0 \\ 0 & 0 & 0 & 0 & G_{31}^{-1} & 0 \\ 0 & 0 & 0 & 0 & 0 & G_{12}^{-1} \end{bmatrix} \tag{1.42}$$

so that the normal strain components are

$$\begin{aligned} \varepsilon_1 &= \frac{\sigma_1}{E_1} - \nu_{21}\frac{\sigma_2}{E_2} - \nu_{31}\frac{\sigma_3}{E_3} \\ \varepsilon_2 &= \frac{\sigma_2}{E_2} - \nu_{12}\frac{\sigma_1}{E_1} - \nu_{32}\frac{\sigma_3}{E_3} \\ \varepsilon_3 &= \frac{\sigma_3}{E_3} - \nu_{13}\frac{\sigma_1}{E_1} - \nu_{23}\frac{\sigma_2}{E_2}. \end{aligned} \tag{1.43}$$

These equations are in line with Eq. (1.10), where the directional Poisson ratio was defined under the condition of uniaxial tension in the i-direction by

$$\nu_{ij} = -\frac{\varepsilon_j}{\varepsilon_i}. \tag{1.44}$$

Finally, we can exploit the symmetry of the compliance matrix S_{ij} to obtain the following very useful relation between the directional Young's moduli and Poisson's ratios:

$$\frac{\nu_{ij}}{E_i} = \frac{\nu_{ji}}{E_j}. \tag{1.45}$$

1.6 Determination of the Full Compliance Matrix

Using the experimental configuration and notation previously introduced in Fig. 1.6, for a specimen under tension along the x_1 direction

$$\sigma_1 = \sigma, \quad \sigma_2 = \sigma_3 = 0, \tag{1.46}$$

$$\varepsilon_1 = \frac{\Delta\ell_1}{\ell_1} = \frac{\sigma}{E_1} \quad \text{and} \quad \varepsilon_2 = \frac{\Delta b_2}{b_2} = -\nu_{12}\frac{\sigma}{E_1}. \tag{1.47}$$

Analogously, for the case of the same material under tension along the x_2 direction

$$\sigma_2 = \sigma, \quad \sigma_1 = \sigma_3 = 0, \tag{1.48}$$

$$\varepsilon_2 = \frac{\Delta\ell_2}{\ell_2} = \frac{\sigma}{E_2} \quad \text{and} \quad \varepsilon_1 = \frac{\Delta b_1}{b_1} = -\nu_{21}\frac{\sigma}{E_2}\,. \tag{1.49}$$

Examining Eqs. (1.48) and (1.49) and their symmetry properties, we notice that for the same uniaxial stress σ, the lateral strains are also the same in both loading directions

$$\frac{\Delta b_1}{b_1} = \frac{\Delta b_2}{b_2}\,. \tag{1.50}$$

Finally, using Eqs. (1.39) and (1.42), the elements of the elastic stiffness matrix can be expressed in terms of engineering constants as

$$\begin{aligned}
C_{11} &= \frac{(1-\nu_{23}\nu_{32})}{N}E_1\\
C_{22} &= \frac{(1-\nu_{13}\nu_{31})}{N}E_2\\
C_{33} &= \frac{(1-\nu_{12}\nu_{21})}{N}E_3\\
C_{12} &= \frac{\nu_{21}+\nu_{31}\nu_{23}}{N}E_1 = \frac{\nu_{12}+\nu_{32}\nu_{13}}{N}E_2,\\
C_{13} &= \frac{\nu_{31}+\nu_{21}\nu_{32}}{N}E_1 = \frac{\nu_{13}+\nu_{12}\nu_{23}}{N}E_3,\\
C_{23} &= \frac{\nu_{32}+\nu_{12}\nu_{31}}{N}E_2 = \frac{\nu_{23}+\nu_{21}\nu_{13}}{N}E_3,\\
C_{44} &= G_{23}, \quad C_{55} = G_{31}, \quad C_{66} = G_{12},
\end{aligned} \tag{1.51}$$

where N is a factor given by

$$N = 1 - \nu_{12}\nu_{21} - \nu_{23}\nu_{32} - \nu_{31}\nu_{13} - 2\nu_{21}\nu_{32}\nu_{13}. \tag{1.52}$$

A further useful relation among the directional Poisson ratios can be easily deduced by generalizing Eq. (1.45)

$$\nu_{12}\nu_{31}\nu_{23} = \nu_{32}\nu_{13}\nu_{21}. \tag{1.53}$$

1.7 Material Coordinate System Transformations

As composite materials are configured in ways to optimize their strength and stiffness in the main directions of likely loading, laminates with plies in several directions are very common. For this reason, one needs to relate elastic properties of different plies in a single coordinate system. This goal can be accomplished by a simple coordinate rotation transforming the material properties to the desired orientation, taking advantage of the fact that finite rotations form a group of orthogonal transformations.

1.7.1 Property matrices in a rotated coordinate system

Suppose a material is loaded in a direction different from any of its principal directions, as indicated in Figure 1.10. In such a case, the effective stiffness matrix C'_{ij} in the rotated (x'_1, x'_2, x'_3) coordinate system can be expressed in terms of the stiffness matrix C_{ij} in the original (*e.g.*, principal) coordinate system (x_1, x_2, x_3). For a vector F_i given in the original coordinate system, the components F'_i in the rotated coordinate system are

$$F'_i = \alpha_{ik} F_k, \tag{1.54}$$

where α_{ik} is an orthogonal tensor whose elements are composed of the direction cosines of the rotation. For a finite rotation of angle ϕ from (x_1, x_2, x_3) to (x'_1, x'_2, x'_3) about the x_3 axis,

$$\begin{aligned}
&\cos(x'_1 x_1) = \cos(x'_2 x_2) = \cos\phi \\
&\cos(x'_1 x_2) = \cos(90° - \phi) = \sin\phi \\
&\cos(x'_2 x_1) = \cos(90° + \phi) = -\sin\phi \\
&\cos(x'_1 x_3) = \cos(x'_2 x_3) = \cos(x'_3 x_1) = \cos(x'_3 x_2) = 0 \\
&\cos(x'_3 x_3) = 1,
\end{aligned} \tag{1.55}$$

where functions such as $\cos(x'_2 x_1)$ denote the cosine of the angle between the rotated x'_2 and the original x_1 axes. We can also write this orthogonal

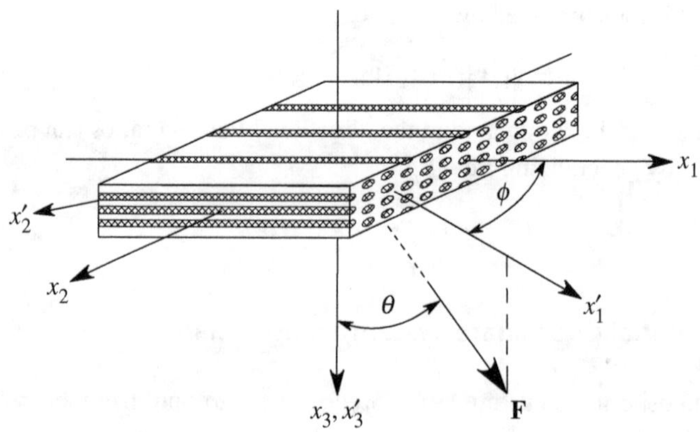

Figure 1.10. Loading of an anisotropic composite plate in a non-principal direction.

rotation tensor in matrix form as follows

$$\alpha_{ij} = \begin{bmatrix} \cos\phi & \sin\phi & 0 \\ -\sin\phi & \cos\phi & 0 \\ 0 & 0 & 1 \end{bmatrix}. \tag{1.56}$$

The transformation tensors for rotations about the x_1 and x_2 axes can be written in a similar manner.

In the case of a second-rank tensor, the rotated tensor can be obtained by rotating each projection of the tensor to be transformed. Thus, for a tensor of the form $A_{k\ell}$, the transformed quantity will be

$$A'_{ij} = \alpha_{ik}\alpha_{j\ell}A_{k\ell}. \tag{1.57}$$

The same rule applies to the stress and strain tensors as well. Similarly, the transformation of the fourth-rank elastic stiffness tensor $c_{ijk\ell}$ will be given by

$$c'_{ijk\ell} = \alpha_{im}\alpha_{jn}\alpha_{kp}\alpha_{\ell q}c_{mnpq}. \tag{1.58}$$

The complexity of this last operation and, especially, the fact that the stiffness matrix rather than the stiffness tensor is the engineering quantity of interest suggest that a simpler way should be sought to apply the transformations. Such a simplification was introduced by Bond [7] in 1943 [8, 9]. His contribution consists of realizing that partial products of the rotation tensors in Eq. (1.58) can be expressed as invariant 6×6 matrices, and the elements of these matrices can be easily tabulated. The argument proceeds by considering first the effective transformation of the stress and strain tensors individually.

For example, a finite rotation applied to the stress tensor σ_{ij} or the stress vector $\boldsymbol{\sigma}_i$ is

$$\sigma'_{ij} = \alpha_{ik}\alpha_{j\ell}\sigma_{k\ell}, \text{ where } i, j, k = 1, 2, 3 \tag{1.59}$$

$$\boldsymbol{\sigma}'_p = M_{pq}\boldsymbol{\sigma}_q, \text{ where } p, q = 1, 2, ..., 6. \tag{1.60}$$

The matrix elements in M_{ik} are all combinations of direction cosines between the original and transformed axes, because of course they are only products of the α_{ik} and $\alpha_{j\ell}$ tensors. This same logic can be applied to the strain tensor ϵ_{ij} or the strain vector $\boldsymbol{\varepsilon}_i$,

$$\epsilon'_{ij} = \alpha_{ik}\alpha_{j\ell}\epsilon_{k\ell}, \text{ where } i, j, k = 1, 2, 3 \tag{1.61}$$

$$\boldsymbol{\varepsilon}'_p = N_{pq}\boldsymbol{\varepsilon}_q, \text{ where } p, q = 1, 2, ..., 6, \tag{1.62}$$

where N_{ik} denotes the transformation matrix for strain, which is closely related to M_{ik}. The matrices M_{ik} and N_{ik} are given in the Appendix following Chapter 1.

Combining Eq. (1.60) above with the constitutive equation, we have

$$\sigma'_i = M_{ik} C_{k\ell} \varepsilon_\ell, \tag{1.63}$$

where ε_ℓ is the strain in the original system. Substituting the Bond transformation for the strain from Eq. (1.62) into Eq. (1.63) yields

$$\sigma'_i = M_{ik} C_{k\ell} N_{\ell j}^{-1} \varepsilon'_j, \tag{1.64}$$

where $N_{\ell j}^{-1}$ is the inverse matrix of $N_{\ell j}$. Equation (1.64) has the desired result, namely that the transformed stress vector σ'_i is written in terms of the transformed strain vector ε'_j. The remaining quantity in Eq. (1.64) must then be the transformed stiffness matrix C'_{ij}. It is given in terms of the original matrix and the Bond matrices as

$$C'_{ij} = M_{ik} C_{k\ell} N_{\ell j}^{-1}. \tag{1.65}$$

From a similar calculation, the transformed compliance matrix is given by

$$S'_{ij} = N_{ik} S_{k\ell} M_{ik}^{-1}. \tag{1.66}$$

We note without proof the fact that the inverse of matrix $\{N\}$ is just the transpose of matrix $\{M\}$, or $\{N\}^{-1} = \{M\}^T$, from which follows also that $\{M\}^{-1} = \{N\}^T$ [8]. Here and everywhere else later, to keep the notation consistent, when a matrix quantity is presented without subscripts, for brevity, we will designate it with curly brackets { } to indicate that it is to be used in ordered matrix multiplication and not simply as a set of indexed elements. We can now write

$$C'_{ij} = M_{ik} C_{k\ell} M_{\ell j}^T \tag{1.67}$$

and

$$S'_{ij} = N_{ik} S_{k\ell} N_{\ell j}^T. \tag{1.68}$$

1.7.2 Rotation about the x_3 axis

Let us consider the special case of rotation about a single axis. This is the case depicted in Fig. 1.10. Using Eq. (1.95) and (1.96) from the Appendix following this chapter, the Bond matrix $\{M\}$ can be written as follows

$$M_{ij} = \begin{bmatrix} \cos^2\phi & \sin^2\phi & 0 & 0 & 0 & 2\cos\phi\sin\phi \\ \sin^2\phi & \cos^2\phi & 0 & 0 & 0 & -2\cos\phi\sin\phi \\ 0 & 0 & 1 & 0 & 0 & 0 \\ 0 & 0 & 0 & \cos\phi & -\sin\phi & 0 \\ 0 & 0 & 0 & \sin\phi & \cos\phi & 0 \\ -\sin\phi\cos\phi & \sin\phi\cos\phi & 0 & 0 & 0 & \cos 2\phi \end{bmatrix}. \tag{1.69}$$

Then, the elastic stiffness matrix for an orthotropic material (as given in Eq. (1.34)) becomes

$$C'_{ij} = \begin{bmatrix} C'_{11} & C'_{12} & C'_{13} & 0 & 0 & C'_{16} \\ C'_{12} & C'_{22} & C'_{23} & 0 & 0 & C'_{26} \\ C'_{13} & C'_{23} & C'_{33} & 0 & 0 & C'_{36} \\ 0 & 0 & 0 & C'_{44} & C'_{45} & 0 \\ 0 & 0 & 0 & C'_{45} & C'_{55} & 0 \\ C'_{16} & C'_{26} & C'_{36} & 0 & 0 & C'_{66} \end{bmatrix}, \qquad (1.70)$$

where the elements C'_{ij} can be calculated from the matrix product given in Eq. (1.67). In rotated coordinate system it takes the form of Eq. (1.33).

1.8 Analysis in a Planar Geometry

As composite materials so often appear in the form of laminates composed of individual planar plies, stress analysis using a 3-D formulation appropriate for anisotropic crystalline solids is sometimes redundant. Instead, a planar formulation is often more suitable and entirely sufficient. Next, we consider constitutive laws and their transformation under the simplifying assumption of planar elasticity. Note that the 3-D formulation will be needed in the following chapters to treat wave phenomena in composites.

1.8.1 Two-dimensional stress–strain relations

Suppose a thin plate exists in a state of plane stress, as shown in Fig. 1.11. In this case, the surfaces of the plate are traction free, *i.e.*, the following three stress components vanish

$$\sigma_3 = \sigma_4 = \sigma_5 = 0. \qquad (1.71)$$

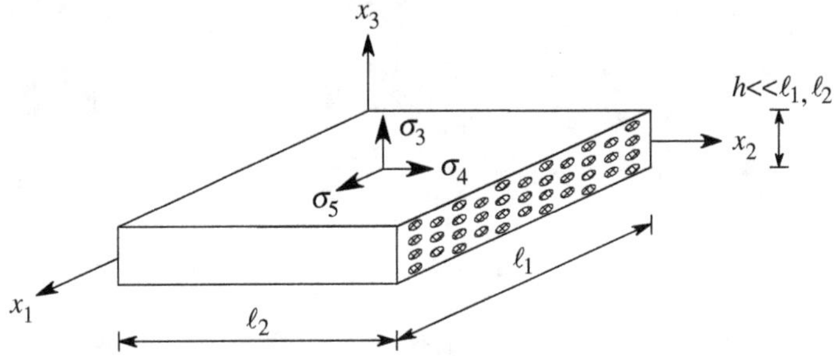

Figure 1.11. Schematic diagram of a thin composite lamina.

Hooke's law can now be expressed as

$$\begin{bmatrix} \varepsilon_1 \\ \varepsilon_2 \\ \varepsilon_3 \\ 0 \\ 0 \\ \varepsilon_6 \end{bmatrix} = \begin{bmatrix} S_{11} & S_{12} & S_{13} & 0 & 0 & 0 \\ S_{12} & S_{22} & S_{23} & 0 & 0 & 0 \\ S_{13} & S_{23} & S_{33} & 0 & 0 & 0 \\ 0 & 0 & 0 & S_{44} & 0 & 0 \\ 0 & 0 & 0 & 0 & S_{55} & 0 \\ 0 & 0 & 0 & 0 & 0 & S_{66} \end{bmatrix} \begin{bmatrix} \sigma_1 \\ \sigma_2 \\ 0 \\ 0 \\ 0 \\ \sigma_6 \end{bmatrix}. \tag{1.72}$$

According to Eqs. (1.43), ε_1, ε_2, and ε_3 in Eq. (1.72) can be written as

$$\begin{aligned} \varepsilon_1 &= \frac{\sigma_1}{E_1} - \nu_{21}\frac{\sigma_2}{E_2} \\ \varepsilon_2 &= \frac{\sigma_2}{E_2} - \nu_{12}\frac{\sigma_1}{E_1} \\ \varepsilon_3 &= -\nu_{13}\frac{\sigma_1}{E_1} - \nu_{23}\frac{\sigma_2}{E_2}. \end{aligned} \tag{1.73}$$

Assuming that the nonzero ε_3 normal strain is calculated separately if needed, Eq. (1.72) can be expressed with a 3 × 3 matrix, *i.e.*, exactly one-half the dimension of the full 3-D case. The foregoing argument also explains why the indices in the 2-D property matrices below run from 1 to 6, with 3, 4, and 5 absent. We find that

$$\begin{bmatrix} \varepsilon_1 \\ \varepsilon_2 \\ \varepsilon_6 \end{bmatrix} = \begin{bmatrix} S_{11} & S_{12} & 0 \\ S_{12} & S_{22} & 0 \\ 0 & 0 & S_{66} \end{bmatrix} \begin{bmatrix} \sigma_1 \\ \sigma_2 \\ \sigma_6 \end{bmatrix}, \tag{1.74}$$

or, after inversion to determine the three-element stress vector

$$\begin{bmatrix} \sigma_1 \\ \sigma_2 \\ \sigma_6 \end{bmatrix} = \begin{bmatrix} Q_{11} & Q_{12} & 0 \\ Q_{12} & Q_{22} & 0 \\ 0 & 0 & Q_{66} \end{bmatrix} \begin{bmatrix} \varepsilon_1 \\ \varepsilon_2 \\ \varepsilon_6 \end{bmatrix}, \tag{1.75}$$

where $Q_{ij} = S_{ij}^{-1}$. Explicit relations between $\{Q\}$ and $\{S\}$ are given by

$$\begin{aligned} Q_{11} &= \frac{S_{22}}{S_{11}S_{22} - S_{12}^2} \\ Q_{22} &= \frac{S_{11}}{S_{11}S_{22} - S_{12}^2} \\ Q_{12} &= -\frac{S_{12}}{S_{11}S_{22} - S_{12}^2} \\ Q_{66} &= \frac{1}{S_{66}}. \end{aligned} \tag{1.76}$$

In terms of the Poisson ratios, shear, and Young's moduli, these terms can be expressed as

$$Q_{11} = \frac{E_1}{1-\nu_{12}\nu_{21}}, \qquad Q_{22} = \frac{E_2}{1-\nu_{12}\nu_{21}},$$
$$Q_{66} = G_{12}, \qquad Q_{12} = \nu_{21}Q_{11} = \nu_{12}Q_{22}. \tag{1.77}$$

Writing Eq. (1.75) in terms of the elements of the elastic stiffness matrix C_{ij}, we obtain

$$\begin{aligned}\sigma_1 &= C_{11}\varepsilon_1 + C_{12}\varepsilon_2 + C_{13}\varepsilon_3\\ \sigma_2 &= C_{12}\varepsilon_1 + C_{22}\varepsilon_2 + C_{23}\varepsilon_3\\ \sigma_3 &= C_{13}\varepsilon_1 + C_{23}\varepsilon_2 + C_{33}\varepsilon_3 = 0.\end{aligned} \tag{1.78}$$

The last equation is a consequence of the plane-stress condition, and from it ε_3 can be obtained as a function of ε_1 and ε_2

$$\varepsilon_3 = -\frac{C_{13}}{C_{33}}\varepsilon_1 - \frac{C_{23}}{C_{33}}\varepsilon_2. \tag{1.79}$$

By substituting ε_3 into Eq. (1.78), the in-plane normal stresses σ_1 and σ_2 can be expressed in terms of the in-plane normal strains ε_1 and ε_2 only as follows

$$\sigma_1 = (C_{11} - \frac{C_{13}^2}{C_{33}})\varepsilon_1 + (C_{12} - \frac{C_{13}C_{23}}{C_{33}})\varepsilon_2 \tag{1.80}$$

and

$$\sigma_2 = (C_{12} - \frac{C_{13}C_{23}}{C_{33}})\varepsilon_1 + (C_{22} - \frac{C_{23}^2}{C_{33}})\varepsilon_2. \tag{1.81}$$

Hence, the 2-D stiffnesses can be written in terms of the full elastic stiffness matrix elements C_{ij},

$$Q_{11} = C_{11} - \frac{C_{13}^2}{C_{33}}, \quad Q_{22} = C_{22} - \frac{C_{23}^2}{C_{33}}$$
$$Q_{12} = Q_{21} = C_{12} - \frac{C_{13}C_{23}}{C_{33}}. \tag{1.82}$$

For isotropic materials, $Q_{11} = Q_{22} = E/(1-\nu^2)$, which leads us to conclude that $E < Q_{11} < C_{11}$, which is of course is a simple consequence of the Poisson effect.

1.8.2 Rotation about the principal axis

We may apply a rotation through an angle ϕ about the principal axis x_3, as we did for the full 3-D case with $\{S\}$ or $\{C\}$. Making use of all the foregoing

3-D analysis in the planar geometry, this transformation reduces to

$$\begin{bmatrix} \sigma'_1 \\ \sigma'_2 \\ \sigma'_6 \end{bmatrix} = \{T\}\{Q\}\{T\}^T \begin{bmatrix} \varepsilon'_1 \\ \varepsilon'_2 \\ \varepsilon'_6 \end{bmatrix}, \tag{1.83}$$

where $\{T\}$ is the 2-D analog of the Bond rotation matrix. This matrix can be obtained from Eq. (1.69) by eliminating the third, fourth, and fifth rows and columns of the matrix that are not used. As the corresponding stress components are identically zero, the elements in those rows and columns are not needed; instead, we have

$$\{T\} = \begin{bmatrix} \cos^2\phi & \sin^2\phi & 2\cos\phi\sin\phi \\ \sin^2\phi & \cos^2\phi & -2\cos\phi\sin\phi \\ -\sin\phi\cos\phi & \sin\phi\cos\phi & \cos^2\phi - \sin^2\phi \end{bmatrix}. \tag{1.84}$$

By analogy with the 3-D analysis for the stiffness and compliance matrices, we conclude that

$$\{Q'\} = \{T\}\{Q\}\{T\}^T \tag{1.85}$$

is the transformed 2-D matrix of $\{Q\}$ after a finite rotation. The elements of the $\{Q'\}$ matrix are given by

$$\begin{aligned}
Q'_{11} &= Q_{11}\cos^4\phi + Q_{22}\sin^4\phi + 2(Q_{12}+2Q_{66})\sin^2\phi\cos^2\phi \\
Q'_{22} &= Q_{11}\sin^4\phi + Q_{22}\cos^4\phi + 2(Q_{12}+2Q_{66})\cos^2\phi\sin^2\phi \\
Q'_{12} &= Q'_{21} = (Q_{11}+Q_{22}-4Q_{66})\sin^2\phi\cos^2\phi + Q_{12}(\sin^4\phi+\cos^4\phi) \\
Q'_{16} &= Q'_{61} = (Q_{12}+2Q_{66}-Q_{11})\cos^3\phi\sin\phi \\
&\quad -(Q_{12}+2Q_{66}-Q_{22})\sin^3\phi\cos\phi \\
Q'_{26} &= Q'_{62} = (Q_{12}-2Q_{66}-Q_{11})\sin^3\phi\cos\phi \\
&\quad -(Q_{12}-2Q_{66}-Q_{22})\cos^3\phi\sin\phi \\
Q'_{66} &= (Q_{11}+Q_{22}-2Q_{12})\sin^2\phi\cos^2\phi + Q_{66}(\cos^2\phi-\sin^2\phi)^2.
\end{aligned} \tag{1.86}$$

In terms of the engineering constants, we have

$$\begin{aligned}
\frac{1}{E'_1} &= \frac{1}{E_1}\cos^4\phi + \left(\frac{1}{G_{12}} - \frac{2\nu_{12}}{E_1}\right)\sin^2\phi\cos^2\phi + \frac{1}{E_2}\sin^4\phi \\
\frac{1}{E'_2} &= \frac{1}{E_2}\cos^4\phi + \left(\frac{1}{G_{12}} - \frac{2\nu_{12}}{E_1}\right)\sin^2\phi\cos^2\phi + \frac{1}{E_1}\sin^4\phi
\end{aligned} \tag{1.87}$$

$$\frac{1}{G'_{12}}=2\left(\frac{2}{E_1}+\frac{1}{E_2}+\frac{4\nu'_{12}}{E_1}-\frac{1}{G_{12}}\right)\sin^2\phi\cos^2\phi+\frac{1}{G_{12}}\left(\sin^4\phi+\cos^4\phi\right)$$

$$\nu'_{12}=\nu_{12}(\sin^4\phi+\cos^4\phi)-\left(1+\frac{E_1}{E_2}-\frac{E_1}{G_{12}}\right)\sin^2\phi\cos^2\phi,$$

where primed quantities on the left-hand side refer to the new rotated coordinate system.

There is also a set of quantities that is invariant under a rotation about the x_3 axis; these quantities are [2, 3]

$$\begin{aligned}
U_1&=\frac{1}{8}(3Q_{11}+3Q_{22}+2Q_{12}+4Q_{66})\\
U_2&=\frac{1}{2}(Q_{11}-Q_{22})\\
U_3&=\frac{1}{8}(Q_{11}+Q_{22}-2Q_{12}-4Q_{66})\\
U_4&=\frac{1}{8}(Q_{11}+Q_{22}+6Q_{12}-4Q_{66})\\
U_5&=\frac{1}{8}(Q_{11}+Q_{22}-2Q_{12}+4Q_{66}).
\end{aligned}\tag{1.88}$$

These quantities are useful in verifying whether the elements of Q'_{ij}, calculated in the rotated system, are correct. As they are invariant under finite rotation, their values should be identical in both coordinate systems. Equations (1.86) can be rewritten in terms of the rotation invariants given by Eqs. (1.88) as follows:

$$\begin{aligned}
Q'_{11}&=U_1+U_2\cos 2\phi+U_3\cos 4\phi\\
Q'_{22}&=U_1-U_2\cos 2\phi+U_3\cos 4\phi\\
Q'_{12}&=U_4-U_3\cos 4\phi\\
Q'_{66}&=U_5-U_3\cos 4\phi\\
Q'_{16}&=-\frac{1}{2}U_2\sin 2\phi-U_3\sin 4\phi\\
Q'_{26}&=-\frac{1}{2}U_2\sin 2\phi+U_3\sin 4\phi.
\end{aligned}\tag{1.89}$$

1.8.3 Boron-epoxy composite: An example

A unidirectional boron-epoxy composite lamina has the following elastic properties [3]:

$$E_1=210.0\ \text{GPa},\quad E_2=21.0\ \text{GPa},$$
$$G_{12}=7.0\ \text{GPa},\quad \nu_{12}=0.3.$$

We can obtain the values of Q_{ij} from Eqs. (1.77). For a specified rotation angle, Q'_{ij} in the rotated coordinate system can be obtained either from Eq. (1.86) or from Eq. (1.89). Figure 1.12 shows the dependence of the "normal" stiffness coefficients (Q'_{11}, Q'_{22}, and Q'_{12}) in a thin unidirectional boron-epoxy composite lamina on the rotation angle ϕ. The curves for Q'_{11} and Q'_{22} are the inverse of each other since through rotation of 90°, Q'_{11} takes the place of Q_{22}, and vice versa. Figure 1.13 shows the angular dependence of the "shear" stiffness coefficients (Q'_{16}, Q'_{26}, and Q'_{66}) in the

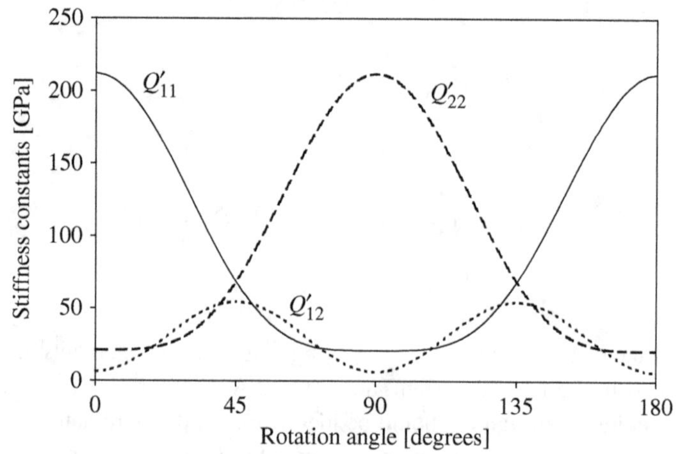

Figure 1.12. Angular dependence of the normal stiffness coefficients in a thin unidirectional boron-epoxy composite lamina.

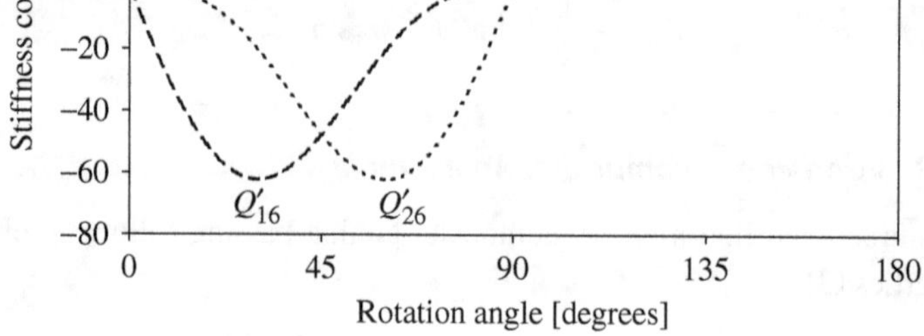

Figure 1.13. Angular dependence of the shear stiffness coefficients in a thin unidirectional boron-epoxy composite lamina.

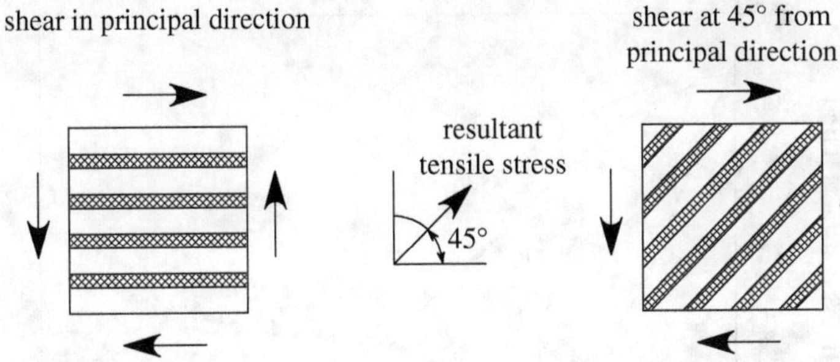

Figure 1.14. Shear loading of a unidirectional composite lamina in the principal material direction and 45° from it.

same lamina. We note that the maxima of Q'_{66} are reached at 45° and 135°. As is shown in Fig. 1.14, when the composite is rotated through these angles, the resultant tensile force from the shear load is applied in the fiber direction; therefore, in this orientation, composite lamina seems to be very stiff. Figures 1.12 and 1.13 demonstrate well the physical meaning of the rotation invariants U_4 and U_5. As one can also see from Eq. (1.89), these invariants are the average values of Q'_{12} and Q'_{66}, respectively, during the rotation. It is also important to notice that Q'_{12} and Q'_{26} are both zero at $\phi=0°$ and 90°. This is best explained by noting that in the principal directions, the shear strain components are decoupled from the normal stresses.

As an additional note on the relationship between the relevant elastic stiffnesses of the same material in different configurations, let us consider the axial stiffness, *i.e.*, σ'_1/ε'_1, of the above-introduced boron-epoxy composite along the x'_1 axis of the rotated x'_1, x'_2, x'_3 coordinate system so that x'_1 subtends an angle of ϕ with respect to the fiber orientation. Assuming that the lateral strains vanish ($\varepsilon'_2=\varepsilon'_3=0$), the axial stiffness is simply C'_{11}. This condition can be reached in a finite-size specimen only by enforcing the restriction on the lateral strain via appropriate boundary conditions. However, we will see in the next chapter that these conditions are inherently satisfied in the case of longitudinal wave propagation in the medium at ultrasonic frequencies; therefore, the relevant axial stiffness is C'_{11}. In the case of plane stress ($\sigma'_3=\sigma'_4=\sigma'_5=0$), we have shown earlier that the relevant axial stiffness is Q'_{11}, and that due to the Poisson effect, $Q'_{11}<C'_{11}$. This condition can be reached in a finite-size specimen also only by enforcing the restriction on the lateral in-plane strain via appropriate boundary conditions. However, we will also see later that such conditions occur in the case of longitudinal wave propagation in a thin plate at ultrasonic frequencies when the thickness of the plate is negligible with respect to the acoustic wavelength, at which point the relevant axial stiffness is equal to Q'_{11}. Finally, in the case of uniaxial tension of a bar ($\sigma'_2=\sigma'_3=\sigma'_4=\sigma'_5=\sigma'_6=0$) the

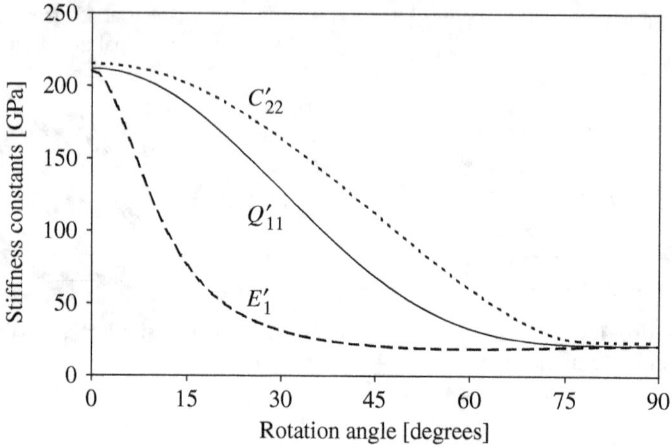

Figure 1.15. Comparison between C'_{11}, Q'_{11}, and E'_1 in a uniaxial fiber reinforced boron-epoxy composite as a function of rotation angle ϕ.

relevant axial stiffness is simply Young's modulus E'_1, which is even lower than Q'_{11}.

Figure 1.15 shows the comparison between C'_{11}, Q'_{11}, and E'_1 in a uniaxial fiber reinforced boron-epoxy composite as a function of rotation angle ϕ. C'_{11} was calculated by the Bond transformation as given in Eqs. (1.67) and (1.69). Q'_{11} was calculated from Eq. (1.86) and was already plotted in Fig. 1.12, while Young's modulus E'_1 was calculated from Eq. (1.87). As expected, these three axial stiffness parameters exhibit similar angular dependence, although their values are significantly different and always satisfy the condition that $C'_{11} > Q'_{11} > E'_1$.

1.9 Experimental Determination of Stiffness

The geometry of composite samples used in experiments to determine their stiffness is critically important to achieving accurate and reproducible results [4]. Unlike the typical "dog bone" shaped samples used in tensile tests of other structural materials, such as metals, composite samples are fabricated to be rectangular in shape with a thickness reinforcement (attached with adhesive) in the region where the sample is held by the grips of the tensile machine, as illustrated in Fig. 1.16. This sample configuration avoids weak points in the composite arising from the presence of broken fibers, thereby enhancing the accuracy of the measurement. Perhaps the most difficult aspect of determining the stiffness of composites is the measurement of the shear modulus. One method is to measure Young's modulus in the sample with fibers oriented at 45° and use the equations of Eq. (1.87) to find

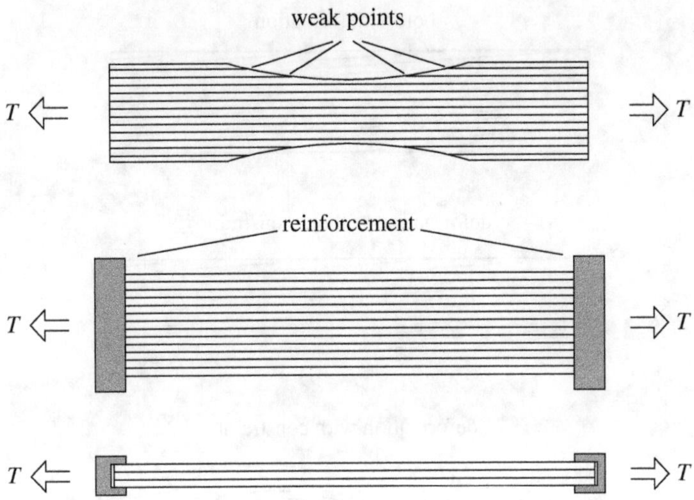

Figure 1.16. Sample geometry for tension testing of unidirectional composite specimens.

the value of G_{12}. Here is how this method works

$$\frac{1}{E_{45^\circ}}=\frac{1}{4}\left(\frac{1}{E_1}+\frac{1}{E_2}+\frac{1}{G_{12}}-\frac{2\nu_{12}}{E_1}\right) \text{ or}$$

$$G_{12}=\left(\frac{4}{E_{45^\circ}}-\frac{1}{E_1}-\frac{1}{E_2}+\frac{2\nu_{12}}{E_1}\right)^{-1}. \qquad (1.90)$$

For isotropic materials, this method yields the well-known relationship between Young's modulus, Poisson's ratio, and the shear modulus

$$G=\left(\frac{4}{E}-\frac{1}{E}-\frac{1}{E}+\frac{2\nu}{E}\right)^{-1}=\frac{E}{2(1+\nu)}. \qquad (1.91)$$

When unidirectional tension is applied in a direction different from $\phi=0^\circ$ or $\phi=90^\circ$, the sample will deform in a way that is mainly determined by the constraint introduced by the tensile machine, as illustrated in Fig. 1.17. Owing to this constraint on specimens loaded at angles other than 0°or 90°with respect to the fiber orientation, the quantity measured is not actually the Young's modulus E'. Young's modulus is defined for a prismatic bar that experiences axial loading only, whereas the sample used in the tensile test is a plate of finite width that also experiences in-plane bending and shear depending on the constraints of the grips. Therefore, the quantity measured in a real tensile test is an effective Young's modulus $E^{*\prime}$. For loading at approximately parallel ($\phi\approx0^\circ$) or normal ($\phi\approx90^\circ$) to the fiber orientation, the difference between the two moduli, E' and $E^{*\prime}$, is relatively small. At other angles, however, the difference might be significant.

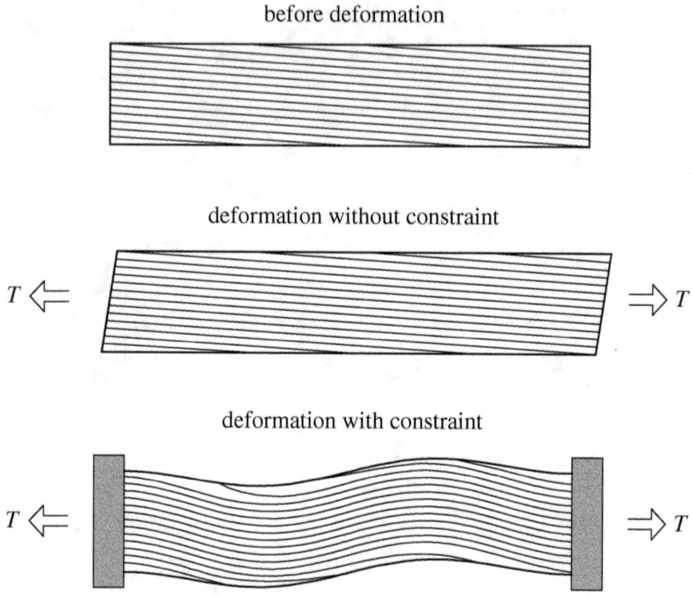

Figure 1.17. Deformation of a composite specimen loaded in a direction different from the fiber orientation.

Empirical evidence suggests that Young's modulus can be estimated from the effective stiffness determined from such tensile tests using the following equation [4]:

$$E' = E^{*\prime}(1-\eta), \tag{1.92}$$

where E' is the value of the true Young's modulus, $E^{*\prime}$ is the measured effective modulus, and η is a correction factor given by

$$\eta = \frac{3(S'_{16})^2/S'_{11}}{3S'_{66}+2S'_{11}(\ell/b)^2}, \tag{1.93}$$

where ℓ is the distance between the grips on the load frame, and b is the width of the sample. From Eq. (1.93), it is clear that the the measured effective modulus is always higher than the true Young's modulus

$$E^{*\prime} > E'. \tag{1.94}$$

This result is a consequence of the constraint introduced by the tensile machine, which makes it more difficult to deform the sample, thereby increasing its apparent stiffness. Figure 1.18 shows the calculated effective modulus E^* of a uniaxial fiber reinforced boron-epoxy composite plate as a function of rotation angle ϕ for different length-to-width ℓ/b ratios. Based on Eq. (1.83), the the rotated compliance matrix $\{S'\}$ can be obtained as the

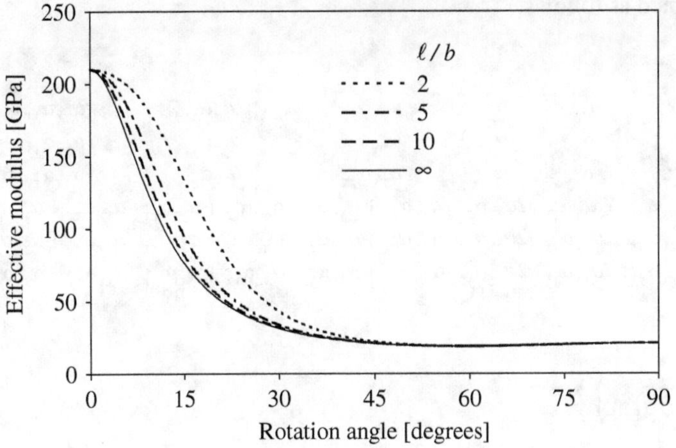

Figure 1.18. Effective modulus $E^{*\prime}$ of a uniaxial fiber reinforced boron-epoxy composite plate as a function of rotation angle ϕ for different length-to-width ℓ/b ratios.

inverse of the rotated stiffness matrix $\{Q'\}$, where the elements of $\{Q'\}$ were listed in Eq. (1.86). As expected, for slender specimens, *i.e.*, for values of $\ell/b > 10$, the correction factor η becomes negligible.

Appendix

The Bond matrices $\{M\}$ and $\{N\}$ for stiffness and compliance matrix transformations are given below. $\{M\}$ is derived by examining the elements of the tensor multiplication in Eq. (1.59) and comparing them with the matrix elements in Eq. (1.60). Here, we use the (x,y,z) notation to designate the coordinate axes.

$$\{M\} =$$

$$\begin{bmatrix} \alpha_{xx}^2 & \alpha_{xy}^2 & \alpha_{xz}^2 & 2\alpha_{xy}\alpha_{xz} & 2\alpha_{xz}\alpha_{xx} & 2\alpha_{xx}\alpha_{xy} \\ \alpha_{yx}^2 & \alpha_{yy}^2 & \alpha_{yz}^2 & 2\alpha_{yy}\alpha_{yz} & 2\alpha_{yz}\alpha_{yx} & 2\alpha_{yx}\alpha_{yy} \\ \alpha_{zx}^2 & \alpha_{zy}^2 & \alpha_{zz}^2 & 2\alpha_{zy}\alpha_{zz} & 2\alpha_{zz}\alpha_{zx} & 2\alpha_{zx}\alpha_{zy} \\ \alpha_{yx}\alpha_{zx} & \alpha_{yy}\alpha_{zy} & \alpha_{yz}\alpha_{zz} & \alpha_{yy}\alpha_{zz}+\alpha_{yz}\alpha_{zy} & \alpha_{yx}\alpha_{zz}+\alpha_{yz}\alpha_{zx} & \alpha_{yy}\alpha_{zx}+\alpha_{yx}\alpha_{zy} \\ \alpha_{zx}\alpha_{xx} & \alpha_{zy}\alpha_{xy} & \alpha_{zz}\alpha_{xz} & \alpha_{xy}\alpha_{zz}+\alpha_{xz}\alpha_{zy} & \alpha_{xz}\alpha_{zx}+\alpha_{xx}\alpha_{zz} & \alpha_{xx}\alpha_{zy}+\alpha_{xy}\alpha_{zx} \\ \alpha_{xx}\alpha_{yx} & \alpha_{xy}\alpha_{yy} & \alpha_{xz}\alpha_{yz} & \alpha_{xy}\alpha_{yz}+\alpha_{xz}\alpha_{yy} & \alpha_{xz}\alpha_{yx}+\alpha_{xx}\alpha_{yz} & \alpha_{xx}\alpha_{yy}+\alpha_{xy}\alpha_{yx} \end{bmatrix}, \tag{1.95}$$

By a similar process using the strain transformation in Eqs. (1.61) and (1.62), the Bond matrix $\{N\}$ for transforming compliance matrices can be obtained as follows

$$\{N\}=$$

$$\begin{bmatrix} \alpha_{xx}^2 & \alpha_{xy}^2 & \alpha_{xz}^2 & \alpha_{xy}\alpha_{xz} & \alpha_{xz}\alpha_{xx} & \alpha_{xx}\alpha_{xy} \\ \alpha_{yx}^2 & \alpha_{yy}^2 & \alpha_{yz}^2 & \alpha_{yy}\alpha_{yz} & \alpha_{yz}\alpha_{yx} & \alpha_{yx}\alpha_{yy} \\ \alpha_{zx}^2 & \alpha_{zy}^2 & \alpha_{zz}^2 & \alpha_{zy}\alpha_{zz} & \alpha_{zz}\alpha_{zx} & \alpha_{zx}\alpha_{zy} \\ 2\alpha_{yx}\alpha_{zx} & 2\alpha_{yy}\alpha_{zy} & 2\alpha_{yz}\alpha_{zz} & \alpha_{yy}\alpha_{zz}+\alpha_{yz}\alpha_{zy} & \alpha_{yx}\alpha_{zz}+\alpha_{yz}\alpha_{zx} & \alpha_{yy}\alpha_{zx}+\alpha_{yx}\alpha_{zy} \\ 2\alpha_{zx}\alpha_{xx} & 2\alpha_{zy}\alpha_{xy} & 2\alpha_{zz}\alpha_{xz} & \alpha_{xy}\alpha_{zz}+\alpha_{xz}\alpha_{zy} & \alpha_{xz}\alpha_{zx}+\alpha_{xx}\alpha_{zz} & \alpha_{xx}\alpha_{zy}+\alpha_{xy}\alpha_{zx} \\ 2\alpha_{xx}\alpha_{yx} & 2\alpha_{xy}\alpha_{yy} & 2\alpha_{xz}\alpha_{yz} & \alpha_{xy}\alpha_{yz}+\alpha_{xz}\alpha_{yy} & \alpha_{xz}\alpha_{yx}+\alpha_{xx}\alpha_{yz} & \alpha_{xx}\alpha_{yy}+\alpha_{xy}\alpha_{yx} \end{bmatrix}. \tag{1.96}$$

Bibliography

1. S. P. Timoshenko and J. N. Goodier, *Theory of Elasticity*, 3rd ed (McGraw-Hill Co, New York, 1970).
2. H. T. Hahn and S. W. Tsai, *Introduction to Composite Materials* (CRC Press, New York, 1980).
3. R. M. Jones, *Mechanics of Composite Materials*, 3rd ed (CRC Press, New York, 1998).
4. I. M. Daniel and O. Ishai, *Engineering Mechanics of Composite Materials* (Oxford University Press, New York, 1994).
5. T. J. Chung, *Continuum Mechanics* (Prentice Hall, Englewood Cliffs, 1988).
6. D. Frederick and T. S. Chang, *Continuum Mechanics* (Allyn and Bacon, Boston, 1965).
7. W. Bond, "The mathematics of the physical properties of crystals," Brit. Sci. Tech. J. **22**, 1–72 (1943).
8. B. A. Auld, *Acoustic Fields and Waves in Solids*, 2nd ed (Krieger Publishing, Malabar, 1990).
9. J. F. Nye, *Physical Properties of Crystals: Their Representation by Tensors and Matrices* (Oxford University Press, New York, 1985).

2

Elastic Waves in Anisotropic Media

In this chapter, we provide a brief introduction to ultrasonic wave propagation in unbounded anisotropic solids with emphases on examples suitable for ultrasonics of composites. Many excellent books are relevant to the subject addressed in this chapter [1–13]. Some of them broadly discuss elastic waves in anisotropic solids [5, 7], [9–13] including waves in layered anisotropic media [7]. In-depth theoretical description of elastic waves in anisotropic media is given in classical texts [12, 13], which have influenced and provided guidance to our treatment of some aspects of the theory. Beautiful visualization of ultrasonic waves in crystals (often obtained by laser excitation) is given in reference [11].

2.1 Equations of Motion

2.1.1 Introduction

The equations of motion for the vibration of an elastic medium are extensions of Newton's second law for particles. Treating the elastic continuum as a collection of particles, each of which is assumed to obey Newton's laws, leads to a particularly straightforward argument. We begin by considering a short segment of a bar with length Δx and cross-sectional area S_0 as is illustrated in Fig. 2.1. The material is assumed to be linear and elastic, and its deformations can be described by constitutive equations derived in the previous chapter. For simplicity, we assume only uniaxial stress in the x-direction of the continuum. The normal strain ϵ in the bar is the result of different displacements at its two ends

$$\epsilon = \frac{\Delta u}{\Delta x}. \tag{2.1}$$

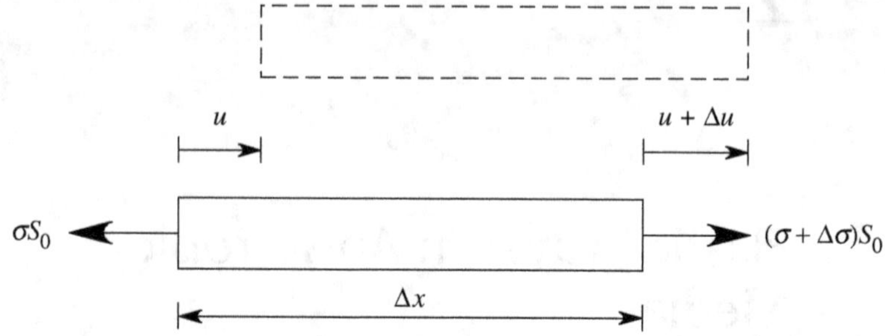

Figure 2.1. Segment of bar under stress.

The normal stress necessary to produce this strain can be calculated from

$$\sigma = E\epsilon, \tag{2.2}$$

where E denotes the stiffness of the bar. The resultant force acting on the segment is

$$\Delta\sigma S_0, \tag{2.3}$$

where σ represents the normal stress on the cross-section. According to D'Alembert's principle, the inertial force in the segment is given by

$$-\rho\Delta x S_0\ddot{u}, \tag{2.4}$$

where ρ is the mass density of the bar, u is the particle displacement in Lagrangian coordinates, and $\ddot{u}$ is the corresponding particle acceleration. From Newton's second law, the equation of motion of the assembly of mass particles (*i.e.*, the continuum) can be obtained by combining Eqs. (2.3) and (2.4)

$$\frac{\Delta\sigma}{\Delta x} - \rho\ddot{u} = 0. \tag{2.5}$$

Finally, combining Eqs. (2.1), (2.2), and (2.5) and taking the $\Delta x \to 0$ limit, leads to the 1-D wave equation in terms of the displacement

$$E\frac{\partial^2 u}{\partial x^2} = \rho\ddot{u}. \tag{2.6}$$

The equation of motion and the wave equation have been obtained under the assumption of 1-D stress in a bar, but the equation shows all the important aspects and dependencies that will appear in the 3-D version. The generalization of this method to the equations of motion of a 3-D elastic solid is shown in the next section.

2.1.2 Differential approach

The static equilibrium of an arbitrary, 3-D solid body occupying volume V that has both surface and body forces requires that the resultant force vector vanishes,

$$\int_{\text{surf}} t_i \, dS + \int_{\text{body}} \rho b_i \, dV = 0, \tag{2.7}$$

where the first integral is evaluated on the surface of the body, and the second over its volume. Here, t_i is the traction vector, given by $t_i = \sigma_{ij} n_j$, and n_j is an outward unit normal on the body surface. The term b_i is the body force per unit mass. Substituting for the traction, and transforming the surface integral to a volume integral by Gauss' theorem yields

$$\int_{\text{body}} \frac{\partial \sigma_{ij}}{\partial x_j} \, dV + \int_{\text{body}} \rho b_i \, dV = 0. \tag{2.8}$$

As the volume V is entirely arbitrary, the integrand itself must vanish, leaving

$$\frac{\partial \sigma_{ij}}{\partial x_j} + \rho b_i = 0, \tag{2.9}$$

and the result is sometimes referred to as *Euler's* first relation. This is a static result, however, and to describe the dynamical motion of a continuum, an inertial term accounting for the particle acceleration must be included, as we saw earlier,

$$\int_{\text{surf}} t_i \, dS + \int_{\text{body}} \rho b_i \, dV - \int_{\text{body}} \rho \ddot{u}_i \, dV = 0, \tag{2.10}$$

where the last term accounts for the inertia effect. The calculation proceeds in a manner identical to the foregoing to yield

$$\frac{\partial \sigma_{ij}}{\partial x_j} + \rho b_i = \rho \ddot{u}_i, \tag{2.11}$$

also called *Cauchy's* equation. If we now allow that the body force b_i vanishes for free, unforced motion, we have

$$\frac{\partial \sigma_{ij}}{\partial x_j} = \rho \ddot{u}_i, \tag{2.12}$$

and this result leads us at once to a dynamical equation of equilibrium for the elastic continuum. For the stress tensor σ_{ij}, substitute its equivalent from the constitutive equations developed in the last chapter,

$$c_{ijk\ell} \frac{\partial \epsilon_{k\ell}}{\partial x_j} = \rho \ddot{u}_i, \tag{2.13}$$

and now insert for the strain $\epsilon_{k\ell}$, its linearized equivalent in terms of the displacement gradient,

$$\epsilon_{k\ell} = \frac{1}{2}\left[\frac{\partial u_k}{\partial x_\ell} + \frac{\partial u_\ell}{\partial x_k}\right]. \tag{2.14}$$

The result is

$$\frac{1}{2}c_{ijk\ell}\frac{\partial}{\partial x_j}\left[\frac{\partial u_k}{\partial x_\ell} + \frac{\partial u_\ell}{\partial x_k}\right] = \rho \ddot{u}_i, \tag{2.15}$$

where evaluation of the vector on the left-hand side needs some thought. The sum over the partial derivatives with subscripts k and ℓ and elements of the stiffness tensor $c_{ijk\ell}$ leads to pairs of identical terms because of the symmetry in the stiffness tensor when exchanging k and ℓ. The result is that the pairs of identical terms sum and cancel the factor of 1/2, leaving

$$c_{ijk\ell}\frac{\partial^2 u_k}{\partial x_\ell \partial x_j} = \rho \ddot{u}_i. \tag{2.16}$$

Equation (2.16) is the equation of motion for an infinite, generally anisotropic linear elastic medium in 3-D.

2.1.3 Integral principle

While economical and quick, the route we have just taken to Eq. (2.16) is not the only way to obtain this equation. Earlier, we have worked from Newton's second law, a differential principle, to obtain the equation of motion of the linear elastic continuum in the absence of forced motion. There is another way to derive this result that proceeds from an integral principle using Hamilton's principle. Its advantages are:

- The system is considered as a whole
- There are no separate forces; instead a single function is sought
- All equations of motion proceed from one unified principle, independent of reference frame and using generalized coordinates.

Our treatment here is neither rigorous nor exhaustive, but rather intended simply as an illustration of the principles involved. For more detailed and rigorous expositions of this subject, there are several excellent sources [1–3].

First, we consider a mechanical continuum characterized by a particular function $\mathcal{L}$, called the Lagrangian density, which is dependent on a "generalized coordinate" q_i and its time and space derivatives $\partial q_i/\partial t$ and $\partial q_i/\partial x_j$. We abbreviate the time derivative as $\dot{q}_i$ and the space derivative as $q_{i,j}$, for the sake of readability in the equations. Then, $\mathcal{L} = \mathcal{L}(q_i, \dot{q}_i, q_{i,j})$. At two particular times t_1 and t_2, the system will be in two particular states

$q_i(t_1)$ and $q_i(t_2)$ corresponding to the values of the generalized coordinates at those times. Let us define a new function S given by

$$S = \int_{t_1}^{t_2} \int_V \mathcal{L}(q_i, \dot{q}_i, q_{i,j})\, dx_j\, dt \tag{2.17}$$

called the *Action*. The spatial integration, necessary to yield a scalar function, extends over coordinates corresponding to the system position at t_1 and t_2. Then, as the mechanical system evolves from time t_1 to t_2, it does so in a way such that function S is minimized. This is the statement of Hamilton's principle, also called the Principle of Least Action. This principle and the function $\mathcal{L}$ replace Newton's second law and the summation of forces.

Imagine that we have a set of generalized coordinates as a function of time $q_i(t)$ that do satisfy Hamilton's principle and make S a minimum. Then, any change in these coordinate values between t_1 to t_2 will lead to an *increase* in S. We may think of the mechanical system and the function S evolving in a 9-D space composed of the three vector variables on which $\mathcal{L}$ depends. If there is more than a single generalized coordinate q, these add to the dimension of the space. One way to increase S is to add a small change or *variation* to q_i,

$$q_i + \delta q_i, \tag{2.18}$$

where δ is an operator resulting in a virtual change of the coordinate. The variational operator δ is identical to the conventional infinitesimal differential d with the following exceptions: 1) δ satisfies all boundary conditions on the virtual change and 2) the variational differential occurs in a span of time equal to zero, not in an infinitesimal time dt.

The first condition above requires that the variation vanish at the endpoints $\delta q_i(t_1) = \delta q_i(t_2) \to 0$. In variational language, the principle states that $\delta S \to 0$. Consider that δS is the result of replacing the q_i and related variables in $[t_1, t_2]$ by their variationally displaced counterparts, $q_i + \delta q_i$, and so on. Therefore,

$$\delta S = S(q_i + \delta q_i, \dot{q}_i + \delta \dot{q}_i, q_{i,j} + \delta q_{i,j}) - S(q_i, \dot{q}_i, q_{i,j}). \tag{2.19}$$

Expanding in a Taylor's series while keeping only first-order terms, we have

$$\begin{aligned} \delta S &= S(q_i, \dot{q}_i, q_{i,j}) + \frac{\partial S}{\partial q_i}\delta q_i + \frac{\partial S}{\partial \dot{q}_i}\delta \dot{q}_i + \frac{\partial S}{\partial q_{i,j}}\delta q_{i,j} - S(q_i, \dot{q}_i, q_{i,j}) \\ &= \frac{\partial S}{\partial q_i}\delta q_i + \frac{\partial S}{\partial \dot{q}_i}\delta \dot{q}_i + \frac{\partial S}{\partial q_{i,j}}\delta q_{i,j}. \end{aligned} \tag{2.20}$$

Applying this result to the integral expression for S we obtain

$$\delta S = \int_{t_1}^{t_2}\int_V \left[\frac{\partial \mathcal{L}}{\partial q_i}\delta q_i + \frac{\partial \mathcal{L}}{\partial \dot{q}_i}\delta \dot{q}_i + \frac{\partial \mathcal{L}}{\partial q_{i,j}}\delta q_{i,j}\right] dx_j\, dt. \tag{2.21}$$

The variational and conventional derivatives can be interchanged [2],

$$\delta \dot{q}_i = \delta \frac{dq_i}{dt} = \frac{d\delta q_i}{dt} \tag{2.22}$$

$$\delta q_{i,j} = \delta \frac{dq_i}{dx_j} = \frac{d\delta q_i}{dx_j}, \tag{2.23}$$

and we now proceed to analyze Eq. (2.21) term by term.

In the second term of Eq. (2.21), we find the presence of an exact differential $d(\delta q_i)$, after interchanging the order of differentiation (see Eq. (2.22)) and canceling the dt terms. Therefore, this term in the integral of Eq. (2.21) can be handled with a parts integration,

$$\int_1^2 \underbrace{\frac{\partial \mathcal{L}}{\partial \dot{q}_i}}_{u} \underbrace{\frac{d(\delta q_i)}{dt} dt}_{dv} = \left.\frac{\partial \mathcal{L}}{\partial \dot{q}_i}\delta q_i\right|_1^2 - \int_1^2 \frac{d}{dt}\left(\frac{\partial \mathcal{L}}{\partial \dot{q}_i}\right)\delta q_i dt. \tag{2.24}$$

By a similar manipulation, the third term in the spatial derivative can be handled the same way, but the operations are a bit more delicate,

$$\int_1^2 \underbrace{\frac{\partial \mathcal{L}}{\partial q_{i,j}}}_{u} \underbrace{\frac{d(\delta q_i)}{dx_j} dx_j}_{dv} = \left.\frac{\partial \mathcal{L}}{\partial q_{i,j}}\delta q_i\right|_1^2 - \int_1^2 \frac{d}{dx_j}\left(\frac{\partial \mathcal{L}}{\partial q_{i,j}}\right)\delta q_i dx_j, \tag{2.25}$$

where we now invoke the volume integration element dx_j instead of the time dt. In both Eqs. (2.24) and (2.25), the first term on the right must vanish according to the boundary conditions on δq_i. Substituting these results into Eq. (2.21) yields

$$\delta S = \int_{t_1}^{t_2}\int_V \left[\frac{\partial \mathcal{L}}{\partial q_i} - \frac{d}{dt}\left(\frac{\partial \mathcal{L}}{\partial \dot{q}_i}\right) - \frac{d}{dx_j}\left(\frac{\partial \mathcal{L}}{\partial q_{i,j}}\right)\right]\delta q_i\, dx_j\, dt \to 0, \tag{2.26}$$

collecting the common term δq_i outside the brackets. In order for the integral to vanish for an arbitrary t_1 and t_2 and arbitrary variations δq_i in the generalized coordinates, we must require the integrand itself to vanish,

$$\frac{\partial \mathcal{L}}{\partial q_i} - \frac{d}{dt}\left(\frac{\partial \mathcal{L}}{\partial \dot{q}_i}\right) - \frac{d}{dx_j}\left(\frac{\partial \mathcal{L}}{\partial q_{i,j}}\right) = 0. \tag{2.27}$$

This vector equation for $i = 1, 2, 3$ is called the *Euler–Lagrange* equation for continua. In the third term, there is an implied sum over j.

It now only remains to express the Lagrangian density function $\mathcal{L}$ in terms of the kinetic $\mathcal{T}$ and elastic strain $\mathcal{U}$ energy densities and to identify the generalized coordinates q_i with the particle displacement u_i,

$$\mathcal{L} = \mathcal{T} - \mathcal{U} \tag{2.28}$$

$$= \frac{1}{2}\rho \dot{u}_i \dot{u}_i - \frac{1}{2}\sigma_{ij}\epsilon_{ij} \tag{2.29}$$

$$= \frac{1}{2}\rho \dot{u}^2 - \frac{1}{2}c_{ijk\ell}\epsilon_{ij}\epsilon_{k\ell}, \tag{2.30}$$

where $\dot{u}$ denotes the magnitude of the particle velocity vector, and the Lagrangian density is a scalar quantity. The second term, or strain energy density, requires more finesse. Here, because of the exchange symmetry in $c_{ijk\ell}$ with $(ij) \leftrightarrow (k\ell)$, either of the two strains summed individually with the stiffness yields the same result. In this sense, ϵ_{ij} is equivalent to $\epsilon_{k\ell}$. A more elegant way, however, to handle this potential energy term is to write the strain energy density in differential form [4]

$$\partial\mathcal{U} = \sigma_{ij}\partial\epsilon_{ij}. \tag{2.31}$$

Then, in the Euler–Lagrange equation, these substitutions for $\mathcal{L}$ in Eq. (2.27) yield

$$-\frac{1}{2}\frac{d}{dt}\left(\frac{\partial}{\partial \dot{u}_i}\rho \dot{u}_i \dot{u}_i\right) + \frac{d}{dx_j}\left(\frac{\partial \mathcal{U}}{\partial u_{i,j}}\right) = 0 \tag{2.32}$$

$$-\frac{d}{dt}(\rho \dot{u}_i) + \frac{d}{dx_j}\left(\sigma_{ij}\frac{\partial \epsilon_{ij}}{\partial u_{i,j}}\right) = 0 \tag{2.33}$$

$$\rho \ddot{u}_i - \frac{d}{dx_j}\left(\frac{1}{2}\sigma_{ij}\frac{\partial}{\partial u_{i,j}}\left[u_{i,j} + u_{j,i}\right]\right) = 0 \tag{2.34}$$

$$\rho \ddot{u}_i - \frac{d}{dx_j}\left(\sigma_{ij}\frac{1}{2}[1+1]\right) = 0. \tag{2.35}$$

The result is clearly

$$\rho \ddot{u}_i = \frac{\partial \sigma_{ij}}{\partial x_j}, \tag{2.36}$$

and this is just the Cauchy equation Eq. (2.12) derived above. The rest of the calculation leading to Eq. (2.16) is identical. While the result is the same, and there are more steps needed to achieve it, it is nonetheless valuable to see the elastic wave equation proceed from such a general principle of nature. The general ideas presented above can also be found in many excellent texts on elastic waves [4–9].

2.2 Plane Wave Propagation in Bulk Materials

The plane harmonic wave solution of Eq. (2.16) can be expressed either in vector notation or in index notation as

$$\mathbf{u}(\mathbf{r}, t) = A_0 \mathbf{p} \exp i(\mathbf{k} \cdot \mathbf{r} - \omega t) \tag{2.37}$$

$$u_j(x_n, t) = A_0 p_j \exp i(k_m x_m - \omega t), \tag{2.38}$$

where A_0 is the displacement amplitude of the wave, $\mathbf{p}$ or p_i is the polarization unit vector with $p = 1$, and ω is the angular frequency. The wave vector $\mathbf{k}$ or k_m is given by

$$\mathbf{k} = k\mathbf{n}, \quad k = \frac{2\pi}{\lambda} \tag{2.39}$$

where $\mathbf{n}$ is a unit vector in the direction of wave propagation, and λ is the wavelength. The phase of the wave θ is given by

$$\theta = \mathbf{k} \cdot \mathbf{r} - \omega t. \tag{2.40}$$

For a point on the wave where the value of the phase is constant

$$\mathbf{k} \cdot \mathbf{r} - \omega t \ = \ \text{constant}. \tag{2.41}$$

Taking differentials gives

$$\mathbf{k} \cdot d\mathbf{r} - \omega dt \ = \ 0\,, \tag{2.42}$$

and rearranging

$$\mathbf{k} \cdot \frac{d\mathbf{r}}{dt} = \omega. \tag{2.43}$$

Now substitute $\mathbf{k}$ from Eq. (2.39) and let ξ be the projection of the wavefront position vector $\mathbf{r}$ on the direction of wave propagation ($\xi = \mathbf{n} \cdot \mathbf{r}$), as shown in Fig. 2.2,

$$k\mathbf{n} \cdot \frac{d\mathbf{r}}{dt} = \omega, \tag{2.44}$$

and the propagation of a point of constant phase proceeds with phase velocity V given by

$$V = \frac{d\xi}{dt} = \frac{\omega}{k} \tag{2.45}$$

or in vectorial form

$$\mathbf{V} = V\mathbf{n} = \frac{\omega}{k}\mathbf{n}. \tag{2.46}$$

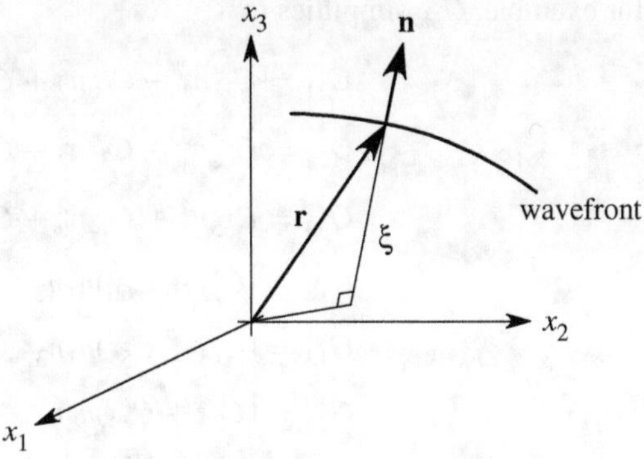

Figure 2.2. Wavefront moving in a rectangular coordinate system.

Substituting the plane harmonic wave solution (Eq. 2.38) into the equation of motion (Eq. 2.16), we have

$$\rho\omega^2 u_i = c_{ij\ell m} k_j k_\ell u_m, \tag{2.47}$$

or

$$c_{ij\ell m} n_j n_\ell p_m - \rho V^2 p_i = 0. \tag{2.48}$$

By using $p_i = \delta_{im} p_m$, the above equation can be rewritten as

$$[c_{ij\ell m} n_j n_\ell - \rho V^2 \delta_{im}]\, p_m = 0. \tag{2.49}$$

In order to have a nontrivial solution for p_m, the determinant of the coefficients for the linear equation Eq. (2.49) must vanish, leading to the following well-known eigenvalue equation, also known as *Christoffel's* equation,

$$| \, c_{ij\ell m} n_j n_\ell - \rho V^2 \delta_{im} \, | = 0. \tag{2.50}$$

The straight braces $|\ldots|$ denote a determinant. An expansion of Eq. (2.50) yields a cubic polynomial in V^2, the phase velocity squared.

The expression in Eq. (2.50) can be simplified considerably by forming the implied sums over j and ℓ in the first term. This operation results in a second-rank tensor, which we call G_{im}. Its elements are given by

$$\begin{aligned} G_{im} &= c_{ij\ell m} n_j n_\ell \\ &= c_{i11m} n_1^2 + c_{i22m} n_2^2 + c_{i33m} n_3^2 + (c_{i12m} + c_{i21m}) n_1 n_2 \\ &\quad + (c_{i13m} + c_{i31m}) n_1 n_3 + (c_{i23m} + c_{i32m}) n_2 n_3 \end{aligned} \tag{2.51}$$

for a generally anisotropic medium. This general result, Eq. (2.51) for G_{im}, can be particularized to specific symmetry cases of interest in the characterization of composite materials. For orthotropic materials, for example, G_{im} simplifies to

$$\begin{aligned}
G_{11} &= C_{11}n_1^2 + C_{66}n_2^2 + C_{55}n_3^2 \\
G_{22} &= C_{66}n_1^2 + C_{22}n_2^2 + C_{44}n_3^2 \\
G_{33} &= C_{55}n_1^2 + C_{44}n_2^2 + C_{33}n_3^2 \\
G_{12} &= (C_{12} + C_{66})n_1 n_2 \\
G_{13} &= (C_{13} + C_{55})n_1 n_3 \\
G_{23} &= (C_{23} + C_{44})n_2 n_3,
\end{aligned} \tag{2.52}$$

where we have written the elastic properties in terms of elements of the stiffness matrix C_{ij} (see Chap. 1).

Continuing with the general anisotropic case, the problem to be solved in Eq. (2.50) can be expressed more conveniently as

$$\begin{vmatrix} G_{11} - \rho V^2 & G_{12} & G_{13} \\ G_{12} & G_{22} - \rho V^2 & G_{23} \\ G_{13} & G_{23} & G_{33} - \rho V^2 \end{vmatrix} = 0. \tag{2.53}$$

There will be three solutions for V^2 in the above equation, and each of these will yield a pair of positive and negative phase velocity values of identical magnitude, pertaining to positive-going and negative-going waves, for a total of six solutions. The implication here is that there are three distinct waves that can propagate along any given direction in the medium. Each of the eigenvalues (velocity squared) corresponds to one type of wave.

Due to the special symmetry properties of the stiffness tensor $c_{ijk\ell}$, the second-rank tensor G_{im} is also symmetric, as can be seen in the following,

$$G_{im} = c_{ij\ell m}n_j n_\ell = c_{\ell mij}n_j n_\ell = c_{m\ell ji}n_j n_\ell = c_{mj\ell i}n_\ell n_j = G_{mi}. \tag{2.54}$$

Expanding the determinant of Eq. (2.53) in a polynomial form, we have

$$\begin{aligned}
&(G_{11} - \rho V^2)(G_{22} - \rho V^2)(G_{33} - \rho V^2) - G_{12}^2(G_{33} - \rho V^2) \\
&\quad - G_{13}^2(G_{22} - \rho V^2) - G_{23}^2(G_{11} - \rho V^2) + 2G_{12}G_{13}G_{23} = 0,
\end{aligned} \tag{2.55}$$

or, collecting terms,

$$(\rho V^2)^3 + a(\rho V^2)^2 + b(\rho V^2) + c = 0, \tag{2.56}$$

where the coefficients a, b, and c are given by

$$\begin{aligned}
a &= -(G_{11} + G_{22} + G_{33}) \\
b &= -(G_{12}^2 + G_{13}^2 + G_{23}^2 - G_{11}G_{22} - G_{11}G_{33} - G_{22}G_{33}) \\
c &= -(G_{11}G_{22}G_{33} + 2G_{12}G_{13}G_{23} - G_{11}G_{23}^2 - G_{22}G_{13}^2 - G_{33}G_{12}^2).
\end{aligned}$$

Equation (2.56) is routinely solved by numerical methods. It is also possible, however, to develop analytical solutions to the cubic equation above that lend more insight into the behavior of the roots. We begin by making a change of variable

$$z = \rho V^2 + \frac{a}{3},$$

so that Eq. (2.56) can be rewritten as

$$z^3 + pz + q = 0, \tag{2.57}$$

where

$$p = b - \frac{a^2}{3} \tag{2.58}$$

and

$$q = c - \frac{ba}{3} + 2\left(\frac{a}{3}\right)^3. \tag{2.59}$$

The roots of Eq. (2.57) are of the form

$$z_{(k)} = 2\sqrt{-\frac{p}{3}}\cos\left(\frac{\theta + 2k\pi}{3}\right), \tag{2.60}$$

with the root index $k = 0, 1, 2$ (in order to avoid confusion, subscripts that do not represent tensorial indices, *i.e.*, the summation convention does not apply to them, will be put in parenthesis). Then, the solutions are

$$\rho V_{(k)}^2 = z_{(k)} - \frac{a}{3}, \tag{2.61}$$

where

$$\theta = \arccos\left[\frac{-q/2}{\sqrt{-(p/3)^3}}\right]. \tag{2.62}$$

For real coefficients in Christoffel's equation, the quantity D given by

$$D = \left(\frac{p}{3}\right)^3 + \left(\frac{q}{2}\right)^2 \tag{2.63}$$

must be negative, and all three roots of $V_{(k)}^2$ must be real and positive, corresponding to true propagating modes. This is so because G_{im} is a positive-definite matrix [12, 13] and follows from the positivity of elastic energy.

2.3 Determination of the Polarization Vectors

Now, we need to determine the polarization vectors p_j of our solution Eq. (2.38). We can accomplish this by using what we have already developed. Returning to Eq. (2.49) and using G_{im} from Eq. (2.51), we find the following relations for the polarization vectors of the solutions corresponding to different phase velocities V that satisfy the Christoffel equations. We have

$$
\begin{aligned}
(G_{11} - \rho V^2)p_1 + G_{12}p_2 + G_{13}p_3 &= 0 \\
G_{12}p_1 + (G_{22} - \rho V^2)p_2 + G_{23}p_3 &= 0 \\
G_{13}p_1 + G_{23}p_2 + (G_{33} - \rho V^2)p_3 &= 0
\end{aligned}
\tag{2.64}
$$

with the condition that

$$p_1^2 + p_2^2 + p_3^2 = 1. \tag{2.65}$$

We first take $p_1 \neq 0$. Solving Eq. (2.64) for p_2 and p_3 gives

$$
\begin{aligned}
p_2 &= -p_1 \frac{(G_{11} - \rho V^2)(G_{33} - \rho V^2) - G_{13}^2}{G_{12}(G_{33} - \rho V^2) - G_{23}G_{13}} \\
p_3 &= -p_1 \frac{(G_{11} - \rho V^2)(G_{22} - \rho V^2) - G_{12}^2}{G_{13}(G_{22} - \rho V^2) - G_{23}G_{12}}.
\end{aligned}
\tag{2.66}
$$

For convenience, we make a change of variables to

$$
\begin{aligned}
A &= (G_{11} - \rho V^2)(G_{22} - \rho V^2) - G_{12}^2 \\
B &= (G_{11} - \rho V^2)(G_{33} - \rho V^2) - G_{13}^2 \\
C &= (G_{22} - \rho V^2)(G_{33} - \rho V^2) - G_{23}^2 \\
x &= (G_{22} - \rho V^2)G_{13} - G_{23}G_{12} \\
y &= (G_{11} - \rho V^2)G_{23} - G_{12}G_{13} \\
z &= (G_{33} - \rho V^2)G_{12} - G_{13}G_{23},
\end{aligned}
\tag{2.67}
$$

yielding

$$
\begin{aligned}
p_1 &= \left(1 + \frac{A^2}{x^2} + \frac{B^2}{z^2}\right)^{-1/2} \\
p_2 &= -\frac{B}{z}\left(1 + \frac{A^2}{x^2} + \frac{B^2}{z^2}\right)^{-1/2} \\
p_3 &= -\frac{A}{x}\left(1 + \frac{A^2}{x^2} + \frac{B^2}{z^2}\right)^{-1/2}.
\end{aligned}
\tag{2.68}
$$

The equations for p_i above are straightforward, but inconvenient for calculations because the denominator can vanish when the displacement vectors are in planes of material symmetry. There is, however, a different formulation that avoids these difficulties. Noting that

$$\frac{B}{z} = -\frac{y}{x}, \quad \frac{A}{x} = -\frac{y}{z}, \quad \frac{C}{x} = -\frac{z}{y}, \tag{2.69}$$
$$AB = y^2, \quad AC = x^2, \quad BC = z^2,$$

we can express the polarizations as

$$p_1 = \sqrt{\frac{|C|}{|A| + |B| + |C|}}$$
$$p_2 = -\mathrm{sign}\left(\frac{z}{C}\right)\sqrt{\frac{|B|}{|A| + |B| + |C|}} \tag{2.70}$$
$$p_3 = -\mathrm{sign}\left(\frac{x}{C}\right)\sqrt{\frac{|A|}{|A| + |B| + |C|}}.$$

Equations (2.70) are useful for computer analysis because the denominator never vanishes identically.

There are always three eigenvectors, $p_{k(1)}$, $p_{k(2)}$, and $p_{k(3)}$, corresponding to the three eigenvalues $V_{(1)}$, $V_{(2)}$, and $V_{(3)}$, respectively. These polarization vectors must satisfy the Christoffel equation Eq. (2.49), that can be re-written with the notation of Eq. (2.51) as follows

$$[G_{ik} - \rho V_{(j)}^2 \delta_{ik}] p_{k(j)} = 0. \tag{2.71}$$

It is very important to realize that

$$p_{k(j)} p_{k(\ell)} = \delta_{j\ell}. \tag{2.72}$$

For $j = \ell$ we see that $p_{k(\ell)} p_{k(\ell)} = p_{(\ell)}^2 = 1$ as it should be since the polarization vectors are defined as unit vectors. More interesting is that for $j \neq \ell$ we see that $p_{k(j)} p_{k(\ell)} = 0$, *i.e.*, the three polarization vectors are mutually orthogonal for all propagation directions. The special importance of the rule of orthogonality calls for a short mathematical derivation and physical explanation. It is sufficient to show only that $\mathbf{p}_{(1)}$ and $\mathbf{p}_{(2)}$ are orthogonal since, by generality, this implies that all three polarization vectors are mutually orthogonal. Specializing Eq. (2.71) for $j = 1$ and $j = 2$ yields

$$G_{ik} p_{k(1)} = \rho V_{(1)}^2 \delta_{ik} p_{k(1)} \tag{2.73}$$

and

$$G_{ik} p_{k(2)} = \rho V_{(2)}^2 \delta_{ik} p_{k(2)}, \tag{2.74}$$

respectively. In order to get the scalar product of the polarization vectors, let us multiply Eq. (2.73) by $p_{i(2)}$ and Eq. (2.74) by $p_{i(1)}$. Subtracting these two equations from each other leads to

$$\rho(V_{(1)}^2 - V_{(2)}^2)\delta_{ik}p_{k(1)}p_{i(2)} = 0, \tag{2.75}$$

where we exploited the symmetry of the Christoffel matrix ($G_{ik} = G_{ki}$). Since the two eigenvalues are necessarily different, $V_{(1)}^2 - V_{(2)}^2 \neq 0$, $p_{i(1)}p_{i(2)} = 0$, which proves the above stated orthogonality between the polarization vectors. The physical significance of the mutual orthogonality of the three polarization vectors for any given wave direction is that it excludes coupling between the modes. Without orthogonality of the polarization directions, these modes could not propagate independently from each other. In this way, mode conversion can occur only as a result of interaction with interfaces, inhomogeneities, etc., when the boundary conditions require the existence of multiple modes.

2.4 Example: Plane Waves in an Orthotropic Material

Suppose a plane wave propagates along the x_1 axis of an orthotropic material (see Chap. 1), as shown in Fig. 2.3. In this case, we have for the propagation unit vector n_j

$$n_1 = 1, \quad n_2 = 0, \quad n_3 = 0, \tag{2.76}$$

and the Christoffel matrix for an orthotropic material Eq. (2.52) is

$$G_{11} = C_{11}, \quad G_{22} = C_{66}, \quad G_{33} = C_{55}, \quad G_{12} = G_{13} = G_{23} = 0. \tag{2.77}$$

Then, Eq. (2.53) becomes

$$(C_{11} - \rho V^2)(C_{66} - \rho V^2)(C_{55} - \rho V^2) = 0, \tag{2.78}$$

and from this equation, we find the velocities as

$$V_{(1)} = \sqrt{\frac{C_{11}}{\rho}}, \quad V_{(2)} = \sqrt{\frac{C_{66}}{\rho}}, \quad V_{(3)} = \sqrt{\frac{C_{55}}{\rho}}. \tag{2.79}$$

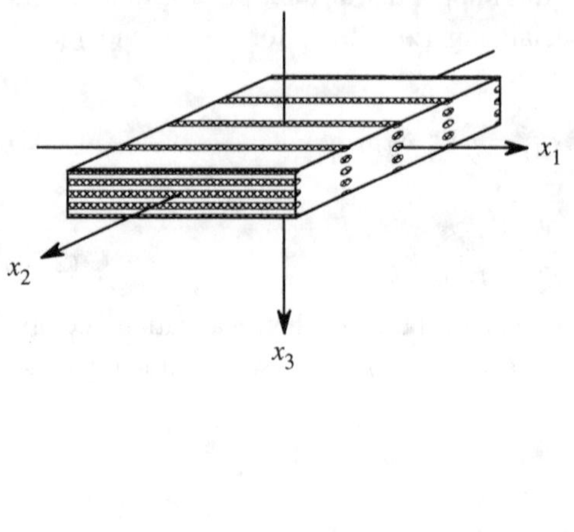

Figure 2.3. Unidirectional composite material of orthotropic symmetry.

Intuition tells us that $V_{(1)}$ is a longitudinal wave polarized along x_1, and $V_{(2)}$ and $V_{(3)}$ are transverse waves polarized along x_2 and x_3, respectively. To test this intuitive hypothesis, we use $V = V_{(1)}$ in Eqs. (2.67) to get

$$A = 0, \quad B = 0, \quad C = (C_{66} - C_{11})(C_{55} - C_{11}), \tag{2.80}$$

and from Eqs. (2.70)

$$p_{1(1)} = 1, \quad p_{2(1)} = p_{3(1)} = 0. \tag{2.81}$$

This result confirms our guess about $V_{(1)}$. For $V_{(2)}$ we have

$$A = 0, \quad B = (C_{11} - C_{66})(C_{55} - C_{66}), \quad C = 0 \tag{2.82}$$

and

$$p_{1(2)} = 0, \quad p_{2(2)} = 1, \quad p_{3(2)} = 0. \tag{2.83}$$

Finally, for $V_{(3)}$ we find

$$A = (C_{11} - C_{55})(C_{66} - C_{55}), \quad B = 0, \quad C = 0 \tag{2.84}$$

and

$$p_{1(3)} = 0, \quad p_{2(3)} = 0, \quad p_{3(3)} = 1. \tag{2.85}$$

These results confirm our initial guess.

Suppose now that the direction of propagation is x_2, then the values for n_i will be

$$n_1 = 0, \quad n_2 = 1, \quad n_3 = 0, \tag{2.86}$$

and for G_{ij} we get

$$G_{11} = C_{66}, \quad G_{22} = C_{22}, \quad G_{33} = C_{44}, \quad \text{and} \quad G_{12} = G_{13} = G_{23} = 0. \tag{2.87}$$

Then, Eq. (2.53) becomes

$$(C_{66} - \rho V^2)(C_{22} - \rho V^2)(C_{44} - \rho V^2) = 0 \tag{2.88}$$

and the roots will be

$$V_{(1)} = \sqrt{\frac{C_{22}}{\rho}}, \quad V_{(2)} = \sqrt{\frac{C_{66}}{\rho}}, \quad V_{(3)} = \sqrt{\frac{C_{44}}{\rho}}. \tag{2.89}$$

Again, our intuition indicates that $V_{(1)}$ is a longitudinal wave polarized along x_2, and $V_{(2)}$ and $V_{(3)}$ are shear waves polarized along x_1 and x_3, respectively,

$$\begin{array}{lll} \text{for } V_{(1)} & \text{for } V_{(2)} & \text{for } V_{(3)} \\ p_{1(1)} = 0 & p_{1(2)} = 1 & p_{1(3)} = 0 \\ p_{2(1)} = 1 & p_{2(2)} = 0 & p_{2(3)} = 0 \\ p_{3(1)} = 0 & p_{3(2)} = 0 & p_{3(3)} = 1. \end{array} \tag{2.90}$$

Finally, taking the direction of propagation as x_3, we have

$$n_1 = 0, \quad n_2 = 0, \quad n_3 = 1, \quad \text{and,} \tag{2.91}$$

$$G_{11} = C_{55}, \quad G_{22} = C_{44}, \quad G_{33} = C_{33}, \quad \text{and} \quad G_{12} = G_{13} = G_{23} = 0. \tag{2.92}$$

These intermediate results lead to

$$(C_{55} - \rho V^2)(C_{44} - \rho V^2)(C_{33} - \rho V^2) = 0, \tag{2.93}$$

and the velocities and the corresponding polarization vectors are

$$V_{(1)} = \sqrt{\frac{C_{55}}{\rho}}, \quad V_{(2)} = \sqrt{\frac{C_{44}}{\rho}}, \quad V_{(3)} = \sqrt{\frac{C_{33}}{\rho}}, \tag{2.94}$$

and

$$\begin{array}{lll} \text{for } V_{(1)} & \text{for } V_{(2)} & \text{for } V_{(3)} \\ p_{1(1)} = 1 & p_{1(2)} = 0 & p_{1(3)} = 0 \\ p_{2(1)} = 0 & p_{2(2)} = 1 & p_{2(3)} = 0 \\ p_{3(1)} = 0 & p_{3(2)} = 0 & p_{3(3)} = 1. \end{array} \tag{2.95}$$

If the velocities are denoted by double subscripts, the first indicating the direction of propagation and the second the direction of polarization, we can write

$$\begin{array}{lll} V_{(11)} = \sqrt{\dfrac{C_{11}}{\rho}}, & V_{(12)} = \sqrt{\dfrac{C_{66}}{\rho}}, & V_{(13)} = \sqrt{\dfrac{C_{55}}{\rho}}, \\ V_{(21)} = \sqrt{\dfrac{C_{66}}{\rho}}, & V_{(22)} = \sqrt{\dfrac{C_{22}}{\rho}}, & V_{(23)} = \sqrt{\dfrac{C_{44}}{\rho}}, \\ V_{(31)} = \sqrt{\dfrac{C_{55}}{\rho}}, & V_{(32)} = \sqrt{\dfrac{C_{44}}{\rho}}, & V_{(33)} = \sqrt{\dfrac{C_{33}}{\rho}}, \end{array} \tag{2.96}$$

where $V_{(11)}$, $V_{(22)}$, and $V_{(33)}$ are longitudinal waves, and $V_{(12)}$, $V_{(13)}$, $V_{(21)}$, $V_{(23)}$, $V_{(31)}$, $V_{(32)}$ are shear waves with velocities:

$$V_{(12)} = V_{(21)}, \quad V_{(13)} = V_{(31)}, \quad V_{(23)} = V_{(32)}, \tag{2.97}$$

which indicates that a shear wave traveling in the i direction and polarized along the j direction $(i, j = 1, 2, 3)$ will have the same velocity as a shear wave traveling along the j direction and polarized along the i direction. In addition, it can be used to check the correctness of the velocity measurements.

This result is very important in the study of composites because it limits the information that can be extracted from measurements of acoustic velocities along the principal directions. Figure 2.4 shows the

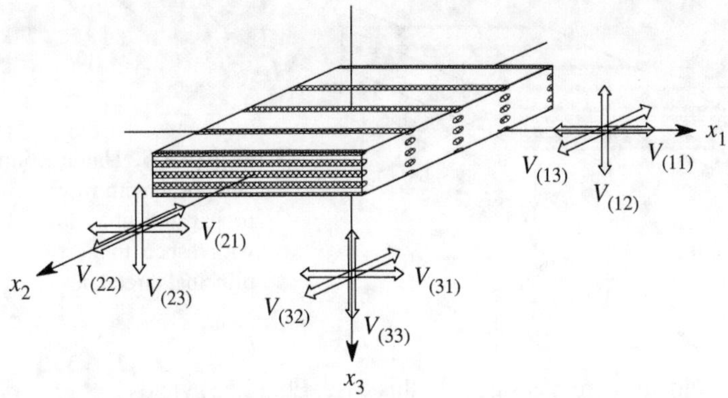

Figure 2.4. Different elastic waves that can be excited along the principal directions in a unidirectional composite and their respective polarization directions.

nine different combinations of polarizations that can exist along the three principal directions x_1, x_2, and x_3. However, Eqs. (2.97) show that only six independent velocities can be found among these nine different combinations, therefore only six independent elastic constants can be determined, namely:

$$C_{11}, \quad C_{22}, \quad C_{33}, \quad C_{44}, \quad C_{55}, \text{ and } C_{66}.$$

To determine the other three elastic constants, C_{12}, C_{13}, and C_{23}, measurements in directions different from the principal directions are needed. These measurements can be made by cutting the material in such a way that the elastic waves can be sent in a direction ϕ with respect to one of the principal directions as shown in Fig. 2.5. Although this technique will provide the necessary information, it cannot always be applied because it is a destructive technique that is mostly limited to laboratory practice.

Suppose now that the angle ϕ of Fig. 2.5 is equal to 45°, *i.e.*, the wave propagates in the [1,1,0] direction. In this case

$$n_1 = \frac{1}{\sqrt{2}}, \quad n_2 = \frac{1}{\sqrt{2}}, \quad n_3 = 0 \quad \text{and} \tag{2.98}$$

$$G_{11} = \frac{1}{2}(C_{11} + C_{66}), \quad G_{22} = \frac{1}{2}(C_{66} + C_{22})$$
$$G_{33} = \frac{1}{2}(C_{55} + C_{44}), \quad G_{12} = \frac{1}{2}(C_{12} + C_{66}). \tag{2.99}$$

As only one new elastic constant C_{12} is involved in this 45° measurement, the other two elastic constants (C_{13} and C_{23}) cannot be obtained even if all

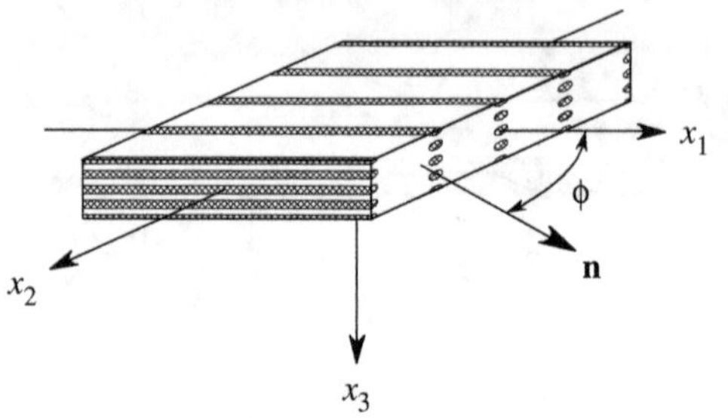

Figure 2.5. Unidirectional composite with wave propagation at an angle ϕ with respect to the x_1 principal direction.

three velocities are measured. In this case, Eq. (2.53) yields

$$\begin{vmatrix} G_{11} - \rho V^2 & G_{12} & 0 \\ G_{12} & G_{22} - \rho V^2 & 0 \\ 0 & 0 & G_{33} - \rho V^2 \end{vmatrix} = 0, \tag{2.100}$$

or

$$(G_{33} - \rho V^2)[(G_{11} - \rho V^2)(G_{22} - \rho V^2) - G_{12}^2] = 0. \tag{2.101}$$

From Eq. (2.101), we immediately have

$$V_{(3)} = \sqrt{\frac{G_{33}}{\rho}} = \sqrt{\frac{C_{55} + C_{44}}{2\rho}}. \tag{2.102}$$

From Eqs. (2.67),

$$A \neq 0 \quad \text{and} \quad B = C = 0, \tag{2.103}$$

so

$$p_{1(3)} = 0, \quad p_{2(3)} = 0, \quad p_{3(3)} = 1, \tag{2.104}$$

which indicates that this is a shear wave with polarization along x_3. The other roots of Eq. (2.101), *i.e.*, $V_{(1)}$ and $V_{(2)}$ are given by

$$(G_{11} - \rho V^2)(G_{22} - \rho V^2) - G_{12}^2 = 0, \tag{2.105}$$

and by applying Eq. (2.67) to (2.68),

$$A = 0, \quad \text{and} \quad B \neq 0, \quad C \neq 0. \tag{2.106}$$

So, for $\ell = 1$ or 2

$$p_{1(\ell)} \neq 0, \quad p_{2(\ell)} \neq 0, \quad p_{3(\ell)} = 0. \tag{2.107}$$

Equation (2.107) verifies the fact that once the polarization direction of the first wave is determined, the polarizations of the other two waves will be normal to that of the first one. Therefore, $V_{(1)}$ and $V_{(2)}$ must be polarized in the (x_1, x_2) plane. C_{11} and C_{22} are known from the waves along axes x_1 and x_2, and for $\ell = 1$ and 2, we obtain

$$p_{1(\ell)}^2 = \frac{1}{B/C+1} = \{[G_{11} - \rho V_{(\ell)}^2]/[G_{22} - \rho V_{(\ell)}^2] + 1\}^{-1}$$

$$p_{2(\ell)}^2 = \frac{1}{C/B+1} = \{[G_{22} - \rho V_{(\ell)}^2]/[G_{11} - \rho V_{(\ell)}^2] + 1\}^{-1}. \quad (2.108)$$

Equations (2.108) indicate that the waves propagating in the (x_1, x_2) plane are neither pure longitudinal nor pure transverse. These kinds of waves are called quasitransverse or quasilongitudinal, depending on which one of the polarization vector components, $p_{1(\ell)}$ or $p_{2(\ell)}$, is greater. However, we shall see that for strongly anisotropic materials, in practice, the difference between quasitransverse and quasilongitudinal is not very clear.

For a cubic material $C_{11} = C_{22}$, so $G_{11} = G_{22}$, and $B = C$. In this case, $p_{1(1)} = p_{2(1)} = 1/\sqrt{2}$, and $p_{1(2)} = -p_{2(2)} = 1/\sqrt{2}$, *i.e.*, all modes are pure as could be expected since the [1,1,0] direction is an axis of symmetry. The velocities of the two waves polarized in the (x_1, x_2) plane can be obtained by solving Eq. (2.105) as

$$V_{(1)} = \sqrt{\frac{C_{11} - C_{12}}{2\rho}} \quad (2.109)$$

and

$$V_{(2)} = \sqrt{\frac{C_{11} + C_{12} + 2C_{44}}{2\rho}}. \quad (2.110)$$

For cubic materials, the velocity of the third mode propagating along the [1,1,0] direction and polarized along the [0,0,1] direction can be obtained from Eq. (2.102) as follows

$$V_{(3)} = \sqrt{\frac{C_{44}}{\rho}}. \quad (2.111)$$

The problem of finding solutions for waves propagating as shown in Fig. 2.6 in the x_1' direction (not along a symmetry axis) can be addressed in a different way which sometimes simplifies the solution. This is achieved by finding the elastic constants in the rotated coordinate system. In this case, the stiffness matrix C_{ij}' in a rotated coordinate system is given by

$$C_{ij}' = \begin{bmatrix} C_{11}' & C_{12}' & C_{13}' & 0 & 0 & C_{16}' \\ C_{12}' & C_{22}' & C_{23}' & 0 & 0 & C_{26}' \\ C_{13}' & C_{23}' & C_{33}' & 0 & 0 & C_{36}' \\ 0 & 0 & 0 & C_{44}' & C_{45}' & 0 \\ 0 & 0 & 0 & C_{45}' & C_{55}' & 0 \\ C_{16}' & C_{26}' & C_{36}' & 0 & 0 & C_{66}' \end{bmatrix}. \quad (2.112)$$

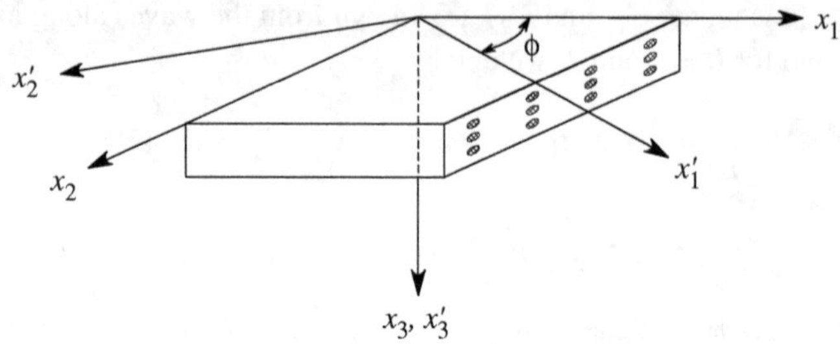

Figure 2.6. Rotation around one of the principal axes in a unidirectional composite material.

Thus, the elements of G_{ij} become

$$G_{k\ell} = C'_{1k\ell 1} n_1^2 = C'_{1k\ell 1} \tag{2.113}$$

or

$$\begin{aligned} G_{11} &= C'_{1111} = C'_{11} \\ G_{22} &= C'_{1221} = C'_{66} \\ G_{33} &= C'_{1331} = C'_{55} \\ G_{23} &= C'_{1231} = C'_{65} = 0 \\ G_{13} &= C'_{1131} = C'_{15} = 0 \\ G_{12} &= C'_{1121} = C'_{16}. \end{aligned}$$

Now Eq. (2.53) can be written as

$$\begin{vmatrix} (C'_{11} - \rho V^2) & C'_{16} & 0 \\ C'_{16} & (C'_{66} - \rho V^2) & 0 \\ 0 & 0 & (C'_{55} - \rho V^2) \end{vmatrix} = 0. \tag{2.114}$$

Immediately, $V_{(1)}$ can be written as

$$V_{(1)} = \sqrt{\frac{C'_{55}}{\rho}}, \tag{2.115}$$

which involves only terms in the $x_1 x_3$ directions; thus this mode is polarized along x'_3, *i.e.*, normal to the propagation direction x'_1.

Finally, for the other two waves the velocities are determined by

$$(\rho V^2)^2 - (C'_{66} + C'_{11})\rho V^2 + (C'_{11} C'_{66} - C'_{16})^2 = 0. \tag{2.116}$$

2.5 Energy Flux and Group Velocity

Understanding the propagation of elastic wave energy is essential in an anisotropic medium because the energy direction differs in general from that of the wave normal, and thus an ultrasonic beam deviates from the wave vector direction. This situation leads to differences in magnitude and direction between the phase and group velocities, knowledge of which is essential for their accurate measurement. Our analysis below is patterned after Fedorov [12], and the interested reader is referred to this source for further information. For acoustic waves propagating in elastic solids, the kinetic $\mathcal{T}$ and strain $\mathcal{U}$ energy densities are given by

$$\mathcal{T} = \frac{1}{2}\rho\dot{u}^2 \tag{2.117}$$

and

$$\mathcal{U} = \frac{1}{2}\sigma_{ik}\epsilon_{ik} = \frac{1}{2}c_{ik\ell m}\epsilon_{ik}\epsilon_{\ell m}, \tag{2.118}$$

respectively. Both $\mathcal{T}$ and $\mathcal{U}$ are scalar quantities. The total energy contained in a volume V is given by

$$E = \int_V (\mathcal{T} + \mathcal{U})dV, \tag{2.119}$$

and the rate of change of energy with respect to time is given by

$$\dot{E} = \frac{dE}{dt} = \int_V \frac{\partial(\mathcal{T} + \mathcal{U})}{\partial t}dV. \tag{2.120}$$

Using Eqs. (2.117) and (2.118) and the chain rule in the derivatives, Eq. (2.120) can be written as

$$\dot{E} = \int_V \left(\rho\dot{u}_i\ddot{u}_i + \frac{\partial\mathcal{U}}{\partial\epsilon_{ik}}\dot{\epsilon}_{ik}\right)dV. \tag{2.121}$$

Analyzing the derivative of the strain energy density, we obtain

$$\begin{aligned}\frac{\partial\mathcal{U}}{\partial\epsilon_{ik}} &= \frac{\partial}{\partial\epsilon_{ik}}\left[\frac{1}{2}c_{nw\ell m}\epsilon_{nw}\epsilon_{\ell m}\right] \\ &= \frac{1}{2}\left[c_{nw\ell m}\frac{\partial\epsilon_{nw}}{\partial\epsilon_{ik}}\epsilon_{\ell m} + c_{nw\ell m}\epsilon_{nw}\frac{\partial\epsilon_{\ell m}}{\partial\epsilon_{ik}}\right] \\ &= c_{ik\ell m}\epsilon_{\ell m} = \sigma_{ik},\end{aligned} \tag{2.122}$$

and we get the following form for the time derivative of the strain energy density

$$\dot{\mathcal{U}} = \frac{1}{2}\sigma_{ik}\left(\frac{\partial\dot{u}_i}{\partial x_k} + \frac{\partial\dot{u}_k}{\partial x_i}\right). \tag{2.123}$$

Exploiting again the symmetry of the stress tensor σ_{ik}, we have

$$\dot{\mathcal{U}} = \sigma_{ik}\frac{\partial \dot{u}_i}{\partial x_k}, \tag{2.124}$$

thus, the strain energy term in Eq. (2.121) is

$$\int_V \dot{\mathcal{U}} dV = \int_V \sigma_{ik}\frac{\partial \dot{u}_i}{\partial x_k} dV. \tag{2.125}$$

By using the fact that

$$\sigma_{ik}\frac{\partial \dot{u}_i}{\partial x_k} = \frac{\partial}{\partial x_k}(\sigma_{ik}\dot{u}_i) - \dot{u}_i\frac{\partial \sigma_{ik}}{\partial x_k}, \tag{2.126}$$

we have

$$\int_V \dot{\mathcal{U}} dV = \int_V \left[\frac{\partial}{\partial x_k}(\sigma_{ik}\dot{u}_i) - \dot{u}_i\frac{\partial \sigma_{ik}}{\partial x_k}\right] dV, \tag{2.127}$$

and from Gauss' theorem we have

$$\int_V \left[\frac{\partial}{\partial x_k}(\sigma_{ik}\dot{u}_i) - \dot{u}_i\frac{\partial \sigma_{ik}}{\partial x_k}\right] dV = -\int_V \dot{u}_i\frac{\partial \sigma_{ik}}{\partial x_k} dV + \oint \sigma_{ik}\dot{u}_i dS_k. \tag{2.128}$$

Inserting these results into Eq. (2.121) yields

$$\dot{E} = \int_V \dot{u}_i\left[\rho\ddot{u}_i - \frac{\partial \sigma_{ik}}{\partial x_k}\right] dV + \oint \sigma_{ik}\dot{u}_i dS_k, \tag{2.129}$$

where the term in brackets of the first integral is zero from the equation of motion

$$\rho\ddot{u}_i - \frac{\partial \sigma_{ik}}{\partial x_k} = 0. \tag{2.130}$$

These manipulations reduce Eq. (2.129) to

$$\dot{E} = \oint \sigma_{ik}\dot{u}_i dS_k, \tag{2.131}$$

as we could have expected.

Equation (2.131) allows us to define the vector of energy flow density **P**, that is often called the acoustic Poynting vector, as follows

$$P_k = -\sigma_{ik}\dot{u}_i. \tag{2.132}$$

The vector **P** is directed toward the local direction of energy transfer. Finally, from Eq. (2.131)

$$\dot{E} + \oint \mathbf{P}\cdot d\mathbf{S} = 0, \tag{2.133}$$

where $d\mathbf{S}$ denotes the elementary surface area vector that is everywhere normal to the surface S which encloses the volume V.

Using the real part of the plane harmonic wave solution in Eq. (2.38), we can write the real displacement, velocity, and displacement gradient distributions as follows

$$u_i = A_0 p_i \cos(k_n x_n - \omega t), \tag{2.134}$$

$$\dot{u}_i = \omega A_0 p_i \sin(k_n x_n - \omega t), \tag{2.135}$$

and

$$\frac{\partial u_i}{\partial x_\ell} = -k_\ell A_0 p_i \sin(k_n x_n - \omega t). \tag{2.136}$$

The kinetic and strain energy densities are

$$\mathcal{T} = \frac{1}{2}\rho \dot{u}^2 = \frac{1}{2}\rho\omega^2 A_0^2 \sin^2(k_n x_n - \omega t) \tag{2.137}$$

and

$$\mathcal{U} = \frac{1}{2} c_{ijk\ell} k_j k_\ell A_0^2 p_i p_k \sin^2(k_n x_n - \omega t), \tag{2.138}$$

respectively, where we exploited the fact that the polarization vector is defined as a unit vector, *i.e.*, $p_i p_i = p^2 = 1$.

To relate kinetic and strain energy densities let us use

$$\rho\omega^2 u_i = c_{ij\ell m} k_j k_\ell u_m. \tag{2.139}$$

Multiplying both sides of the equation by u_i, we get

$$\rho\omega^2 u^2 = c_{ij\ell m} k_j k_\ell u_m u_i, \tag{2.140}$$

where we substituted $u_i u_i = u^2$. Alternatively, we can divide both sides of this equation by u^2 to obtain the following equation

$$\rho\omega^2 = c_{ij\ell m} k_j k_\ell p_i p_m, \tag{2.141}$$

which leads to the conclusion that the kinetic and strain energy densities of a propagating plane wave are always equal

$$\mathcal{T} = \mathcal{U}. \tag{2.142}$$

At this point in our derivation, we should take a short excursion and provide a physical interpretation of this very simple, but not entirely trivial, conclusion. The equality of the kinetic and strain energy densities in a propagating plane wave seems to be in stark contrast with one's expectations based on vibrations of simple mechanical systems such as one made of a spring and a mass. Then, the strain energy of the spring is alternatingly transformed into the kinetic energy of the mass and vice versa, *i.e.*, the sum of the kinetic and strain energies is constant. In other words, when one of them is increasing, the other one is decreasing. Of course, this is because the combined energy is not going anywhere; so this case is analogous to standing waves. In contrast, propagating waves exhibit spatial and temporal invariances requiring that the kinetic and potential components maintain the same balance everywhere and all the time. In terms of energy density, this means that the two components must be equal, *i.e.*, they rise and fall together everywhere and all the time. Why they are exactly equal is a consequence of the Reciprocity Theorem, which requires that the transfer efficiency between two coupled forms in which energy can exist must be the same in both directions. Under these conditions, balance is feasible only if the kinetic and strain energies are equal. One should note that Eq. (2.142) is also valid for plane transient waves.

The total energy density $\mathcal{E}$ of an acoustic plane wave can be written as follows

$$\mathcal{E} = \rho\omega^2 A_0^2 \sin^2(k_n x_n - \omega t). \tag{2.143}$$

The time-averaged total energy density $\bar{\mathcal{E}}$ and energy flow density vector $\bar{\mathbf{P}}$ are given by

$$\bar{\mathcal{E}} = \frac{1}{2}\rho\omega^2 A_0^2 \tag{2.144}$$

and

$$\bar{P}_i = \frac{1}{2}\omega c_{ij\ell m} k_\ell A_0^2 p_m p_j, \tag{2.145}$$

where p_i denotes the polarization unit vector. (Equation (2.145) is easily obtainable by substituting Eq. (2.135) and Eq. (2.136) into Eq. (2.132)). Finally, the energy flow or *group* velocity $\mathbf{V}_\mathrm{g}$ (or in index notation $V_{\mathrm{g}i}$) is defined by

$$V_{\mathrm{g}i} = \frac{\bar{P}_i}{\bar{\mathcal{E}}} = \frac{\frac{1}{2}\omega c_{ij\ell m} k_\ell A_0^2 p_m p_j}{\frac{1}{2}\rho\omega^2 A_0^2} = \frac{c_{ij\ell m} k_\ell p_m p_j}{\rho\omega}. \tag{2.146}$$

The group velocity vector $V_{\mathrm{g}i}$ indicates the velocity of acoustic energy (wavefront) propagation in the material, and it is parallel to the direction of the energy flux vector. In contrast, the previously introduced phase velocity V is the velocity of the propagation of the wave in a direction normal to the wavefront, as in Fig. 2.7.

2.6 Relation between the Phase and Group Velocities

Let us consider the relationship between the phase velocity V and the group or energy velocity $\mathbf{V}_\mathrm{g}$. Based on the previous definition of the scalar phase velocity V, we can introduce a phase velocity vector $\mathbf{V}$ so that $V = |\mathbf{V}|$, and $\mathbf{V}$ lies in the direction of the wave propagation $\mathbf{n}$. The wave vector k_j can be

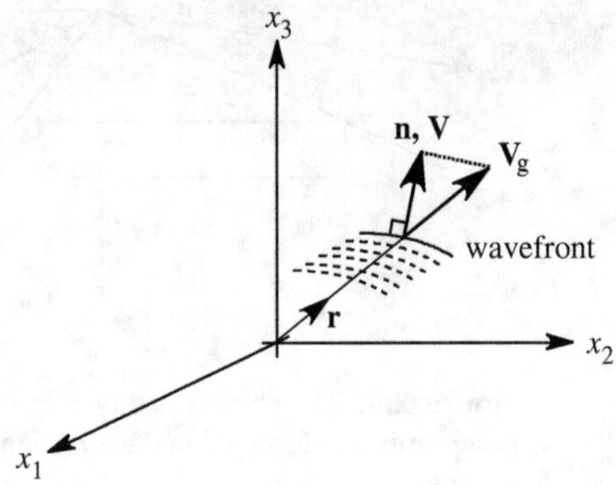

Figure 2.7. Group and phase velocities. The phase velocity vector **V** is parallel to the normal **n** of the wave front, while the group velocity $\mathbf{V}_g$ represents the propagation of the acoustic energy and it is parallel to the beam orientation.

written in terms of the the angular frequency ω, the phase velocity V, and the wave unit vector n_j as follows

$$k_j = \frac{\omega}{V} n_j. \tag{2.147}$$

Then, we can re-write Eq. (2.146) in terms of the phase velocity as follows

$$V_{gi} = \frac{c_{ij\ell m} n_\ell p_m p_j}{\rho V}. \tag{2.148}$$

This relation is illustrated in Fig. 2.7. We can see from the figure that **V** is in the direction of **k** and **n**. The direction of $\mathbf{V}_g$ is the actual direction of the acoustic beam.

A standard definition of the group velocity is given by

$$V_{gi} = \frac{\partial \omega}{\partial k_i}, \tag{2.149}$$

while the phase velocity is given by

$$V = \frac{\omega}{k}. \tag{2.150}$$

Again, from the equation of motion we have

$$\omega^2 = \frac{c_{ij\ell m}}{\rho} k_j k_\ell p_i p_m. \tag{2.151}$$

By differentiating the above equation with respect to k_ℓ, we can obtain the following expression for the group velocity $V_{g\ell}$

$$V_{g\ell} = \frac{c_{ij\ell m}}{\omega \rho} k_j p_i p_m. \tag{2.152}$$

The above equation, which is identical to Eq. (2.146), shows that the group velocity is actually the same as the energy flow velocity. To find the relationship between $\mathbf{V}_g$ and **V**, a vector **Q** is defined here as follows

$$Q_i = c_{im\ell j} n_j p_\ell p_m. \tag{2.153}$$

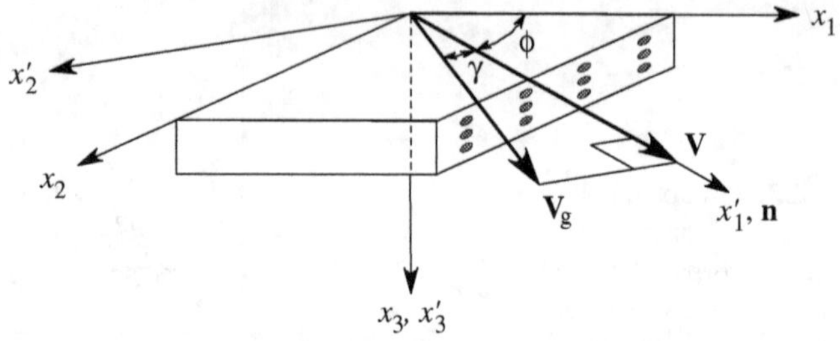

Figure 2.8. The phase velocity vector **V** is the projection of the group velocity vector $\mathbf{V}_g$ to the direction of wave normal **n**.

From the equation of motion in the form of Eq. (2.48), after multiplying both sides by p_i, we get

$$c_{ij\ell m}n_jn_\ell p_mp_i - \rho V^2 = 0. \tag{2.154}$$

Using Eq. (2.153), we have

$$Q_in_i = \rho V^2. \tag{2.155}$$

From Eq. (2.152), Q_i can be also written as

$$Q_i = V_{gi}\rho V. \tag{2.156}$$

Finally, from Eqs. (2.155) and (2.156), we obtain

$$V_{gi}n_i = V \tag{2.157}$$

or

$$V_g \mid \cos\gamma \mid = V, \tag{2.158}$$

where γ is the angle between $\mathbf{V}_g$ and **V**. This last equation indicates that **V** is the projection of $\mathbf{V}_g$ in the direction of wave propagation as shown in Fig. 2.8. Therefore, the magnitude V is always smaller than, or equal to, V_g. It is also convenient to write V_g in terms of a matrix D_{ij} as

$$V_{gi} = \frac{D_{ij}n_j}{\rho V}, \tag{2.159}$$

where the matrix is given by

$$D_{ij} = C_{im\ell j}p_mp_\ell. \tag{2.160}$$

As an analogy we note that the group and phase velocities are also different in dispersive media (dispersion occurs when velocity depends on frequency). When an ultrasonic pulse propagates in such media, each frequency, or spectral, component propagates with a different speed. The resulting speed of the pulse envelope, or its group velocity, will be different from the phase velocity; depending on dispersion, the group velocity can be larger or smaller than the phase velocity. We consider in this chapter a solid without dispersion, but with anisotropy. The direction of the phase velocity depends, however, on angle. As the ultrasonic beam is composed of plane waves, each with its own wave vector orientation, those plane waves will have different speeds. The resulting beam, or group, velocity will be different from the phase velocity (equal to, or larger than, the phase speed). In later chapters, both effects, geometrical dispersion and anisotropy, will be considered.

2.7 Measurement of the Group Velocity

The group velocity can be determined experimentally using the configuration shown in Fig. 2.9. The receiver can be moved along the back surface of the sample with respect to the transmitter until the signal reaches a maximum. The angle γ and the distance ℓ can be determined from the known thickness h of the specimen and the off-set between the transmitting and receiving transducers. The group velocity will be given by

$$V_g = \frac{\ell}{t}, \tag{2.161}$$

where t is the measured propagation time. In contrast, the phase velocity will be given by

$$V = \frac{h}{t}, \tag{2.162}$$

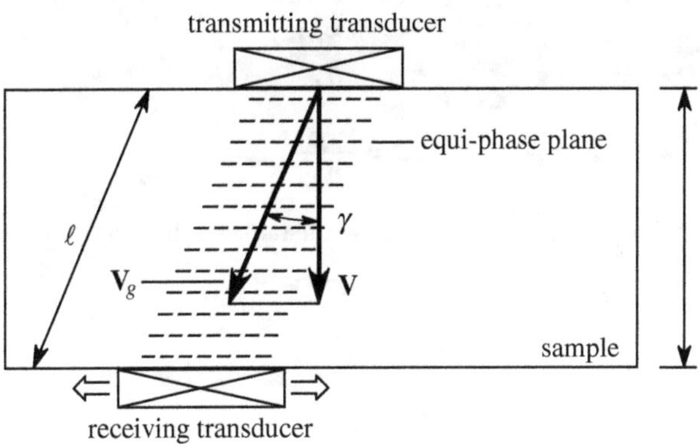

Figure 2.9. Experimental determination of the group velocity $\mathbf{V}_g$.

and the ratio between them is

$$\frac{V_{\text{g}}}{V} = \frac{\ell}{h} = \frac{1}{\cos\gamma}, \tag{2.163}$$

i.e., $V = V_{\text{g}} \cos\gamma$.

2.8 Examples

Consider a wave traveling along the x'_1 axis of a rotated coordinate system in a plate made of composite material as shown in Fig. 2.10. In this rotated system, the matrix of elastic constants for an orthotropic composite becomes populated like the one shown in Eq. (2.112). The direction cosines in the rotated coordinate system are

$$n_1 = 1, \quad n_2 = 0, \quad n_3 = 0. \tag{2.164}$$

According to Eq. (2.152), the group velocity components in this case are given by

$$V_{\text{g}i} = \frac{c'_{im\ell 1} p_\ell p_m}{\rho V}. \tag{2.165}$$

Suppose that one of the waves is polarized along the x'_3 axis. Then, the components of the group velocity vector will be given by

$$V_{\text{g}i} = \frac{c'_{i331}}{\rho V}, \tag{2.166}$$

with $i = 1, 2, 3$, *i.e.*,

$$\begin{aligned} V_{\text{g}1} &= \frac{c'_{1331}}{\rho V} \\ V_{\text{g}2} &= \frac{c'_{2331}}{\rho V} \\ V_{\text{g}3} &= \frac{c'_{3331}}{\rho V}. \end{aligned} \tag{2.167}$$

Using matrix notation for the stiffness constants of the material, we have

$$\begin{aligned} V_{\text{g}1} &= \frac{C'_{55}}{\rho V} \\ V_{\text{g}2} &= \frac{C'_{45}}{\rho V} \\ V_{\text{g}3} &= \frac{C'_{35}}{\rho V}. \end{aligned} \tag{2.168}$$

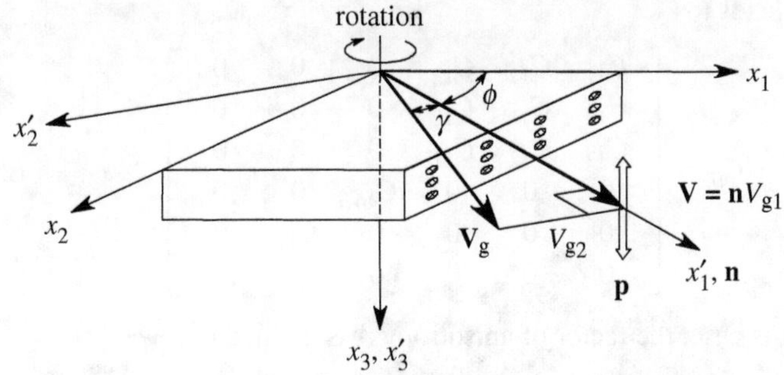

Figure 2.10. Ultrasonic wave traveling along the x_1' direction in a composite material with polarization along the $x_3 = x_3'$ direction.

The vector component V_{g3} equals zero because $C_{35}' = 0$ (See Eq. (2.112)). Therefore, the group velocity lies in the (x_1', x_2') plane. According to Fig. 2.9, the components V_{g1} and V_{g2} are given by

$$V_{g1} = V_g \cos\gamma = \frac{C_{55}'}{\rho V}$$
$$V_{g2} = V_g \sin\gamma = \frac{C_{45}'}{\rho V}. \tag{2.169}$$

By combining these two equations, the angle γ can be found from

$$\tan\gamma = \frac{C_{45}'}{C_{55}'}. \tag{2.170}$$

From this equation, it can be concluded that, in this case, the angle of deviation between the energy flow vector (or group velocity direction) and the wave vector (or phase velocity direction) depends only on the rotated elastic constants of the material.

Although composite materials rarely exhibit cubic symmetry, it is very enlightening to consider this particular case as an example. For materials with cubic symmetry, the elastic constants have the following relations

$$C_{11} = C_{22} = C_{33}$$
$$C_{12} = C_{21} = C_{13} = C_{31} = C_{23} = C_{32} \tag{2.171}$$
$$C_{44} = C_{55} = C_{66},$$

which leads to

$$C_{ij} = \begin{bmatrix} C_{11} & C_{12} & C_{12} & 0 & 0 & 0 \\ C_{12} & C_{11} & C_{12} & 0 & 0 & 0 \\ C_{12} & C_{12} & C_{11} & 0 & 0 & 0 \\ 0 & 0 & 0 & C_{44} & 0 & 0 \\ 0 & 0 & 0 & 0 & C_{44} & 0 \\ 0 & 0 & 0 & 0 & 0 & C_{44} \end{bmatrix}. \tag{2.172}$$

In this case, the factor of anisotropy A is defined by

$$A = \frac{2C_{44}}{C_{11} - C_{12}}. \tag{2.173}$$

If $A = 1$ the material is isotropic. For a given propagation direction n_i, Eq. (2.52) simplifies to

$$\begin{aligned} G_{11} &= n_1^2 C_{11} + (n_2^2 + n_3^2) C_{44} \\ G_{22} &= n_2^2 C_{11} + (n_1^2 + n_3^2) C_{44} \\ G_{33} &= n_3^2 C_{11} + (n_1^2 + n_2^2) C_{44} \\ G_{12} &= G_{21} = n_1 n_2 (C_{12} + C_{44}) \\ G_{23} &= G_{32} = n_2 n_3 (C_{12} + C_{44}) \\ G_{13} &= G_{31} = n_1 n_3 (C_{12} + C_{44}). \end{aligned} \tag{2.174}$$

If $\phi = 45°$, $n_1^2 = n_2^2 = 1/2$, $n_3 = 0$ and

$$G_{11} = G_{22} = G_{12} = \frac{C_{12} + C_{44}}{2}, \quad G_{33} = C_{44}, \quad G_{13} = G_{23} = 0, \tag{2.175}$$

which leads to

$$V_{(3)} = \sqrt{\frac{C_{44}}{\rho}} \tag{2.176}$$

and

$$\begin{vmatrix} G_{11} - \rho V^2 & G_{12} \\ G_{12} & G_{11} - \rho V^2 \end{vmatrix} = 0, \tag{2.177}$$

or

$$G_{11} - \rho V^2 = \pm G_{12}. \tag{2.178}$$

The roots of this equation are given by

$$V_{(1)} = \sqrt{\frac{C_{11} + C_{12} + 2C_{44}}{2\rho}} \tag{2.179}$$

and

$$V_{(2)} = \sqrt{\frac{C_{11} - C_{12}}{2\rho}}. \tag{2.180}$$

The polarization vector for $V_{(1)}$ is given as

$$p_{1(1)} = p_{2(1)} = \frac{1}{\sqrt{2}}, \quad p_{3(1)} = 0, \tag{2.181}$$

which indicates that the wave propagating with the phase velocity $V_{(1)}$ is pure longitudinal. The polarization vector for $V_{(2)}$ is given as

$$p_{1(2)} = -p_{2(2)} = \frac{1}{\sqrt{2}}, \quad p_{3(2)} = 0, \tag{2.182}$$

which indicates that the wave with phase velocity $V_{(2)}$ is pure transverse.

Finally, let us consider a wave propagating along the body diagonal of the cube, as indicated in Fig. 2.11. In crystallographic notation, this direction is called the [1,1,1] direction. It can be shown that along this direction, there are three propagating waves: a pure longitudinal wave and two pure shear waves, whose velocities are given by

$$V_{(1)} = \sqrt{\frac{C_{11} + 2C_{12} + 4C_{44}}{3\rho}} \tag{2.183}$$

and

$$V_{(2)} = V_{(3)} = \sqrt{\frac{C_{11} - C_{12} + C_{44}}{3\rho}}. \tag{2.184}$$

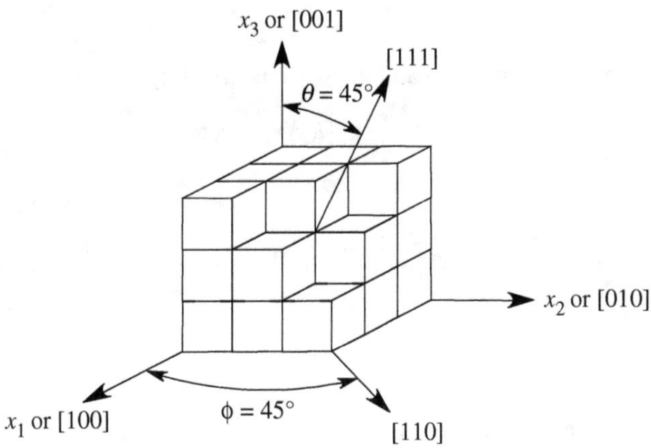

Figure 2.11. Acoustic waves traveling along symmetry directions in a cubic material.

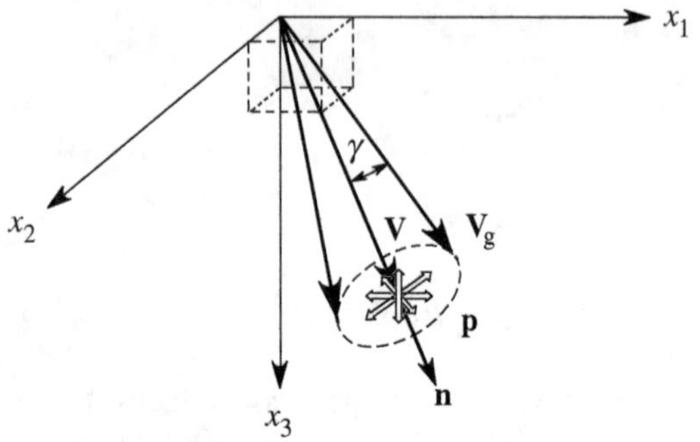

Figure 2.12. Conical refraction of the group velocity for an acoustic wave propagating along the body diagonal [1,1,1] in a cubic material.

Equation (2.184) signifies that there is no birefringence in the direction of the body diagonal in a medium of cubic symmetry, *i.e.*, the velocity of the transverse wave is independent of its polarization. Such directions are called acoustic axes of the material, in which the polarization for shear waves can be arbitrary. The angle by which $\mathbf{V}_g$ deviates from the acoustic axis is given by

$$\gamma = \arctan \frac{C_{11} - C_{12} - 2C_{44}}{\sqrt{2}(C_{11} - C_{12} + C_{44})}, \tag{2.185}$$

which indicates that the energy flux or the group velocity of waves propagating along an acoustical axis is directed along the surface of a symmetric cone of angle γ as shown in Fig. 2.12. When the polarization of the transducer changes, the energy beam rotates along this conical surface.

2.9 Phase Velocity, Group Velocity, and Slowness

2.9.1 Slowness

Suppose the displacement of a plane wave is given by

$$u_j = A_0 p_j \exp[i(k_\ell x_\ell - \omega t)], \tag{2.186}$$

where

$$k_\ell = \frac{\omega n_\ell}{V}. \tag{2.187}$$

Then the quantity

$$m_\ell = \frac{n_\ell}{V} \tag{2.188}$$

is defined as the slowness vector **m**, and can be also written as

$$m_\ell = \frac{k_\ell}{\omega}. \tag{2.189}$$

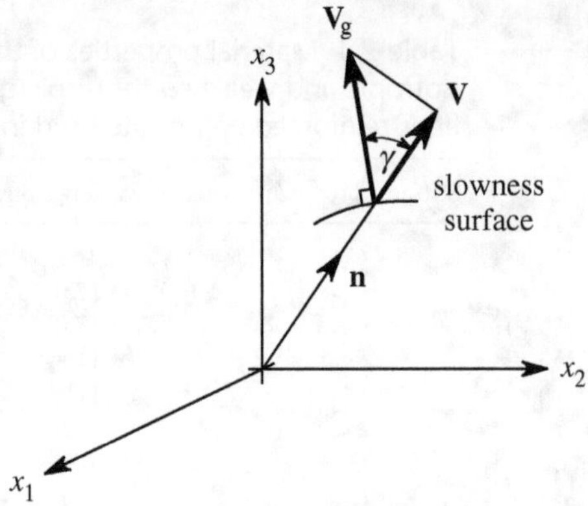

Figure 2.13. The relative orientation of the slowness surface and the group velocity vector. Group velocity $\mathbf{V}_g$ is normal to slowness surface.

The loci of the slowness vectors is called the slowness surface, while the loci of the phase velocity vectors is called the phase velocity surface. The relation between slowness surface, group velocity, and phase velocity is shown in Fig. 2.13. As will be shown in Section 4.6 of Chapter 6, a group velocity vector is always orthogonal to the slowness surface as shown in the figure.

Finally, relations between $\mathbf{V}_g$ and $\mathbf{m}$ can be obtained from

$$\mathbf{V}_g \cdot \mathbf{n} = V_g \cos\gamma = V \tag{2.190}$$

or

$$\mathbf{V}_g \cdot \mathbf{m} = 1. \tag{2.191}$$

2.9.2 Examples for a graphite-epoxy composite

To illustrate the general trends of the slowness surfaces in typical composite materials, we are going to present some numerical results for a graphite fiber-reinforced epoxy matrix composite. The material properties of the transversely isotropic composite are listed in Table 2.1 along with those of a weakly orthotropic composite to be used in following examples. Figure 2.14 shows the slowness surfaces for the quasilongitudinal $V_{(1)}$, fast quasishear $V_{(2)}$, and slow quasishear $V_{(3)}$ waves in a transversely isotropic graphite fiber-reinforced epoxy matrix composite. For definition of the angle ϕ, refer to Figs. 2.11 and 2.21; fibers are along the x_1 axis ($\phi = 0°$, $\theta = 90°$). At $\phi = \pm 90°$, this condition corresponds to the (x_2, x_3) plane of isotropy; therefore, the slownesses of all three modes are independent of θ the polar angle. In comparison, $\phi = 0°$ (Fig. 2.10) and $\phi = 180°$ correspond to the (x_1, x_3) plane that exhibits the strongest degree of anisotropy; therefore, the slownesses of all three modes are highly dependent on θ. Quantitative information can be better illustrated using polar diagrams corresponding

Table 2.1 Material properties of the transversely isotropic and weakly orthotropic graphite fiber-reinforced composite used in our illustrations.

Stiffness [10^9 N/m^2]	Transversely isotropic	Slightly orthotropic
C_{11}	162	162
C_{22}	17.0	15.3
C_{33}	17.0	18.7
C_{12}	11.8	10.6
C_{13}	11.8	13.0
C_{23}	8.2	8.2
C_{44}	4.4	4.4
C_{55}	8.0	8.8
C_{66}	8.0	7.2
ρ[10^3 kg/m^3]	1.61	1.61

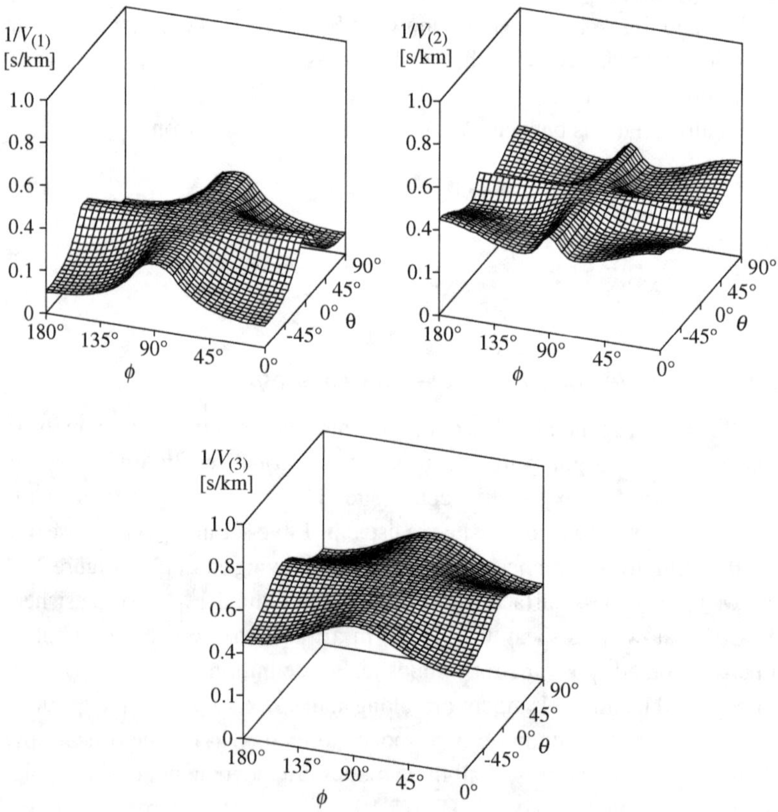

Figure 2.14. Slowness surfaces for the quasilongitudinal $V_{(1)}$, fast quasishear $V_{(2)}$, and slow quasishear $V_{(3)}$ waves in a transversely isotropic composite.

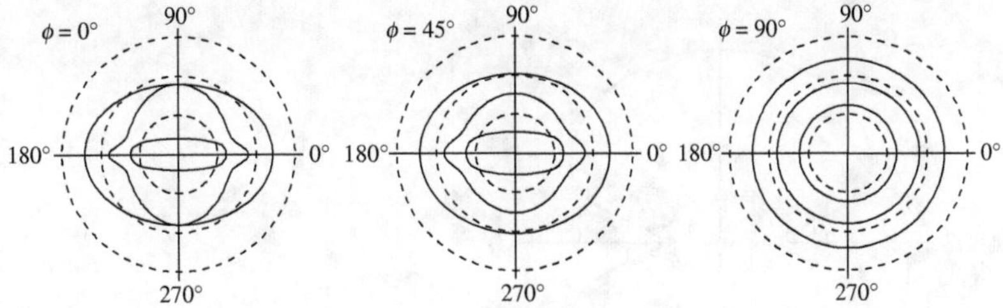

Figure 2.15. Polar diagrams of the slowness curves for all three modes in the same graphite fiber-reinforced epoxy matrix composite at $\phi = 0°$, $\phi = 45°$, and $\phi = 90°$.

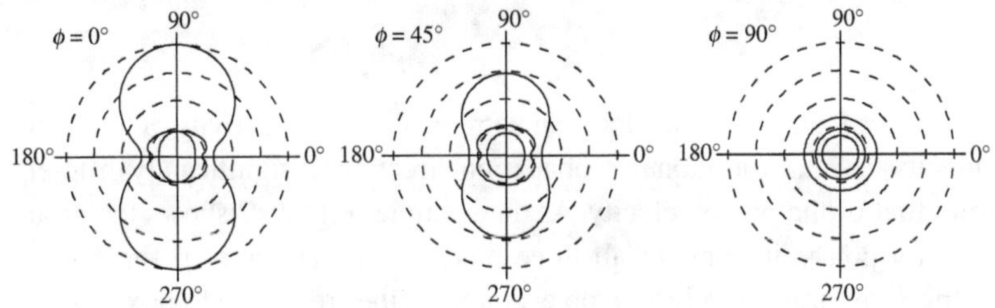

Figure 2.16. Polar diagrams of the phase velocity distributions for all three modes in the same graphite fiber-reinforced epoxy matrix composite at $\phi = 0°$, $\phi = 45°$, and $\phi = 90°$.

to constant values of azimuthal angles. As an example, Fig. 2.15 shows the polar diagrams of the slowness curves for all three modes in the same graphite fiber-reinforced epoxy matrix composite at $\phi = 0°$, $\phi = 45°$, and $\phi = 90°$. Finally, the same information can also be presented by polar diagrams of the phase velocity distributions. Figure 2.16 shows such polar diagrams of the phase velocity versus the polar angle θ for the same three constant values of the azimuthal angle ϕ. The quasilongitudinal mode reaches its highest velocity of approximately $V_1 \approx 10$ km/s along the fiber direction ($\phi = 0°, \theta = 90°$), which represents the absolute maximum for the velocity of any wave in this material. In comparison, its minimum is only $V_1 \approx 3.25$ km/s when it propagates in any direction within the plane of isotropy ($\phi = 90°$). The absolute minimum for the velocity of any wave in this material is reached by the slow quasitransverse mode also in the plane of isotropy at $V_3 \approx 1.65$ km/s. Owing to symmetry requirements, true modes always represent local minima or maxima in the phase velocity. Consequently, the absolute lowest and highest velocities always belong to true modes.

Although the minima and maxima of the group velocity necessarily coincide with the corresponding minima and maxima of the phase velocity

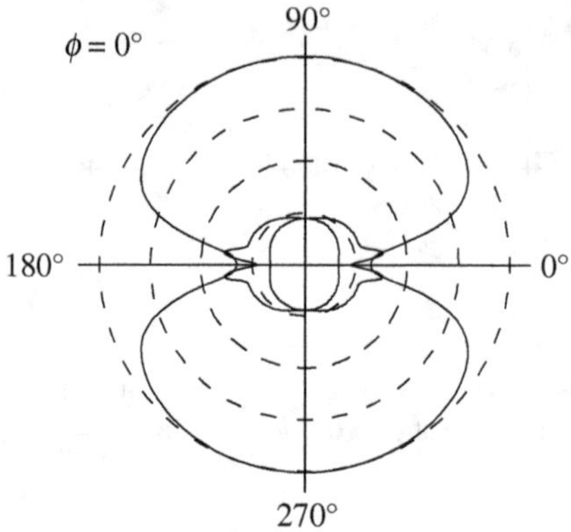

Figure 2.17. Group velocity polar diagrams as functions of θ for $\phi = 0°$ in a graphite fiber-reinforced epoxy matrix composite (2.5 km/s per division).

in symmetry directions, the local anisotropy, *i.e.*, the sensitivity of the group velocity to small variation in propagation direction, is significantly different from that of the phase velocity. As an example, Fig. 2.17 shows the group velocity polar diagrams of all three modes as functions of θ for $\phi = 0°$ in the same transversely isotropic graphite fiber-reinforced epoxy matrix composite (2.5 km/s per division). In well over a $\pm 45°$ range centered around the fiber direction ($\theta = 90°$), the group velocity of the quasilongitudinal wave barely changes. Since the same absolute variation occurs over a much smaller range centered around $\theta = 0°$ as the one exhibited by the phase velocity, the rate of change in the vicinity of the plane of isotropy is much higher for the group velocity than for the phase velocity. In the same region, the fast quasitransverse wave exhibits an even sharper variation as a result of which it actually exceeds the velocity of the quasilongitudinal wave.

Another important parameter that we can use to characterize the anisotropic nature of acoustic wave propagation in a composite material is the degree of misalignment between the group and phase propagation directions. Figure 2.18 shows the previously defined skewing angle γ as a function of the polar angle θ for $\phi = 0°$. As one can expect in a highly anisotropic material like this uniaxial composite, the misalignment could be very large, in excess of $\gamma = 50°$, but it vanishes in both symmetry directions, *i.e.*, along the fibers ($\theta = 90°$) and normal to them ($\theta = 0°$) in the plane of isotropy. As in this case the direction of wave propagation always remains within one of the planes of symmetry (x_1, x_3), the group velocity is also always contained in this plane, and the skewing orientation can be fully described with a single angle γ that is either positive or negative depending upon whether the group velocity bends toward the direction of higher or lower polar angles, respectively. For the quasilongitudinal and slow quasishear waves the dependence of the skewing angle on the polar angle is a relatively simple function of odd symmetry relative to the fiber direction ($\theta = 90°$).

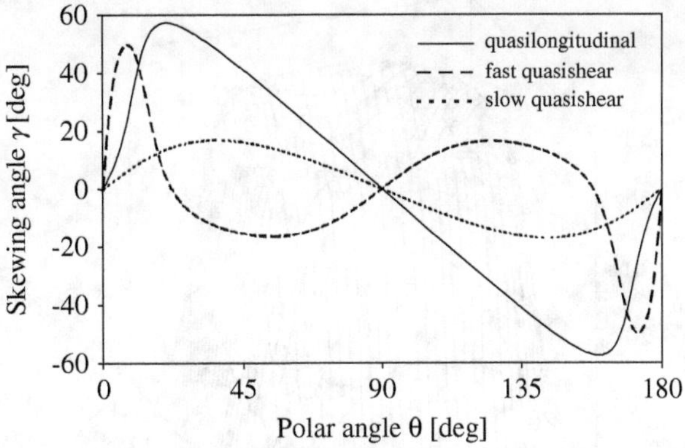

Figure 2.18. Skewing angle γ between the group and phase propagation directions versus the polar angle θ for $\phi = 0°$ in a transversely isotropic graphite fiber-reinforced epoxy matrix composite.

The group velocity vector of these two modes always bends from the wave direction toward the fiber orientation. The behavior of the fast quasishear mode is somewhat more complicated. In the vicinity of the plane of isotropy ($\theta = 0°$), the group velocity vector of the fast quasishear mode also bends toward the fiber orientation, but when the wave propagation direction is inclined by more than $\theta \approx 20°$ with respect to the plane of isotropy, the group velocity bends away from the fiber direction toward this plane of symmetry.

Generally, the polarization angle $\delta_{(\ell)}$ with respect to the wave propagation direction can be defined as a positive angle calculated from

$$\delta_{(\ell)} = \arccos(p_{i(\ell)} n_i), \tag{2.192}$$

where $\ell = 1, 2, 3$ indicates the three modes. In transversely isotropic materials, there exists a symmetry plane associated with any direction of wave propagation (see later in Sect. 2.9.4). Two modes, namely the quasilongitudinal and one of the quasishear waves, are always polarized in this plane of symmetry, while the third mode is a true transverse wave polarized normal to it. Therefore, the degree of misalignment between the propagation direction and the polarization vector of the quasilongitudinal wave alone is sufficient to fully characterize the polarization of both quasi modes.

As an example, Fig. 2.19 shows the polarization skewing angle δ for the quasilongitudinal mode in the same transversely isotropic graphite fiber-reinforced epoxy matrix composite we used in our previous illustrations. In certain directions, the polarization vector of the quasilongitudinal wave is misaligned by more than $\delta = 45°$ from the wave propagation direction, *i.e.*, the polarization is actually more transverse than longitudinal. The edges of this plot correspond to the (x_2, x_3) plane of symmetry ($\phi = \pm 90°$ and any θ) and the x_3 axis ($\theta = 0°$ or $\phi = 180°$ and any ϕ), therefore the polarization skewing angle is identically zero around the edges. In addition, there is a

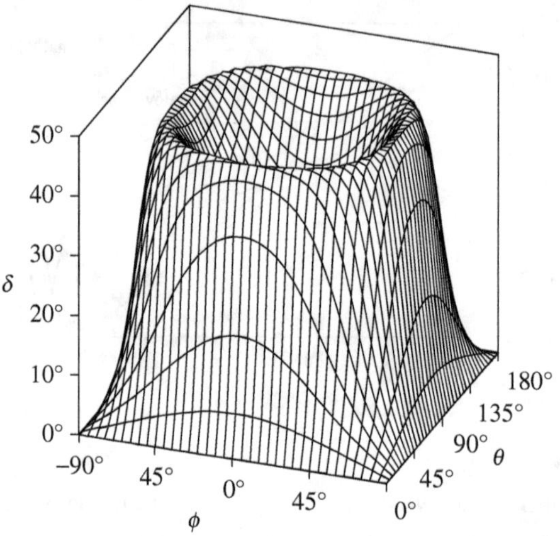

Figure 2.19. Skewing angle δ between the polarization vector of the quasilongitudinal mode and the propagation direction in a transversely isotropic graphite fiber-reinforced epoxy matrix composite.

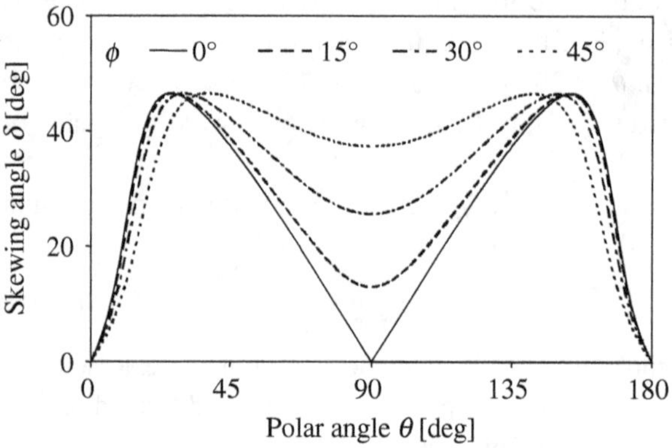

Figure 2.20. Skewing angle δ between the polarization vector of the quasilongitudinal mode and the propagation direction versus the polar angle θ for different values of ϕ.

sharp zero at the center corresponding to the fiber direction ($\phi = 0°$ and $\theta = 90°$) where the polarization skewing angle also vanishes. This effect is further illustrated in Fig. 2.20 where we plotted δ as a function of the polar angle θ for different values of ϕ. This figure well demonstrates that even a relatively small deviation of the quasilongitudinal wave direction from the fiber orientation will result in very substantial transverse displacements because of the particularly low stiffness of the unidirectional composite normal to the reinforcement.

2.9.3 Phase and group velocities in symmetry planes of orthotropic materials

As we discussed in the previous sections, the group velocity $\mathbf{V}_g$ of a plane wave with known phase velocity V and polarization direction $\mathbf{p}$ can be

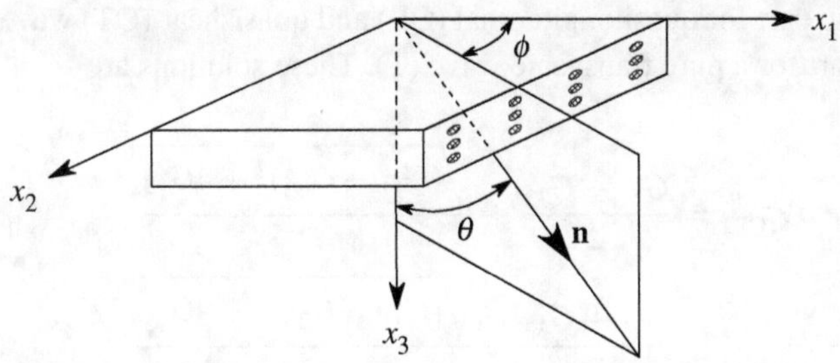

Figure 2.21. Unidirectional composite with wave propagation in a general direction.

determined from Eq. (2.148) as follows

$$V_{gi} = \frac{c_{ijlm} n_\ell p_m p_j}{\rho V}. \tag{2.193}$$

We also saw that the phase velocity is the projection of the group velocity vector onto the wave normal **n**

$$\mathbf{V}_g \cdot \mathbf{n} = V. \tag{2.194}$$

It is often desirable to have explicit analytical equations for the group velocity. Kim [14] derived closed-form analytical formulas for the group velocities of quasilongitudinal and quasitransverse waves in symmetry planes of orthotropic materials. Below, we present the resulting formulas for the group velocities using a different approach [15, 16] which makes them more convenient for subsequent use in the inversion method to be described in the subsequent chapter and briefly outlined in this section.

Let us consider wave propagation in the (x_1, x_3) symmetry plane ($\phi = 0°$) of an orthotropic material as shown in Fig. 2.21. In this case, the elements of the Christoffel matrix (previously given by Eq. (2.53)) have the following form

$$\begin{vmatrix} G_{11} - \rho V^2 & 0 & G_{13} \\ 0 & G_{22} - \rho V^2 & 0 \\ G_{13} & 0 & G_{33} - \rho V^2 \end{vmatrix} = 0, \tag{2.195}$$

where, from Eq. (2.52) we can write that

$$G_{11} = C_{11} \sin^2\theta + C_{55} \cos^2\theta; \qquad G_{33} = C_{55} \sin^2\theta + C_{33} \cos^2\theta;$$
$$G_{13} = (C_{13} + C_{55}) \sin\theta \cos\theta; \qquad G_{22} = C_{66} \sin^2\theta + C_{44} \cos^2\theta. \tag{2.196}$$

The angle θ is between the direction of propagation and the x_3 axis as shown in Fig. 2.21. In the plane of symmetry, Eq. (2.195) decouples into a quadratic part for quasilongitudinal (QL) and quasishear (QT) waves and a linear part for a pure transverse wave (T). These solutions are

$$\rho V_{(\mathrm{QL})}^2 = \frac{G_{11}+G_{33}}{2} + \frac{\sqrt{(G_{11}-G_{33})^2+4G_{13}^2}}{2}$$
$$\rho V_{(\mathrm{QT})}^2 = \frac{G_{11}+G_{33}}{2} - \frac{\sqrt{(G_{11}-G_{33})^2+4G_{13}^2}}{2} \tag{2.197}$$
$$\rho V_{(\mathrm{T})}^2 = G_{22}.$$

For the quasilongitudinal (ℓ = QL) and quasishear (ℓ = QT) waves, the polarization vector can be calculated from

$$p_{1(\ell)} = \frac{1}{\sqrt{1+\xi_{(\ell)}^2}}, \quad p_{2(\ell)} = 0, \quad p_{3(\ell)} = \frac{\xi_{(\ell)}^2}{\sqrt{1+\xi_{(\ell)}^2}}, \tag{2.198}$$

where

$$\xi_{(\ell)} = \frac{\rho V_{(\ell)}^2 - G_{11}}{G_{33}}. \tag{2.199}$$

For the pure transverse wave, the polarization vector is simply

$$p_{1(\mathrm{T})} = 0, \quad p_{2(\mathrm{T})} = 1, \quad p_{3(\mathrm{T})} = 0. \tag{2.200}$$

Along the symmetry axes, the quasilongitudinal and quasitransverse modes also become pure modes that are polarized along the propagation direction and normal to it, respectively. For $\theta = 0°$, $\mathbf{p}_{(\mathrm{QL})} = [0, 0, 1]$ and $\mathbf{p}_{(\mathrm{QT})} = [1, 0, 0]$. For $\theta = 90°$, $\mathbf{p}_{(\mathrm{QL})} = [1, 0, 0]$ and $\mathbf{p}_{(\mathrm{QT})} = [0, 0, 1]$.

Owing to symmetry, for all three waves propagating in the (x_1, x_3) plane the group velocity vectors also lie in the same plane. The nonzero components of the group velocity vectors for the quasilongitudinal ($\ell = \mathrm{QL}$) and quasitransverse ($\ell = \mathrm{QT}$) waves are

$$V_{g1(\ell)} = \frac{(C_{11}p_{1(\ell)}^2 + C_{55}p_{3(\ell)}^2)\sin\theta + (C_{13}+C_{55})p_{1(\ell)}p_{3(\ell)}\cos\theta}{\rho V_{(\ell)}}$$
$$V_{g3(\ell)} = \frac{(C_{13}+C_{55})p_{1(\ell)}p_{3(\ell)}\sin\theta + (C_{55}p_{1(\ell)}^2 + C_{33}p_{3(\ell)}^2)\cos\theta}{\rho V_{(\ell)}}, \tag{2.201}$$

where $\mathbf{p}_{(\ell)}$ is the polarization vector as given in Eq. (2.198). For the pure transverse wave, the nonzero group velocity components are

$$V_{g1(T)} = \frac{C_{66} \sin\theta}{\rho V_{(T)}}$$

$$V_{g3(T)} = \frac{C_{44} \cos\theta}{\rho V_{(T)}}. \tag{2.202}$$

The group velocity vector $\mathbf{V}_g$ subtends an angle of $\zeta = \theta + \gamma$ with respect to the x_3 axis. This angle equals

$$\zeta = \arctan\left(V_{g1}/V_{g3}\right) \quad \text{when } V_{g3} \neq 0, \tag{2.203}$$

so that Eq. (2.194) yields

$$V_g \cos(\zeta - \theta) = V_g \cos(\gamma) = V. \tag{2.204}$$

For better understanding of the above relations, the reader is encouraged to make a 2-D drawing for the symmetry plane (x_1, x_3) ($\phi = 0°$), showing the group and phase velocity vectors.

The quasilongitudinal and quasitransverse phase and group velocities depend on four elastic constants, namely C_{11}, C_{33}, C_{13}, and C_{55}, and the velocity of the pure transverse wave depends on C_{44} and C_{66}. Wave propagation in the (x_2, x_3) symmetry plane is described by equations similar to Eqs. (2.195–2.204) with the index substitution $1 \leftrightarrow 2$ and $55 \leftrightarrow 44$. From group or phase velocity measurements in the (x_2, x_3) plane, it is possible to determine C_{22}, C_{33}, C_{23}, and C_{44}. From combined measurements in the (x_1, x_3) and (x_2, x_3) symmetry planes, one can determine seven of the nine elastic constants of an orthotropic material, *i.e.*, all except C_{12} and C_{66}, which can be determined from measurements in a non-symmetry plane assuming that the other elastic constants are already known, as will be shown in the next section.

2.9.4 Phase and group velocities in non-symmetry planes of orthotropic and transversely isotropic materials

In the case of wave propagation in a non-symmetry plane of orthotropic and transversely isotropic materials, the cubic Christoffel equation (Eq. 2.56) cannot be decoupled into lower-order terms as was done for a plane of symmetry in the previous section. However, if the material is transversely isotropic, there exists a symmetry plane associated with any direction of wave propagation. If we consider the rotated coordinate system associated with this plane, the Christoffel equation can be decoupled into linear and quadratic parts. This factorization is well known in crystal acoustics [12, 13] for hexagonal symmetry. In this section, we review a simple factorization method and analytical formulas for phase and group velocities

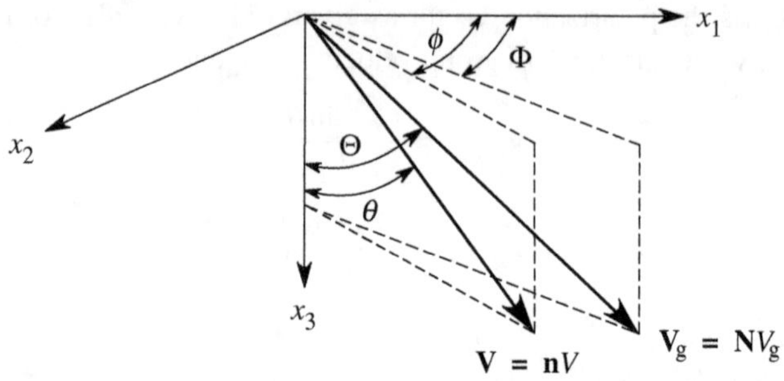

Figure 2.22. Phase **n** and group **N** velocity directions in a Cartesian coordinate system.

and polarization vectors obtained in [15, 16]. We also show that the exact equations for transversely isotropic materials also provide a very good approximation for non-symmetry planes in unidirectional orthotropic composites.

Consider a transversely isotropic material with its plane of symmetry coinciding with the (x_2, x_3) plane of a Cartesian coordinate system. For a wave propagating in an arbitrary direction **n** as shown in Fig. 2.22, the wave vector is characterized by angular coordinates ϕ and θ as follows

$$\mathbf{n} = [\cos\phi \sin\theta, \sin\phi \sin\theta, \cos\theta], \tag{2.205}$$

and the energy propagation direction is given by

$$\mathbf{N} = [\cos\Phi \sin\Theta, \sin\Phi \sin\Theta, \cos\Theta]. \tag{2.206}$$

Generally, the plane determined by the x_3 axis and the **n** vector is not a symmetry plane. To simplify the otherwise rather complicated problem, let us introduce the rotated coordinate system (x_1', x_2', x_3') as shown in Fig. 2.23 so that the (x_1', x_3') plane of the rotated coordinate system contains both the axis of material symmetry x_1 and the wave normal **n**, which is aligned with the x_3' axis of the rotated coordinate system. Due to symmetry, the group velocity vector **N** also lies in the (x_1', x_3') plane. The solution of the Christoffel equation for the phase velocities V and their associated polarization unit vectors **p** in the rotated coordinate system (x_1', x_2', x_3') were given by Chu *et al.* as follows [15].

The wave speeds of the quasimodes are

$$\begin{aligned} \rho V_{(QL)}^2 &= C_{33}' + \frac{1}{2}\left(C_{33}' - C_{55}'\right) f(\xi^2) \\ \rho V_{(QT)}^2 &= C_{55}' - \frac{1}{2}(C_{33}' - C_{55}') f(\xi^2) \\ \rho V_{(T)}^2 &= C_{44}', \end{aligned} \tag{2.207}$$

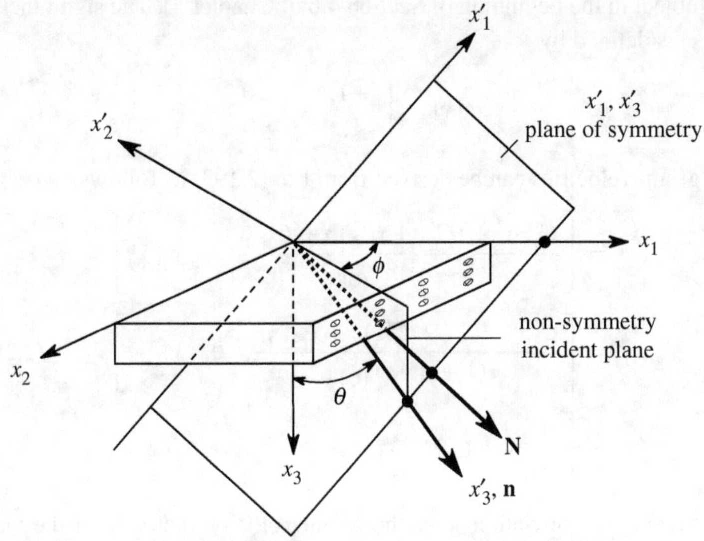

Figure 2.23. Plane of symmetry (x'_1, x'_3) for an arbitrary direction **n** of wave propagation in a transversely isotropic material. The plane of isotropy coincides with the (x_2, x_3) plane.

and

$$\mathbf{p}_{(QL)} = \left[\frac{\eta}{\sqrt{1+\eta^2}},\ 0,\ \frac{1}{\sqrt{1+\eta^2}}\right]$$
$$\mathbf{p}_{(QT)} = \left[\frac{1}{\sqrt{1+\eta^2}},\ 0,\ \frac{-\eta}{\sqrt{1+\eta^2}}\right] \tag{2.208}$$
$$\mathbf{p}_{(T)} = [0,\ 1,\ 0],$$

where the function f, polarization factor ξ, and polarization component η are defined as

$$f(\xi^2) = \text{sgn}(C'_{33} - C'_{55})\sqrt{1+\xi^2} - 1, \tag{2.209}$$

$$\xi = \frac{2C'_{53}}{|C'_{33} - C'_{55}|}, \tag{2.210}$$

$$\eta = \frac{\xi}{\text{sgn}(C'_{33} - C'_{55}) + \sqrt{1+\xi^2}}, \tag{2.211}$$

and C'_{ij} are elastic constants in the rotated (x'_1, x'_2, x'_3) coordinate system. Note that one of the transverse modes (third one in Eq. (2.207)) is a "true" (pure) mode, and it is always the slower of the two modes (see also discussion on the subject in the beginning of Section 4.6 of Chapter 4). The sign function sgn(x) is defined by

$$\operatorname{sgn}(x) = \begin{cases} -1, & x < 0 \\ 1, & x \geq 0 \end{cases} . \tag{2.212}$$

The group velocities can be derived from Eq. (2.193) as follows

$$\begin{aligned}
\mathbf{V}_{\mathrm{g(QL)}} &= \left[\frac{C'_{15}\eta^2 + (C'_{13} + C'_{55})\eta + C'_{35}}{(1+\eta^2)\rho V_{\mathrm{(QL)}}},\ 0,\ V_{\mathrm{(QL)}} \right] \\
\mathbf{V}_{\mathrm{g(QT)}} &= \left[\frac{C'_{15} - (C'_{13} + C'_{55})\eta + C'_{35}\eta^2}{(1+\eta^2)\rho V_{\mathrm{(QT)}}},\ 0,\ V_{\mathrm{(QT)}} \right] \\
\mathbf{V}_{\mathrm{g(T)}} &= \left[\frac{C'_{46}}{\rho V_{\mathrm{(T)}}},\ 0,\ \frac{C'_{44}}{\rho V_{\mathrm{(T)}}} \right],
\end{aligned} \tag{2.213}$$

and the angular deviation γ of the group velocity vector from the wave vector is

$$\gamma = \arctan\left(\frac{V_{\mathrm{g1}}}{V_{\mathrm{g3}}}\right). \tag{2.214}$$

In Eq. (2.214), the V_{g} projections are in the primed (rotated) coordinate system shown in Fig. 2.23; θ' is zero in this system, while the original θ is not.

When the material is slightly anisotropic in the (x_2, x_3) plane, Eqs. (2.207–2.213) can still be used as approximations, if exact equations for the elastic constants C'_{ij} (for explicit formulas for C'_{ij}, see [15, 16]) of the orthotropic material in the rotated coordinate system are used. The group velocity vector in this case will deviate slightly from the (x'_1, x'_3) plane. The difference between group and phase velocity directions can be characterized by in-plane α and out-of-plane β components

$$\alpha = \arctan\left(\frac{N'_1}{N'_3}\right), \quad \beta = \arcsin(N'_2), \tag{2.215}$$

where $\mathbf{N} = [N'_1, N'_2, N'_3]$ is the group velocity unit vector in the rotated coordinate system. We call such materials weakly orthotropic. A range of materials, including uniaxial composites might exhibit such a property.

To illustrate the accuracy of the above-described approximation, let us consider [16] the phase and group velocities for a slightly orthotropic graphite fiber-reinforced epoxy matrix composite. The material properties of this hypothetical orthotropic composite were obtained by slightly offsetting the corresponding elastic constants of a typical transversely isotropic composite. The desired modest orthotropy was produced by decreasing C_{22}, C_{12}, and C_{66} by 10 percent and increasing C_{33}, C_{13}, and C_{55} by the

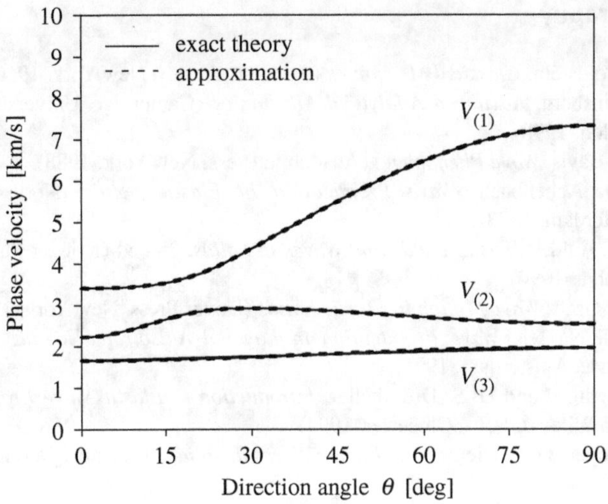

Figure 2.24. Phase velocity versus the polar angle θ for $\phi = 15°$ in a weakly orthotropic material.

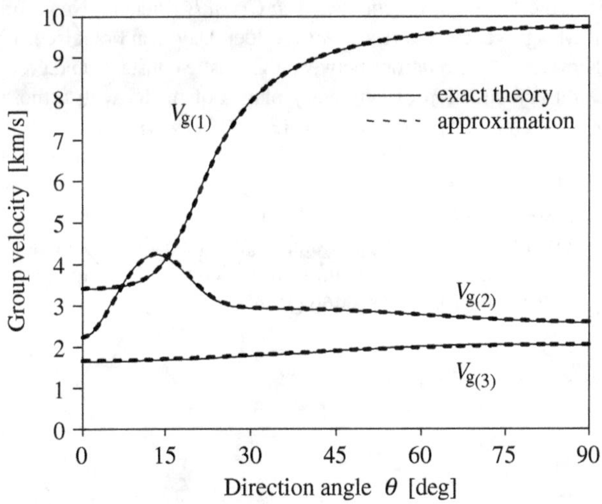

Figure 2.25. Group velocity versus the polar angle θ for $\phi = 15°$ in a weakly orthotropic material.

same amount so that $C_{22} \approx 0.8C_{33}$, $C_{12} \approx 0.8C_{13}$, and $C_{66} \approx 0.8C_{55}$. The material properties of both transversely isotropic and slightly orthotropic composites were listed earlier in Table 2.1. Figures 2.24 and 2.25 show the comparison between the exact and approximate phase and group velocities, respectively, for this slightly orthotropic composite. As one can see, the agreement between the exact and approximate solutions is very good.

Bibliography

1. H. Goldstein, *Classical Mechanics* (Addison-Wesley, New York, 1950).
2. J. Ginsberg, *Advanced Analytical Mechanics* (Cambridge University Press, London, 1998).
3. J. L. Davis, *Wave Phenomena* (Academic Press, New York, 1988).
4. J. D. Achenbach, *Wave Propagation in Elastic Solids* (North-Holland, Amsterdam, 1973).
5. B. A. Auld, *Acoustic Fields and Waves in Solids*, 2nd ed (Krieger Publishing, Malabar, 1990).
6. K. F. Graff, *Wave Motion in Elastic Solids* (Dover Press, New York, 1990).
7. A. H. Nayfeh, *Wave Propagation in Layered Anisotropic Media* (Elsevier Science, Amsterdam, 1995).
8. A. Bedford and D. S. Drumheller, *Introduction to Elastic Wave Propagation* (John Wiley & Sons, Chicester, 1994).
9. D. Royer and E. Dieulesaint, *Elastic Waves in Solids I* (Springer-Verlag, Berlin, 2000).
10. J. L. Rose, *Ultrasonic Waves in Solid Media* (Cambridge University Press, London, 1999).
11. J. P. Wolfe, *Imaging Phonons: Acoustic Wave Propagation in Solids* (Cambridge University Press, London, 1998).
12. F. I. Fedorov, *Theory of Elastic Waves in Crystals* (Plenum, New York, 1968).
13. M. J. P. Musgrave, *Crystal Acoustics* (Holden Day, San Francisco, 1970).
14. K. Y. Kim, "Analytic relations between the elastic constants and group velocity in an arbitrary direction of symmetry planes of media with orthorhombic or higher symmetry," Phys. Rev. **B49**, 3713–3724 (1994).
15. Y. C. Chu, A. D. Degtyar, and S. I. Rokhlin, "On determination of orthotropic material moduli from ultrasonic velocity data in non-symmetry planes," J. Acoust. Soc. Am. **95**, 3191–3203 (1994).
16. A. D. Degtyar and S. I. Rokhlin, "Comparison of elastic constant determination in anisotropic materials from ultrasonic phase and group velocity data," J. Acoust. Soc. Am. **102**, 3458–3466 (1997).

3

Bulk Ultrasonic Techniques for Evaluation of Elastic Properties

Currently, the design of most composite components is based on stiffness, and therefore methods for static measurement of stiffness are in wide use. The disadvantages of these methods lie in their destructive nature (the samples must be cut from parts of different orientations), in the difficulty of measuring shear properties, and in the need for extra care when measuring Young's modulus in off-axis directions. Ultrasonic methods are more accurate and have higher spatial resolution than static measurements. As we showed in Chapter 2, by measuring ultrasonic velocities in several predefined directions, all elastic constants can be determined. The generic method described there is also destructive, however, requiring cutting numerous samples with appropriate fiber orientation.

Specialized nondestructive methods for determining the elastic moduli of composite materials are more powerful and they can be applied to composite coupons before, during, and after strength or fatigue testing. It is important to have a fast and inexpensive technique to estimate input parameters for composite design. It is even more important to have a technique to evaluate composites during service to verify that the manufactured elastic stiffnesses match those assumed in the design.

Several methods that utilize bulk ultrasonic waves for measurement of composite elastic constants are considered in this chapter. By bulk wave methods, we mean quasilongitudinal and quasitransverse ultrasonic wave velocity measurement methods that are applicable when the sample thickness h is larger than both the ultrasonic pulse space length τV and the wavelength λ (τ is the ultrasonic pulse length in time, and V is the wave speed). Other methods, which are applicable in the range $h < \tau V$ and which account for wave interference with the boundaries of the specimen, will be considered in the following chapters.

3.1 Bulk Wave Refraction Method for Phase Velocity Measurement

3.1.1 Introduction

The most promising way to evaluate composite elastic properties non-destructively is to measure ultrasonic velocities in different directions in the composite material and reconstruct the elastic constants from these values using some kind of an inversion technique. One possible method has been suggested by Markham in the 1970s [1], who used ultrasonic waves obliquely incident from water onto a composite plate to measure ultrasonic velocities in various directions and evaluated the results to determine elastic constants. Using Markham's technique, Smith [2] was able to determine a set of five independent elastic constants. Similar experiments were conducted by Gieske and Allred [3] and a set of nine independent elastic constants was determined. Later, this technique was utilized with greater precision [4, 5] for planes of symmetry in unidirectional graphite-epoxy composite plates. In addition to velocity measurements, Hosten *et al.* [5] have determined the tensor of attenuation factors, which leads to better comparison between their theoretical and experimental results.

However, as recognized by Gieske and Allred [3], the time of flight measured by Markham's technique yields the group velocity, whereas straightforward determination of the elastic constants requires phase velocity data. Since these two velocities deviate in general from each other in anisotropic materials, the difference between the measured results (group velocity) and those used for calculation of the elastic constants (phase velocity) raises serious questions about the utility of the technique. Pearson and Murri [4] have shown that, in the plane of symmetry of transversely isotropic composites, for Markham's arrangement of transducers, the group velocity measurement gives the correct results for phase velocity. This result has been extended to arbitrary directions in generally anisotropic materials by Rokhlin and Wang [6].

A further development of this technique has been suggested by Rokhlin and Wang [6–8], who have developed an improved double through-transmission method of measuring ultrasonic bulk velocities in arbitrary directions in orthotropic materials [6, 8]. It has been shown [7] that a plane wave approximation used for the data analyses for the method used in [6, 8] is sufficient; it is demonstrated that a phase correction for the transmitted transverse wave is recommended for more accurate measurements. Kline and Chen [9] have developed a method that uses, instead of transmitted waves, waves reflected from the back side of the composite plate, requiring access to one side of the sample only. Here, we will follow a description of the method given in [8].

3.1.2 Delay times for phase and group velocities in a solid layer of general anisotropy

Let us consider a generally anisotropic composite plate arbitrarily oriented relative to the incident plane. For a fluid-coupled anisotropic plate, at a given angle of oblique incidence, three elastic waves may be excited with different velocities, one quasilongitudinal and two quasitransverse. In general, the group velocities of these three waves are different from the phase velocities not only in magnitude but also in direction. Figure 3.1 illustrates this difference between the three refracted waves at oblique incidence. The directions of the wave vectors are shown by dashed lines, and the projections of the group velocities on the incident plane are shown by solid lines. All three phase velocity vectors lie in the incident plane (plane of the figure). However, the group velocity vectors do not necessarily lie in the incident plane unless incidence occurs in a plane of symmetry.

To measure the ultrasonic velocity, one measures the acoustic signal time delay compared to the length of the acoustic path. The velocity determined in this way is the group velocity. To measure the time delay, one scans the field with an ultrasonic transducer as shown in Fig. 3.1 to find the position at which

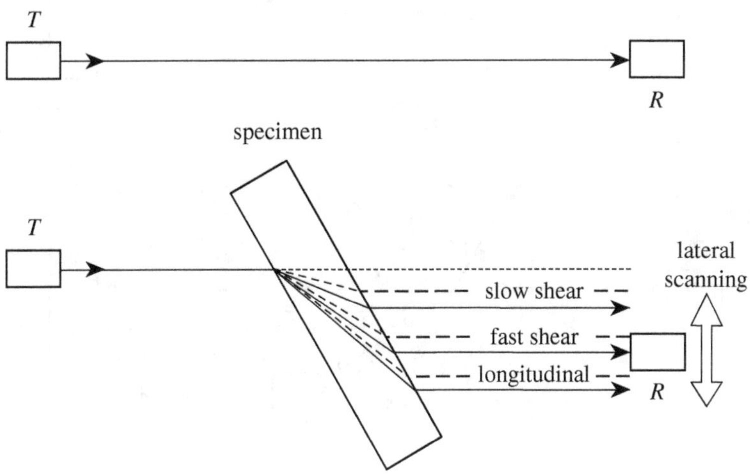

Figure 3.1. General concept of wave velocity measurements in an anisotropic plate. The reference measurement is shown at the top of the figure. The receiving transducers should be positioned in the center of the refracted ultrasonic beam (dashed lines indicate the virtual path of the phase vector).

the ultrasonic signal is maximum. When the group velocity deviates from the incident plane (plane of the figure in Fig. 3.1), one scans additionally in a plane perpendicular to the incident plane. Thus, the refraction angle for the group velocity (ultrasonic beam) can be measured. To find the length of the acoustic path for the phase velocity, one needs the phase velocity, refraction angle. While the refraction angles of the phase velocities are directly related to the angle of incidence through Snell's law, they cannot be calculated immediately since the phase velocity in the material is unknown.

The simple pitch–catch method shown in Fig. 3.1 is rendered rather troublesome by the need for lateral scanning to locate the three transmitted compressional beams in the fluid that correspond to the three different refracted waves in the anisotropic specimen. This scanning can be avoided in pulse-echo mode if the separate receiver is replaced by a plane reflector normally aligned to a single ultrasonic transducer that is used in both transmission and reception, as shown in Fig. 3.2. This method is particularly useful to study Lamb modes in thin plates when spatial and temporal separation of the different modes is even more difficult [10]. As noted in the Introduction, the spatial pulse length is selected to be smaller than the sample

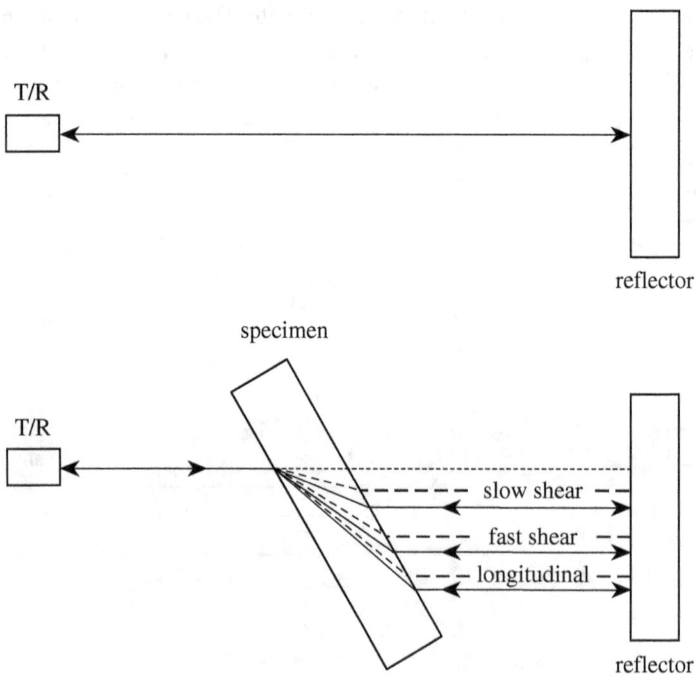

Figure 3.2. Single-transducer immersion testing of an anisotropic composite plate in pulse-echo mode using a plane reflector (dashed lines indicate the virtual path of the phase vector).

thickness, and thus there is no interference of multiple reflections in the sample. Therefore, the pulse propagation in the sample may be considered equivalent to that in an unbounded composite medium. Moreover, in this method of measurement, only the first transmitted signal is used for measurements. Assuming that the opposite surfaces of the specimen are parallel to each other, the transmitted beams remain parallel to the direction of the incident beam, therefore they hit the reflector at normal incidence. Regardless of the degree and direction of the lateral displacement of these transmitted beams, upon reflection from the reflector, owing to reciprocity, they exactly retrace their respective paths through the specimen back to the ultrasonic transducer.

Figure 3.3 illustrates the difference between the refracted wave and refracted ray directions in a unidirectional composite plate when the plane of incidence is rotated from the fiber direction. There are two coordinate systems in Fig. 3.3, primed and unprimed. The unprimed coordinate system is based on the principal material directions in the composite, *i.e.*, x_1 is along the fiber direction, x_2 is normal to the fiber direction in the plane of the sample, and x_3 is normal to the plane of the sample. The primed coordinate system is based on the ultrasonic plane of incidence. In the example shown in Fig. 3.3, it is rotated through an angle ϕ from the nonprimed system around

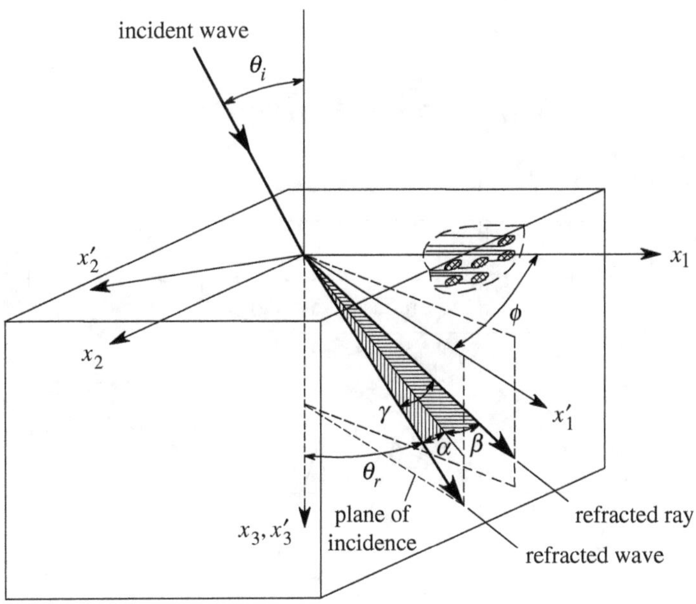

Figure 3.3. Illustration of the different refracted wave and refracted ray directions in a unidirectional composite plate. The plane of incidence is rotated from the fiber direction around the $x_3 = x_3'$ axis by angle ϕ.

axis x_3, *i.e.*, the incident plane (x_1', x_3') subtends angle ϕ from the fiber orientation (axis x_1). For an arbitrary refracted wave with wave vector **k**, the ray direction deviates from the elastic wave vector by an angle γ. This deviation angle has, in general, both an incident-plane (x_1', x_3') component α and an out-of-incident-plane component β. The three angles are related by

$$\cos\gamma = \cos\alpha\,\cos\beta. \tag{3.1}$$

Let us consider an ultrasonic wave incident from fluid onto a generally anisotropic plate at an arbitrary angle. As shown in Fig. 3.4, the incident wave vector $\mathbf{k}_i$ impinges on the plate at point O at an angle θ_i relative to the plate normal. For an arbitrary refracted wave in the anisotropic plate, the phase velocity vector **V** lies in the plane of incidence (defined by the incident wave vector and the plate normal), but its group velocity, $\mathbf{V}_g$, deviates from the wave vector direction by an angle γ.

The path of the phase velocity vector crosses the back surface of the plate at point B and the path of the group velocity vector crosses at point A. The wavefront normal of the refracted wave is directed along the phase velocity vector, and the time of propagation in the plate from point O on the front surface to point B on the back surface is

$$t_p = \frac{\overline{OB}}{V}. \tag{3.2}$$

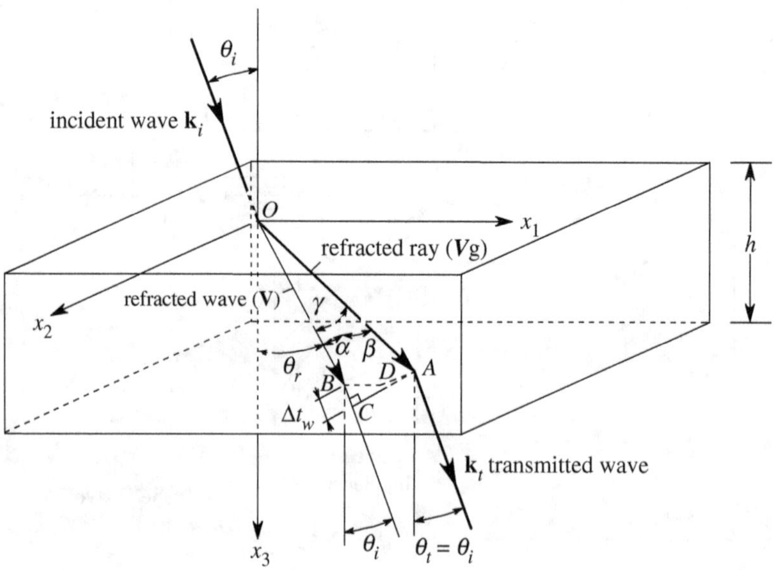

Figure 3.4. Illustration of difference in acoustic paths for propagation with phase **V** and group $\mathbf{V}_g$ velocity.

The acoustic energy propagates along the group velocity direction and reaches the back surface at time

$$t_g = \frac{\overline{OA}}{V_g}, \tag{3.3}$$

which is just the time to pass through the plate. Time-of-flight measurements with ultrasonic pulses directly on the back of the plate give t_g. Since the elastic constants are directly related to the phase velocity V through the Christoffel equation (see Section 2 of the previous chapter), it is the phase velocity that we need to determine the elastic constants. (How to employ the group velocity for this purpose is discussed later in Section 3.5.) Clearly, relations must be found to connect the variables described by Eqs. (3.2) and (3.3).

From geometrical considerations, it is not difficult to see that the distances $\overline{OB}$ and $\overline{OA}$ are given by

$$\overline{OB} = \frac{h}{\cos\theta_r} \tag{3.4}$$

$$\overline{OA} = \frac{h}{\cos(\alpha+\theta_r)\cos\beta}, \tag{3.5}$$

where h is the thickness of the plate and θ_r is the refraction angle. Substituting Eqs. (3.4) and (3.5) into Eqs. (3.2) and (3.3) yields

$$t_p = \frac{h}{V\cos\theta_r} \tag{3.6}$$

$$t_g = \frac{h}{V_g\cos(\alpha+\theta_r)\cos\beta}. \tag{3.7}$$

One can see immediately that there is a difference in time between the energy propagation time t_g and the virtual (phase velocity) wave propagation time t_p in the plate, and that the difference is

$$\Delta t_s = t_g - t_p = \frac{h}{V_g}\left(\frac{1}{\cos\beta\cos(\alpha+\theta_r)} - \frac{1}{\cos\gamma\cos\theta_r}\right), \tag{3.8}$$

where we exploited that $V = V_g\cos\gamma$. Equation (3.8) can be further simplified by taking into account Eq. (3.1) and Snell's law

$$\frac{\sin\theta_r}{V} = \frac{\sin\theta_i}{V_f}, \tag{3.9}$$

where V_f is the velocity in the coupling medium, which is known. Substituting Eqs. (3.1) and (3.9) into Eq. (3.8) yields [6]:

$$\Delta t_s = \frac{h\sin\theta_i\sin\alpha}{V_f\cos(\theta_r+\alpha)\cos\theta_r}. \tag{3.10}$$

From Eq. (3.10), one can see that the difference between the energy propagation time (from the group velocity) and the virtual wave propagation time (from the phase velocity) is a function of the group velocity deviation angle, but only through its out-of-incident-plane component α. When this component vanishes, *i.e.*, when the plane defined by the refracted wave vector and the normal to the incidence plane (normal to the x_2' axis in Fig. 3.3) coincides with a plane of symmetry, we have $\alpha = 0$ and hence $\Delta t_s = 0$. This condition is enforced by symmetry. If this orthogonal to the incident plane is the plane of symmetry (the refraction wave vector and wave normal are in this plane), the group velocity vector must also be in this plane; otherwise, the symmetry will be violated, although the deviation angle β is nonzero if the incident plane itself is a non-symmetry plane. That is, in this circumstance $t_{\rm g} = t_{\rm p}$ and there is no difference between the times for the acoustic energy and the phase front to pass through the anisotropic plate. This condition will be satisfied at critical angles (see Fig. 3.19) if the surface of the sample is the plane of symmetry (Section 3.32). As for isotropic materials the group and phase velocities are equal, it is clear that for isotropic materials this condition is always satisfied. The same result, *i.e.*, $\Delta t_s = 0$, is also valid when the incident beam is normal to the plate, because in this case $\sin \theta_i = 0$; an example of this case is shown in Fig. 2.9.

In all other cases, the difference Δt_s is nonzero. It should be noted that Δt_s can be either positive or negative. Positive Δt_s corresponds to positive α, as is illustrated in Fig. 3.4. Negative α occurs when the angle between the plate normal and the projection of the group velocity on the incidence plane is less than the refraction angle calculated from Snell's law.

After the refracted wave passes through the plate, it converts at the rear plate surface back into a compressional acoustic wave in the coupling fluid. The acoustic wave continues to propagate parallel to its original direction as illustrated in Fig. 3.5. On the left side of the figure, a sample is shown inserted between the transmitting and receiving transducers. The reference path in the coupling medium is shown separately on the right side of Fig. 3.5. The specimen is sectioned so that the particular cross-section of the sample that coincides with the incident plane is shown as the front surface. Due to the difference in direction between the group velocity and phase velocity, the ultrasonic beam, which follows the group velocity direction, and the wave normal emerge from the plate at different locations (points A and B, respectively). Point D in Fig. 3.5 represents the projection of point A onto the incident plane. Since the front and back surfaces of the sample are perpendicular to the incident plane, points A and D are at the same distance from the plane of the receiver. The wave travels an additional distance $\overline{BC}$ in the fluid along the phase velocity path (note that $\overline{BC}$ is parallel to the original incident wave vector $\mathbf{k}_i$) to match its wavefront advance. To travel through this additional distance $\overline{BC}$, an

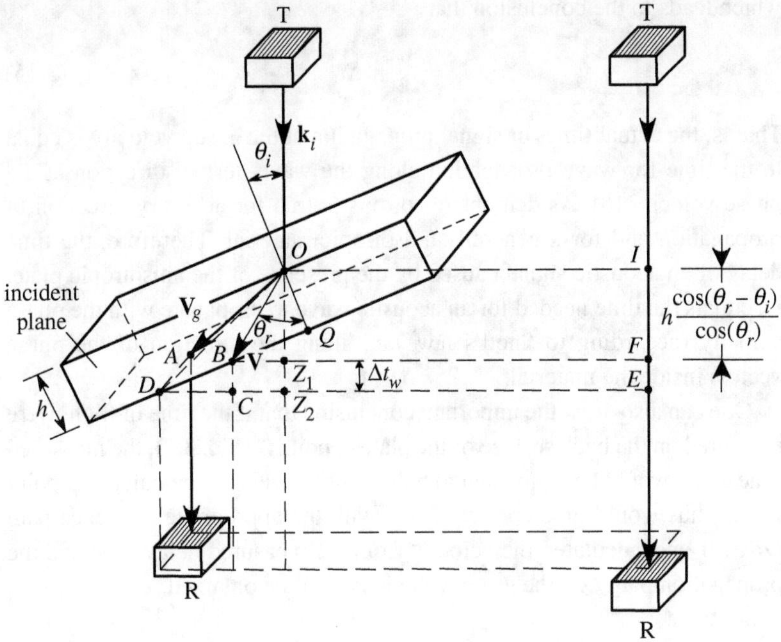

Figure 3.5. Schematic of transducer positions and directions of different acoustic paths. Note the need for 2-D scanning to position the receiving transducer correctly. Both the actual acoustic path (refracted ray, *i.e.*, $\mathbf{V}_g$) and the imaginary acoustic path (refracted wave, *i.e.*, $\mathbf{V}$) are illustrated.

additional time

$$\Delta t_w = \frac{\overline{BC}}{V_f} \tag{3.11}$$

is required. From the triangles OQB, OZ_1B, OQD, and OZ_2D shown in Fig. 3.5, one finds that:

$$\begin{aligned}\overline{BC} = OZ_2 - OZ_1 &= \frac{h\cos(\theta_r - \theta_i + \alpha)}{\cos(\theta_r + \alpha)} - \frac{h\cos(\theta_r - \theta_i)}{\cos\theta_r} \\ &= \frac{h\sin\theta_i \sin\alpha}{\cos(\theta_r + \alpha)\cos\theta_r}.\end{aligned} \tag{3.12}$$

Substituting Eq. (3.12) into Eq. (3.11) yields for the additional time delay [6]:

$$\Delta t_w = \frac{h\sin\theta_i \sin\alpha}{V_f\cos(\theta_r + \alpha)\cos\theta_r}. \tag{3.13}$$

Comparing Eqs. (3.10) and (3.13), one sees that

$$\Delta t_s = \Delta t_w, \tag{3.14}$$

which leads to the conclusion that

$$t_g = t_p + \Delta t_w. \tag{3.15}$$

That is, the actual time for signal propagation at the group velocity is equal to the time for wave propagation along the wave-normal direction at the phase velocity [6]. As demonstrated, this is true for arbitrary direction of propagation and for a generally anisotropic material. Therefore, the time delay of an acoustic signal caused by the presence of the anisotropic plate, t_g, equals the time needed for an acoustic wave to propagate with the phase velocity (according to Snell's law, *i.e.*, along the direction of the phase vector) inside the material.

One can also draw the important conclusion that if the time of flight were measured on the back surface of the plate at point A (Fig. 3.5), the measured time delay would be equivalent to that for phase velocity calculated at point C and thus would give correct results with the appropriate reference path OBC. If one calculated the velocity from the measured delay time and the propagation path OA, the group velocity would be obtained.

3.1.3 Phase velocity measurement

If we subtract the overall time of flight $t(\theta_i)$ along the path in the coupling medium from the time of flight t_0 along the reference path, as shown in Fig. 3.5, we get

$$\Delta t = t_0 - t(\theta_i) = h\left[\frac{\cos(\theta_r - \theta_i)}{V_f \cos\theta_r} - \frac{1}{V(\theta_i)\cos\theta_r}\right]. \tag{3.16}$$

The time shift Δt can be measured. Thus, for a given angle of incidence θ_i there are only two unknowns in Eq. (3.16), namely V and θ_r. These two unknowns are related through Snell's law, Eq. (3.9), therefore by substituting $\sin\theta_r$ and $\cos\theta_r$ in Eq. (3.16) using Eq. (3.9) we can obtain the following analytical expression for the phase velocity V

$$V(\theta_r) = \left[\frac{1}{V_f^2} - \frac{2\Delta t\cos\theta_i}{hV_f} + \frac{\Delta t^2}{h^2}\right]^{-\frac{1}{2}}. \tag{3.17}$$

Equation (3.17) was obtained by a different method in [4] for a plane of symmetry in transversely isotropic materials. In equation (3.17), the incident angle θ_i, the plate thickness h, and the velocity in the immersion fluid V_f are known. The time shift Δt is measured at the given incident angle. Thus, the phase velocity $V(\theta_r)$ can be calculated from Eq. (3.17). Finally, because V_f, θ_i, and $V(\theta_r)$ are known, the angle θ_r can be found from Snell's law (3.9) using $V(\theta_r)$ as follows

$$\theta_r = \arcsin\left[\frac{V(\theta_r)\sin\theta_i}{V_f}\right]. \tag{3.18}$$

3.1.4 Double through-transmission phase velocity measurement

The method in its standard configuration as originally suggested by Markham (see Fig. 3.1) has a serious disadvantage affecting precision, namely when the sample is rotated to change the incident angle, the position of the transmitted beam changes and the receiving transducer should be shifted to the central axis of the transmitted beam. The exact position of the transducer is crucial for precise time delay measurements. Localization of the beam and transducer positioning must be done anew at each angle step. An especially difficult situation arises when the incident plane is not a symmetry plane and, as shown in Fig. 3.5, the receiving transducers should also be translated in the out-of-plane direction. Therefore, to find the transducer position one must seek a signal maximum in two dimensions.

The double through-transmission method, previously illustrated in Fig. 3.2, is free from this disadvantage. This technique can be readily adapted to precision phase velocity measurements in composite samples, as shown in Fig. 3.6. Instead of a separate receiving transducer, a large flat reflector is used, which is oriented normal to the axis of the transducer. The back-reflected wave travels along exactly the same ray path as the incident wave in the opposite direction and arrives at the transducer that operates in pulse-echo mode. The same procedure is used in the reference channel. When the incidence angle is changed by tilting the specimen, the position of the incident beam on the back reflector changes. However, this change does not affect the position of the double-transmitted beam arriving at the receiver, thus eliminating the necessity of moving either the reflector or the transmitter/receiver. To calculate the velocity from the double-transmission time delay, one should substitute $\Delta t/2$ for Δt on the right side of Eq. (3.17), which yields the form [6]:

$$V(\theta_r) = \left[\frac{1}{V_f^2} - \frac{\Delta t\cos\theta_i}{hV_f} + \frac{\Delta t^2}{4h^2}\right]^{-\frac{1}{2}}. \tag{3.19}$$

The modified method has the additional advantage that, since a double path in the material is used, generally greater sensitivity can be achieved.

3.1.5 Self-reference method

The technique described in previous sections uses a long immersion fluid path as reference. This approach has several disadvantages. First and most

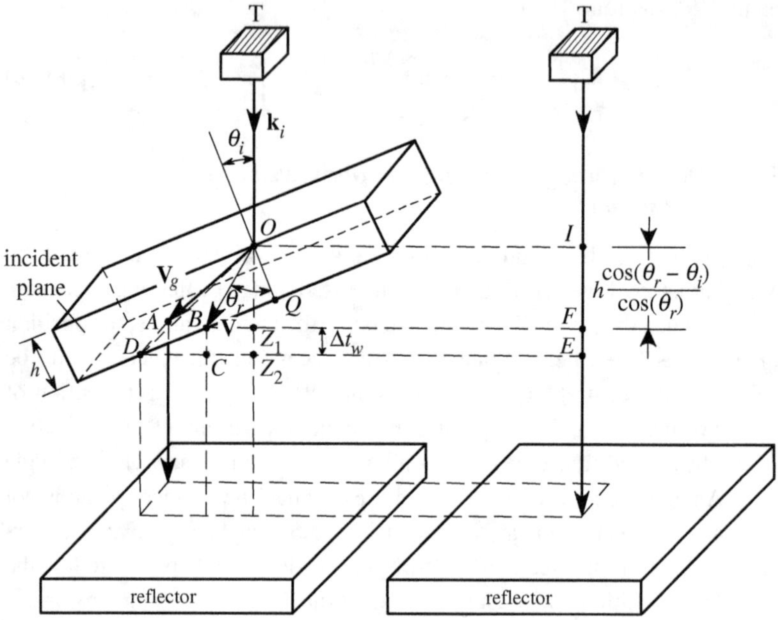

Figure 3.6. Schematic diagram of the double-transmission bulk wave method using a normally aligned reflector.

important, a high degree of parallelism of the sample surfaces is crucial to obtain the desired acoustic path. Even slight non-parallelism of the sample surfaces will produce a small change of the refraction angle in the immersion fluid, leading to a change of the acoustic path in the fluid from the reference path. Second, if the measured sample is highly attenuating, the transmitted signal changes shape compared to the reference, making it difficult to overlap the measured signal with the reference pulse. This effect complicates measurements and introduces inaccuracy. Moreover, sample removal for reference measurement affects precision; because of the usually rather long acoustic path between the transducer and the back reflector, the time-of-flight difference (the time of flight in the sample) cannot be measured very precisely, especially in the presence of temperature variations in the fluid bath.

The self-reference bulk wave method [11,12] shown in Fig. 3.7 improves measurement precision by taking the velocity measured through the sample at normal incidence as the reference. To derive the equation that can be used to calculate the phase velocity from the measurements based on this method, let us calculate the differential quantity $1/V^2(\theta) - 1/V^2(0)$ from Eq. (3.19) and take into account that Δt here is

$$\Delta t = t_0 - t(\theta) = t_0 - t(0) + t(0) - t(\theta) = \Delta t_0 + \Delta t_\theta\,, \tag{3.20}$$

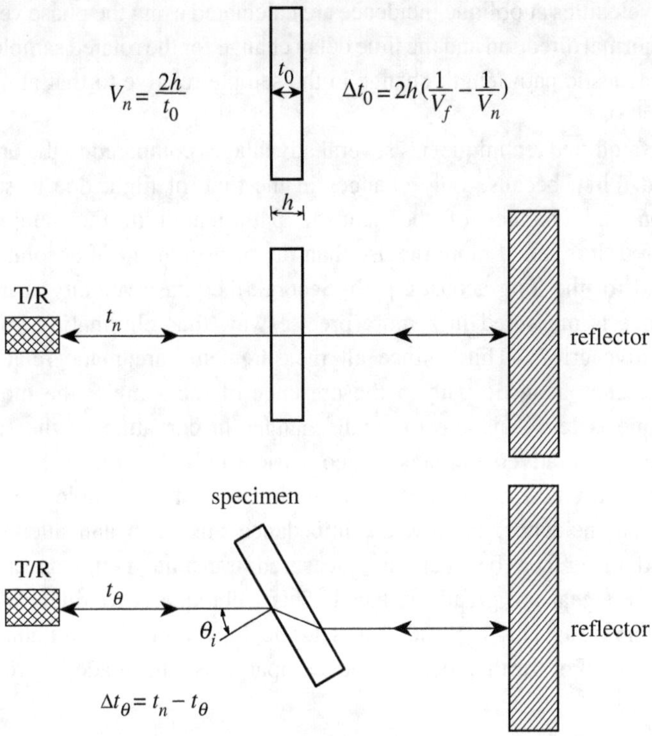

Figure 3.7. Schematic diagram of the self-reference double-transmission bulk wave method.

where $\Delta t_0 = t_0 - t_n = 2h(1/V_f - 1/V_n)$, and $\Delta t_\theta = t_n - t_\theta$ is the difference in the overall time-of-flight measurements between normal incidence $t_n = t(0)$ and arbitrary oblique incidence $t_\theta = t(\theta_i)$ at incidence angle θ_i. For phase velocity calculations at refraction angle θ_r in the material, we now have for the double-transmission case

$$V(\theta_r) = \left[\frac{1}{V_n^2} + \frac{\Delta t_0 - (\Delta t_0 + \Delta t_\theta)\cos\theta_i}{hV_f} + \frac{\Delta t_\theta(2\Delta t_0 + \Delta t_\theta)}{4h^2} \right]^{-\frac{1}{2}}. \tag{3.21}$$

Here V_n is the phase velocity in the sample measured at normal incidence, V_f is the sound speed in water, h is the thickness of the sample, and the refraction angle θ_r can be calculated using Eq. (3.18). In the case of single-transmission experiments, one can still use Eq. (3.21), except that the half-thickness of the plate should be substituted for h to maintain consistency with the reduced delay times measured in this arrangement. The phase velocity at normal incidence is measured with high precision by overlapping multiply reflected signals from the front and back surfaces of the sample.

Phase velocities at oblique incidence are calculated using the phase velocity in the normal direction and the time delay change for the rotated sample (due to the acoustic path length change in the sample relative to that at normal incidence).

The modified technique has several advantages compared to the original method. First, because only changes in the time of flight due to sample rotation and changes of the acoustic path length in the sample are measured, it is much more precise than the measurement of absolute time of flight for the long acoustic path. Second, the phase velocity at normal incidence is measured in a more precise way, thus eliminating possible cumulative errors. Third, since all measurements are made relative to the reference acoustic path in the presence of the sample, the modified technique is less sensitive to small changes in curvature of the sample surface, as the adverse effects of geometrical imperfection of the sample on the ultrasonic measurements are minimized during sample alignment at normal incidence. Finally, the impedance mismatch and attenuation-induced difference between the measured transmitted signals and the reference signal are greatly reduced. This technique is routinely used in laboratory tests [11, 12] and found to be highly precise, reliable, and convenient. For further details and comparisons, the reader is referred to [13].

3.1.6 Multiple reflection method

Due to the typically high acoustic impedance difference between the plate and the surrounding fluid, multiple internal reflections inherently arise that are limited only by intrinsic attenuation in the plate itself. Until now, only the first transmitted signal through the plate has been used for velocity measurements. Using geometrical acoustics interpretation, we consider in the following discussion how these internal reflections can contribute to the total transmitted signal and, ultimately, can be exploited to accomplish the velocity measurement.

A) Transmission mode

A simplified geometrical representation of ultrasonic wave transmission at oblique incidence through the plate is shown in Fig. 3.8, where arrows indicate the direction of the wave vector. Besides the previously mentioned impedance mismatch between the plate and the surrounding fluid and the intrinsic ultrasonic attenuation in the plate, the number of reverberations or repeated transmissions in the detected pulse train is determined by the diameter of the transducer D and the lateral skipping distance $s = 2h \tan \theta_r \cos \theta_i$. The latter increases with plate thickness as well as with increasing angle of incidence and sound velocity in the plate (because

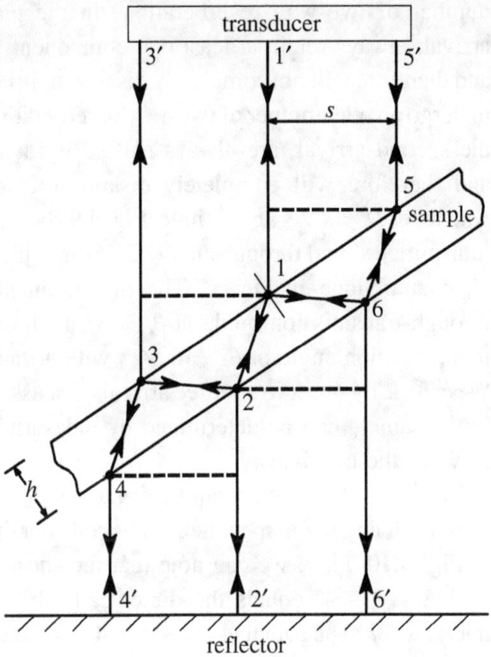

Figure 3.8. A schematic diagram of ultrasonic wave transmission at oblique incidence through a plate.

of Snell's law). In Fig. 3.8, we assumed that the transducer diameter is larger than the lateral skipping distance ($D \approx 2s$). The fastest path to the reflector and back to the transducer is without any reverberation within the plate, for example, along the central path $1' - 1 - 2 - 2' - 2 - 1 - 1'$ (other paths such as $3' - 3 - 4 - 4' - 4 - 3 - 3'$ and $5' - 5 - 6 - 6' - 6 - 5 - 5'$ also contribute to the first arrival in the double-transmitted signal). This signal will be followed by a series of lagging arrivals corresponding to an increasing number of internal reflections within the plate. The second arrival consists of signals that have gone through one full reverberation in the plate. Such paths are, for example, $1' - 1 - 2 - 3 - 4 - 4' - 4 - 3 - 3'$ on the left and $1' - 1 - 2 - 2' - 2 - 1 - 6 - 5 - 5'$ on the right (again, other paths originating from $3'$ and $5'$ also contribute to the second arrival). It should be mentioned that these signals are laterally shifted by a distance of s. The third arrival will be formed by signals that have gone through two full reverberations in the plate, for example, $1' - 1 - 2 - 3 - 4 - 4' - 4 - 3 - 2 - 1 - 1'$ (numerous other paths also contribute to the third arrival). Depending on whether the lateral shifts are added or subtracted, these rays are either not shifted at all or shifted by a distance of $2s$ by the time they return to the transducer.

As the lateral skipping distance increases with respect to the transducer diameter, certain multiple reflections miss the receiver and therefore vanish from the double-transmitted signal. Those pulses that undergo an even number of two-way reverberations in the plate, for example, the third arrival, always retain at least one component with no overall lateral shift and therefore will not completely disappear. In comparison, those pulses that undergo an odd number of two-way reverberations in the plate, for example, the second arrival, are always shifted by at least one skipping distance and therefore will completely disappear from the received pulse train whenever $D < s$. As an example, Fig. 3.9 shows a typical ultrasonic double-transmitted signal through an $h = 2.25$-mm thick graphite/epoxy composite plate at oblique incidence. The measurement was taken in the double through-transmission mode at $\theta_i = 3°$ angle of incidence (corresponding to a refraction angle of $\theta_r = 6.54°$) with a transducer of 12 mm diameter. We can easily observe all three arrivals discussed above. The phase velocity in the sample can be determined by measuring the time delay difference between these arrivals.

To derive the relationship between the measured time delay and sought sound velocity in the specimen, let us consider the schematic diagram shown in Fig. 3.10. First, we can note that the shortest path with multiple skips $(1 - 2 - 3 - 4)$ equals the direct path $(1 - 2 - 7)$ in a plate of triple thickness $3h$. The actual $(1 - 2 - 3 - 4 - 4')$ and equivalent $(1 - 2 - 7 - 7')$ acoustical paths differ by an additional leg of ℓ in the fluid, which results in

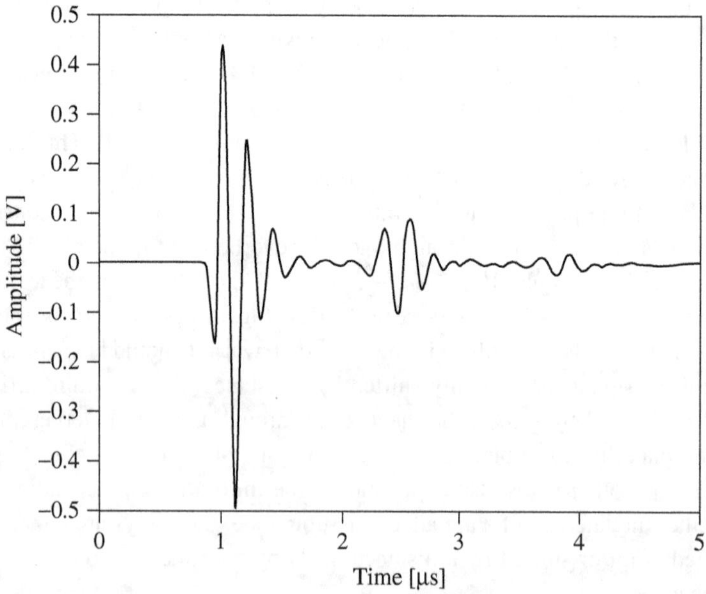

Figure 3.9. Typical ultrasonic double-transmitted signal through a graphite/epoxy composite plate at oblique incidence ($\theta_i = 3°$, $\theta_r = 6.54°$, $h = 2.25$ mm).

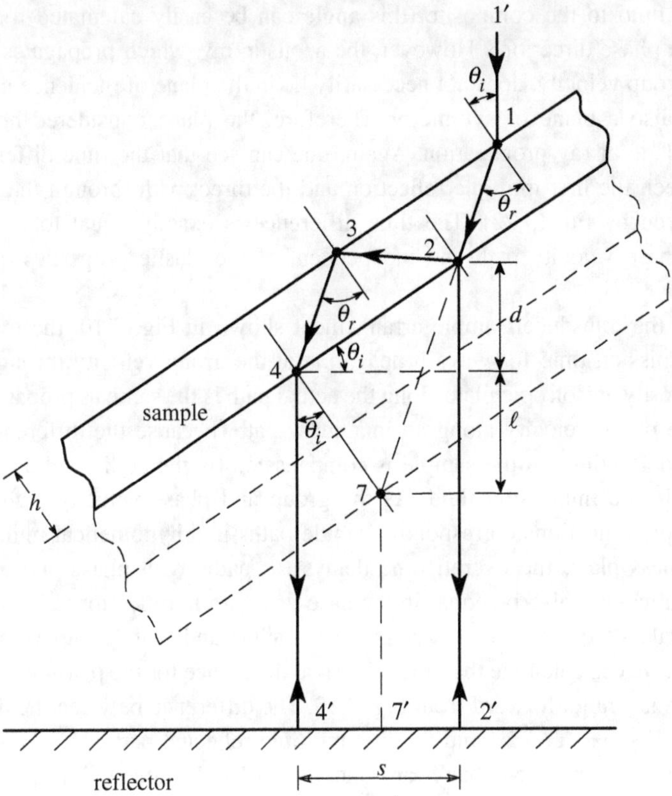

Figure 3.10. A schematic diagram of ultrasonic wave transmission at oblique incidence through a plate.

an additional time delay Δt_ℓ,

$$\ell = 2h\cos\theta_i$$

$$\Delta t_\ell = 2h\cos\theta_i / V_f \tag{3.22}$$

$$t_{1-4'} = t_{1-7'} + \Delta t_\ell\,.$$

One can see that the additional time delay Δt_ℓ depends on the incident angle θ_i, the plate thickness h, and the velocity in the fluid V_f, but it is independent of the wave velocity in the plate. Before continuing with the determination of the velocity in the specimen, let us address the important difference between phase and group velocity paths. One can make exactly the same argument for the energy path (path with group velocity) as for the wave path. In a schematic sense, we can still use Fig. 3.10, but now we have to assume that the directions shown correspond to the group velocity or ray directions, and angle θ_r denotes the refraction angle for energy flow from fluid to the composite (this angle can be easily calculated for any given phase direction). However, the acoustic ray, which propagates with the group velocity, does not necessarily lie in the plane of incidence unless it is also a plane of symmetry. Therefore, the plane considered here is the plane of ray propagation. Again, one can see that the time difference between the first multiple reflection and the direct path through the plate is given by Eq. (3.22). The time difference is exactly equal to that for the phase velocity path and independent of the elastic properties of the plate.

In the transducer/sample arrangement shown in Fig. 3.10, the overall transmission time for wave propagation at the group velocity through an arbitrarily anisotropic plate along the actual path is the same as propagation at the phase velocity along an imaginary path, because the difference in time delay through the sample is compensated by the additional acoustic path in the immersion fluid. As the group and phase velocity paths for multiple reflections correspond to single paths in a hypothetical multiple-thickness plate, the overall time delays for paths with phase and group velocities will also be equal in the case of multiple reflections. Owing to equivalence of the imaginary (phase velocity) and real (group velocity) paths, we can calculate the time-of-arrival difference for the phase velocity, which is straightforward from Fig. 3.10. The difference between the direct path ($1' - 1 - 2 - 2'$) and the first multiple-reflected path ($1' - 1 - 2 - 3 - 4 - 4'$) will be a backward and forward reflection within the plate ($2 - 3 - 4$), supplemented by an additional length of d of the direct path $2 - 2'$ versus the corresponding shorter leg $4 - 4'$ of the multiple-reflected path. The time difference between the two total paths ($1' \rightarrow 2'$) and ($1' \rightarrow 4'$) is

$$\Delta t = t_{1'-4'} - t_{1'-2'} = \frac{2h}{V \cos \theta_r} - \frac{2h \tan \theta_r \sin \theta_i}{V_f} = \frac{2h \cos \theta_r}{V}. \quad (3.23)$$

After substituting the refraction angle θ_r with the incident angle θ_i using Snell's law, we can obtain the expression for the phase velocity V in terms of the time delay Δt between the first and second arrival in the (one-way) through-transmitted signal:

$$V = \frac{2hV_f}{\sqrt{V_f^2 \Delta t^2 + 4h^2 \sin^2 \theta_i}}. \quad (3.24)$$

Let us now return to Fig. 3.8. The time difference measured in the double through-transmission mode for paths $1' - 2' - 1'$ and $1' - 4' - 3'$ (or $1' - 2' - 5'$) occurs due to one full (back and forth) bouncing of the wave in the plate. Therefore, the time difference between the first and second arrivals in the double-transmitted pulse train can be directly substituted in Eq. (3.24) for the phase velocity calculation. The third arrival is related to the path $1' - 4' - 1'$ and corresponds to two full (back and forth) rebounds of the wave in the plate (one bounce in the forward direction when wave propagates from the transducer to the reflector and one in the backward direction). Therefore, one should take half of the delay time between the first and third arrivals when substituting into Eq. (3.24). Special attention should be paid to identifying the order of any given multiple reflection in the received signal because, as we have mentioned above, the true second arrival might disappear when the lateral skipping distance exceeds the transducer diameter and the third arrival appears to follow the first one without any discernible signal between them.

It is useful to calculate the previously introduced skipping distance s between points $2'$ and $4'$ in Fig. 3.10, *i.e.*, the distance between the centers of the first and second transmitted beams. One can easily see that

$$s = 2h \tan\theta_r \cos\theta_i, \tag{3.25}$$

where θ_r can be readily expressed from θ_i using Snell's law. Figure 3.11 shows the normalized skipping distance s/h as a function of the refraction angle θ_r for different values of phase velocity ratio V/V_f. Due to practical limitations, most measurements are made in the range below 60°, where the normalized skipping distance is rather insensitive to the sound velocity in the plate and its value changes from zero at normal incidence to $s/h \approx 3$ at $\theta_r = 60°$.

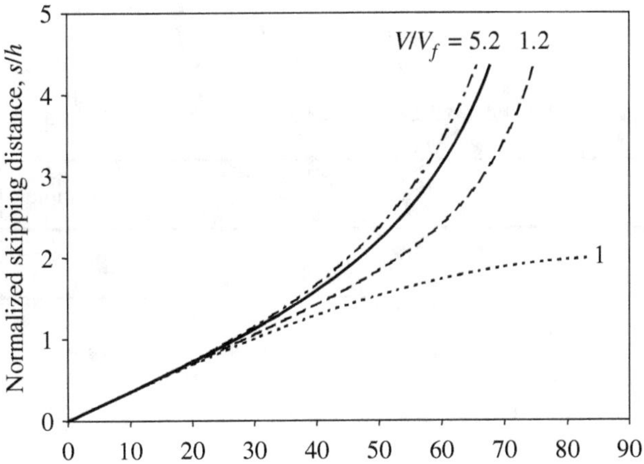

Figure 3.11. Normalized skipping distance s/h as a function of the refraction angle θ_r for different values of phase velocity ratio V/V_f.

B) Reflection mode

Arguments similar to those above are also valid for the reflection mode. Figure 3.12 shows a schematic diagram of the phase and group velocity paths in pulse-echo reflection mode using a plane reflector. The imaginary (phase) time delay of the reflected wave from the back wall of the specimen $(0-1-2-3-3')$ is equal to that of the transmission path from the mirror image of the transducer through a double-thickness plate $(6-7-2-3-3')$. The same results can be applied for the real (group) time delays following the ray paths of the back wall reflection $(0-1-4-5-5')$ and the through-transmission $(6-7-4-5-5')$. As discussed earlier, the total time delays for the actual path with the group velocity and the imaginary path with the phase velocity are identical since the additional time delay in the fluid exactly compensates the time difference in the plate [6]. The phase velocity can be calculated from the measured time delay between the first $(0-1-1')$ and second $(0-1-2-3-3')$ arrivals using Eq. (3.24).

If a reflector is used and the wave transmits through the plate not once but twice, however, special care must be taken. As we have mentioned above, the true second arrival is still lagging by the same amount (given by Eq. (3.23)) as in the case of single transmission, though it is laterally shifted

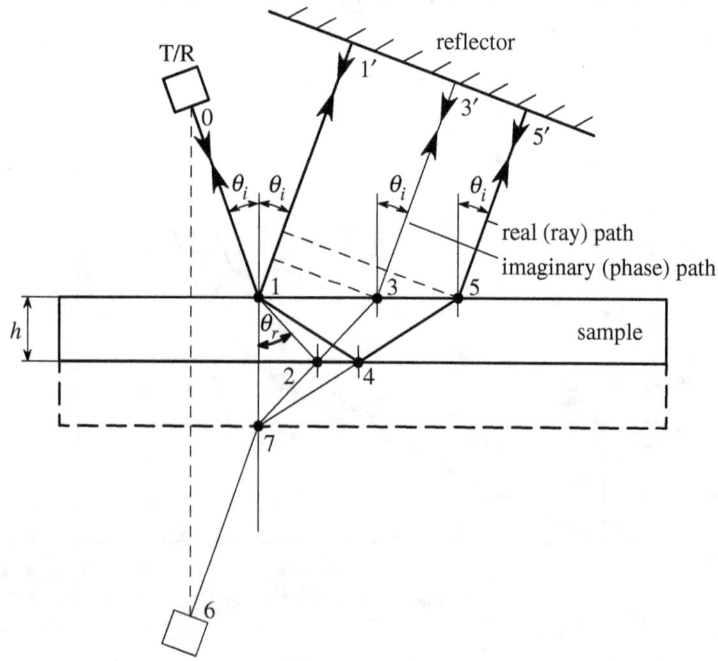

Figure 3.12. A schematic diagram of the phase and group velocity paths in pulse-echo reflection mode using a plane reflector.

by a distance s. Whenever this beam falls beyond the receiving aperture of the transducer, the observed "second" arrival will actually be delayed by the same amount twice, therefore only half of the measured time delay should be substituted for Δt in Eq. (3.24) like before.

Of course, the measurement does not necessarily have to be done in the two-way pulse-echo mode shown in Fig. 3.12. Moreover, the conventional one-way pitch–catch mode using separate transducers for transmission and reception works in a similar way. A small receiving transducer should be used to find the location of the acoustic beam reflected from the back wall of the plate by moving the transducer parallel to the surface of the specimen, as shown in Fig. 3.13. The acoustic path in the fluid will be the same for different positions of the receiver, and the changing time delay will be only due to signal delay in the plate. Unfortunately, the measured time delay will be for the group velocity since no compensation between group and phase velocity paths is produced by the additional acoustic path in the fluid. Thus, the simple equation of $\Delta t = 2h/V \cos\theta_r$ cannot be used together with Snell's law for phase velocity calculations. In this case, the refraction angle Θ_r for energy propagation in

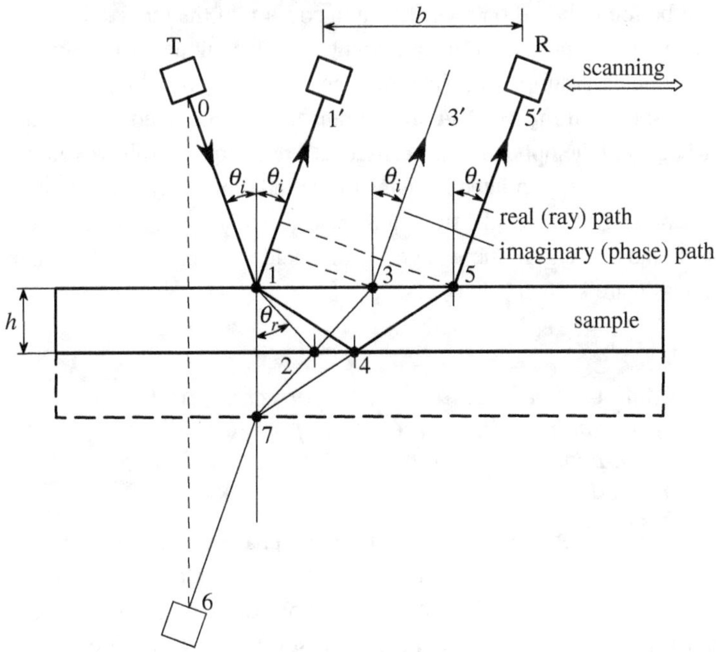

Figure 3.13. A schematic diagram of the phase and group velocity paths in pitch–catch reflection mode using separate transducers for transmission and reception.

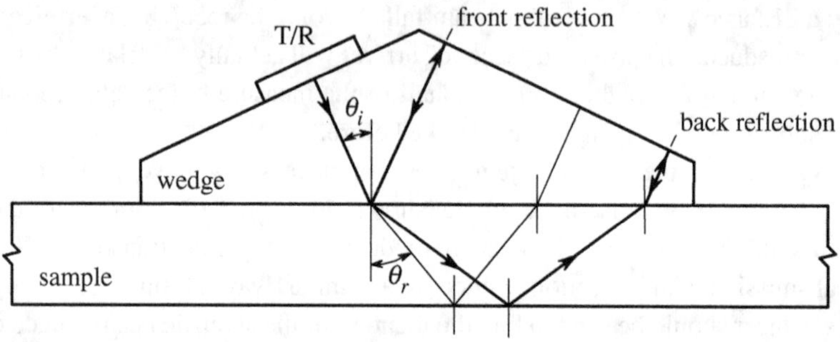

Figure 3.14. Contact angle-beam transducer utilizing a solid wedge for oblique-incidence velocity measurements in pulse-echo mode.

the plate should be used instead of the refraction angle θ_r for the phase direction, which is directly related to the phase velocity V via Christoffel's equation.

The group velocity V_g can be determined from the lateral shift b of the receiver between the front and back wall reflections using $V_g = (4h^2 + b^2)^{1/2}/\Delta t$ [9]. This procedure will introduce some loss of precision and measurement complications because the optimum locations of the receiver should be found based on recording a maximum of the received ultrasonic signal. It is not necessary to use immersion technique to perform these measurements. A simple contact arrangement utilizing a solid wedge of fixed angle is shown in Fig. 3.14. A single transducer is mounted on one face of the wedge and the opposite face is used as a reflector. The phase velocity in the sample can be calculated from the measured time delay between the first and second arrivals, *i.e.*, the front and back wall reflections, using Eq. (3.24) if the sound velocity in the wedge V_f, the incident angle θ_i, and the sample thickness h are known.

3.1.7 Comments on accuracy, phase correction, and applicability of the plane wave approximation in angle-beam, self-referenced, through-transmission method

In the previous sections, we have discussed phase velocity measurements by the double through-transmission method. The elastic stiffnesses are obtained by inverting the Christoffel equation using a set of phase velocity data acquired from sound propagation in several different directions. For a material satisfying orthotropic symmetry, for example, seven of the nine elastic stiffnesses can be found from measurements in two symmetry planes; the solution is independent of the initial guesses and is only slightly affected by data scatter. The other two stiffnesses can be found from measurements in non-symmetry planes.

The accuracy of elastic constant determination depends on the precision of the velocity measurement. As we have discussed above, the self-referenced, double through-transmission method of phase velocity measurement is robust and simple to use. When properly implemented, this method can provide velocity measurements with precisions of 0.1–0.3% [14, 15]. The determination of velocity from time delays uses a simple equation obtained for arbitrary anisotropy from plane wave geometrical acoustic analysis. The phase velocity measurement based on these geometric considerations, however, neglects two effects: 1) the diffraction effect due to the limited size of the transducer; and 2) the signal distortion effect owing to phase change at interfaces between dissimilar materials, *i.e.*, the interface between the test sample and the immersion fluid. The same situation occurs in different implementations of immersion methods for phase or group velocity measurement in through-transmission or reflection modes and also for angle-beam velocity measurements using contact wedge transducers, where the wedge material replaces the immersion fluid. The phase correction for the self-reference, double through-transmission method has been briefly discussed by Lavrentyev and Rokhlin [16] and by Hollman and Fortunko [15] who applied phase/slope analysis for the phase correction in this method.

Both the phase and beam effects on the velocity measurement at oblique incidence, as discussed in this chapter, have been investigated in detail by Wang *et al.* [7], where a time-domain numerical beam model analysis has been performed. It was shown that the phase shift is zero before the first critical angle, and thus one may only need to correct the phase for transverse waves after the first critical angle. For an ultrasonic time-domain signal with a Gaussian shape this phase shift effect can be very accurately corrected using the plane transverse wave transmission coefficient. It was also shown that for accurate results, a zero crossing in the middle of the signal should be used for time delay measurements.

Usually, the time delay change is measured from the shift of a zero crossing of the corresponding ultrasonic signal. The phase shift changes the zero crossings, introducing an additional time delay. This effective time delay for a harmonic or Gaussian-shape signal can be calculated as

$$\Delta t_{\theta_i}^{P} = k\frac{\varphi}{2\pi f}, \tag{3.26}$$

where φ is the plane wave phase shift angle, and f is the ultrasonic wave frequency. The factor k is the number of fluid–solid and solid–fluid interfaces the wave is passing through. For example, $k = 4$ for the double through-transmission method and $k = 2$ in the single through-transmission measurement. Equation (3.26) may be used to account for the phase change at the interface and to correct experimentally measured time delays. For an orthotropic material in a plane of symmetry, the phase shift can be calculated using the analytical equations given in [7], and for arbitrary orientation of the incident plane, it can be calculated by methods described in Chapter 4.

To calculate the phase change at the interface, one needs to know the elastic properties of the solid, which are *a priori* unknown before the measurement. So, for practical implementation of the correction one can use an iterative approach. At first the wave velocity is obtained from the measured time delays without phase correction, and the elastic constants of the sample are determined in a first approximation. Based on the elastic constants found, the transverse wave phase shift is calculated. The calculated phase shift is used to correct the experimentally measured time delay data. (The corrected time delay equals the directly measured time delay plus the time delay $\Delta t^P_{\theta_i}$ introduced by the phase shift.) Then, the time delay and phase velocity are calculated again. The procedure can be repeated to improve the precision of the velocity measurement using iteration; the first iteration, however, is usually satisfactory. Finally, the sample elastic constants are determined from the corrected time delay data.

If the phase correction is not accounted for, one introduces error into the measurement. One can estimate the error ε introduced, owing to transmission through interfaces at angle θ_i, and normalize it by the time of flight t_n through the sample at normal incidence,

$$\varepsilon < \frac{\Delta t^P_{\theta_i}}{t_n} = \frac{\varphi V_n}{\pi f h}. \tag{3.27}$$

As the time of flight corresponding to the transverse wave propagation at oblique incidence through the sample is usually larger than the time of flight t_n at normal incidence, the phase-induced error estimated by Eq. (3.27) is the upper bound of the actual phase-induced error at arbitrary incident angle.

The interface-induced intrinsic phase change of the transmitted signal is especially large and depends strongly on the incident angle in the vicinity of the critical angles. An additional error near the first critical angle comes from distortion in the transmitted ultrasonic pulse. When the incidence plane is close to the fiber direction, the refracted longitudinal wave speed and transmitted beam depend very strongly on the incident angle, thereby distorting the ultrasonic beam. The relative error in reconstructed elastic constants from velocity data is much smaller than that of the maximum velocity error due to interface-induced phase delay or the beam distortion near critical angles. This is the result of the least squares minimization at all measured angles, and therefore the deviation of the velocity from the actual velocity in a small range of angle does not produce a drastic effect on the elastic constants determined. The phase correction is very simple to implement, however, and is recommended for all measurements.

Concerning beam effects, it was found [7] that the plane wave approximation can be used with high precision and the beam effect can be neglected a few degrees away from the critical angles. It is shown that in the far field, the finite width of the transmitter has an effect of less than 0.2% on the time delay measurement in the double through-transmission method and affects only the amplitude of the transmitted signal. Therefore, the plane wave phase correction is sufficient for the precision required from most measurements. Use of phase correction and, if needed, the ultrasonic beam model instead of the plane wave approximation are important when improved accuracy of angle beam velocity measurements is needed, for example, for stress determination from velocity measurements [14]. The conclusions presented in this section are also valid for measurements in multilayered composites where, as in Chapter 7, one can apply Floquet waves, instead of bulk waves, for lamina stiffness measurements [17].

3.2 Examples of Bulk Wave Refraction Velocity Measurements

3.2.1 Experimental apparatus

Figure 3.15 shows a schematic diagram of the experimental apparatus that implements an ultrasonic goniometer. It is distinguished by its ability to make simultaneous measurements of both doubly transmitted (through the sample) and doubly reflected (from the sample) ultrasonic signals using a single ultrasonic transducer. A plane sample is mounted inside an aluminum cylindrical shell that serves as an acoustic reflector. The axis of the cylinder lies in the frontal plane of the specimen that can be rotated around this axis to change the angle of incidence at a given orientation of the specimen. Alternatively, the specimen can be rotated in its own plane to study the dependence of the sound velocity on orientation (azimuthal angle) at a given angle of incidence (polar angle).

In transmission mode, the ultrasonic wave is transmitted through the sample, reflected back from the plane reflector, and then returned to the transducer after a second transmission through the specimen. In reflection mode, the incident and the specularly reflected ultrasonic beams both propagate radially inside the cylinder. After being reflected from the cylinder wall, the ultrasonic signal returns to the sample along the same path, reflects a second time from its surface, and finally returns to the receiving transducer. As opposed to the specularly reflected signal from the sample front surface, the second and subsequent arrivals reflect from the back wall of the specimen and therefore are laterally shifted. In the previously described plane reflector

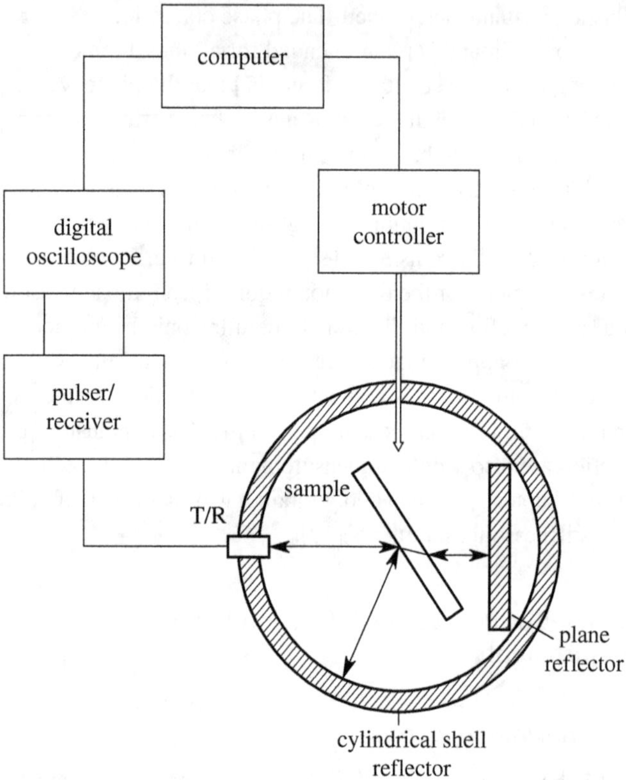

Figure 3.15. Schematic of the experimental apparatus. During measurement, the sample is rotated around the cylinder axis that lies in the frontal plane of the sample.

arrangement (see Figs. 3.12 and 3.14) these waves also strike the reflector at normal incidence, and therefore retrace their own path to the transducer. In the arrangement shown in Fig. 3.15, back wall reflections reach the surface of the cylindrical reflector at slightly off-normal incidence, and therefore they cannot exactly retrace their paths back to the transducer. This inherent misalignment limits the feasibility of this technique to relatively thin plates with respect to the radius of the cylinder. The cylindrical reflector has the unique advantage over the plane reflector arrangement that the angle of incidence can be easily adjusted without the need to adjust the reflector angle.

3.2.2 Double through-transmission method for graphite/epoxy composites

Figures 3.16–3.18 show a couple of examples of experimental phase velocity versus refraction angle curves measured in a unidirectional graphite/epoxy composite plate. These measurements were made by the double-transmission technique outlined earlier. The sample is mounted on a two-axis rotation table as shown in Fig. 3.15. One rotation table rotates the sample to change the angle of incidence using a computer-controlled DC motor. Angular resolution and repeatability of 0.01° can be achieved. The other table rotates the sample in its own plane so that different orientations of the incident plane relative to the fiber direction may be selected.

The ultrasonic signal transmitted through the sample is reflected by the plane reflector and returned to the transducer after the second through-transmission. The received signal is digitized by a digital oscilloscope, averaged, and collected by the computer as a function of the angle of rotation, *i.e.*, angle of incidence. When necessary, the detected ultrasonic signals are analyzed in the frequency domain. Then, both signal and spectrum are fed to the computer through an IEEE interface. All the measurements are conducted in a water tank that is temperature stabilized at 29.8 ± 0.05°C. The following experiments were done on a 2.1-mm thick unidirectional graphite/epoxy composite plate at 15 MHz.

The measurements have been performed for several different azimuthal angles ϕ between the fiber reinforcement and the plane of incidence. For a selected aziumthal angle ϕ, the sample is rotated around the axis perpendicular to the plane of incidence by an angle θ_i as shown in Fig. 3.3. At each set of angles, the time of flight is measured. The ultrasonic phase velocity is calculated from the measured time of flight using Eq. (3.19) or Eq. (3.21).

The phase velocity in the plane of fibers ($\phi = 0°$) is shown in Fig. 3.16 as a function of the refraction angle θ_r. The measured data are shown as discrete points and the solid lines represent theoretical values calculated from the reconstructed elastic constants (the reconstruction of elastic constants is discussed later). Note that there is no experimental data for the slow transverse wave V_{st}, since the incident plane coincides with a plane of symmetry, *i.e.*, the (x_1, x_3) plane. Therefore, only two waves, namely the quasilongitudinal (V_ℓ) and fast quasitransverse (V_{ft}) waves, can be excited and measured in the sample. The fast quasitransverse wave exhibits pure shear vertical (SV) polarization in this incident plane. The third wave (V_{st}) has a pure shear horizontal (SH) character and therefore cannot couple to the fluid.

As the pure SH wave is polarized perpendicular to the fiber direction while the polarization vector of the SV wave has a component along the fiber direction, the SH wave is slower than the SV wave. It should be mentioned that at the critical angles (*i.e.*, $\theta_r = 90°$), all three waves propagate along the fiber direction and the two shear velocities reach the same value. Since the SV wave is polarized in the plane of incidence (x_1, x_3) and the SH wave is polarized in the plane of the surface (x_1, x_2), the equality of the

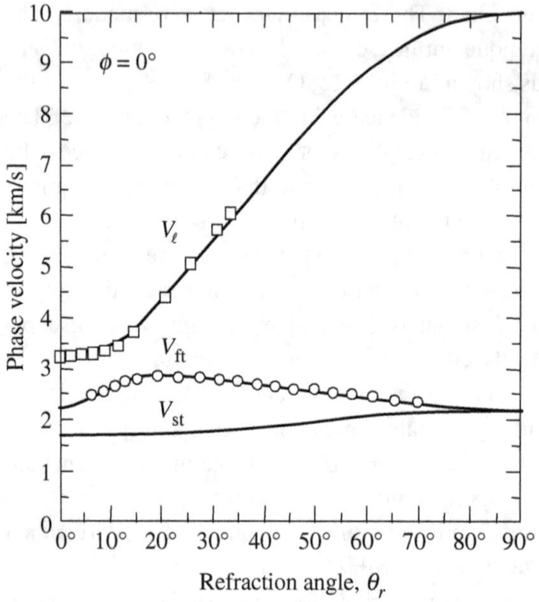

Figure 3.16. Phase velocity versus refraction angle for the three propagating modes in a unidirectional graphite/epoxy composite plate. The incident plane coincides with the fiber direction ($\phi = 0°$), symbols are experimental data, lines show calculated velocity. Note that the slow transverse wave V_{st} has SH character and cannot be excited in the sample.

two shear velocities actually implies that the particular composite sample is transversely isotropic in planes normal to the fiber direction.

The phase velocities of the three propagating modes in the $\phi = 45°$ incident plane are shown as functions of the refraction angle in Fig. 3.17. As the incident plane is not a plane of symmetry, all three elastic waves can be excited in the solid at oblique incidence. The composite is more compliant in this plane than in the previously considered $\phi = 0°$ plane that includes the fiber direction; therefore, all three velocities are somewhat lower than before, though their general dependence on the refraction angle is quite similar.

Finally, the results of the measurements and calculations for the plane orthogonal to the fiber direction ($\phi = 90°$) are shown in Fig. 3.18. As in this case the plane of incidence coincides with the (x_2, x_3) plane, which is also a plane of symmetry, only two of the three waves can be excited in the solid and measured experimentally. The fast transverse wave (V_{ft}) for $\phi = 90°$ has pure SH character, *i.e.*, it is polarized along the fiber direction normally to the plane of incidence. Such a wave cannot be coupled to the fluid and therefore will not be excited by an incident compressional wave from the fluid (when using a sheet beam or line source). On the other hand, the slow transverse wave (V_{st}) for $\phi = 90°$ is polarized in the (x_2, x_3) plane; therefore, it exhibits pure SV character and can be excited and measured experimentally, as shown in Fig. 3.18.

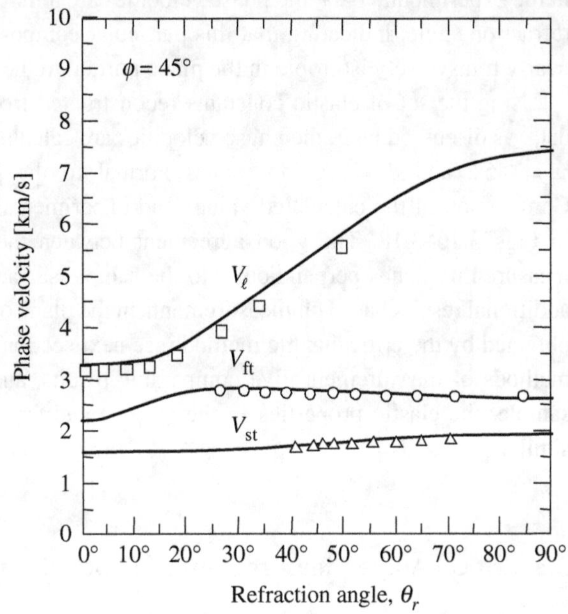

Figure 3.17. The same as Fig. 3.16 except that the incident plane subtends a $\phi = 45°$ azimuthal angle with the fiber direction. Note that both transverse waves have a perceivable vertical polarization component that facilitates their experimental observation.

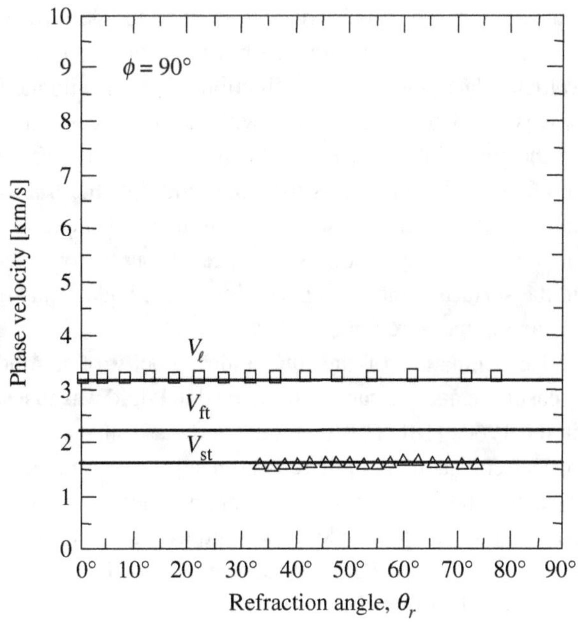

Figure 3.18. The same as Fig. 3.16 except that the incident plane subtends a $\phi = 90°$ azimuthal angle with the fiber direction. Note that the fast transverse wave V_{ft} has SH character and cannot be excited in the sample.

It is interesting to note here that at $\phi = 90°$, except for some minor scatter in the experimental data, the phase velocities are almost independent of the refraction angle, indicating that this particular composite sample is in fact nearly transversely isotropic in the plane normal to the fibers.

Using the set of elastic constants reconstructed from the experimental data, as discussed later, the phase velocities are calculated here for $\phi = 0°$, $\phi = 45°$, and $\phi = 90°$ in planes normal to the plane of the plate. Comparison of the calculated values and experimental results was shown in Figs. 3.16–3.18. The good agreement between the experimental data measured in planes perpendicular to the sample surface (this section) and additional results based on measurements in the plane of the sample surface obtained by the critical angle method (see next section) indicates that both methods of measurement give comparable results, and for this particular sample, the elastic properties in the (x_1, x_2) and (x_1, x_3) planes are very similar.

3.3 Critical Angle Measurement of Elastic Constants

3.3.1 Introduction

The measurements described in the previous sections of this chapter have been performed in planes perpendicular to the surface of the sample. As one can see from the velocity data presented in Figs. 3.16–3.18, it is difficult to perform measurements close to the critical angles. For example, longitudinal velocity data close to fiber direction are not available. Data for transverse waves can also not be obtained with the wave vector oriented nearly parallel to the fiber direction, nor with the wave vector nearly perpendicular to the fiber direction, unless the finite width of the transducer is taken fully into account. (This question is examined in Chapter 8.) The critical angle method described in this section can be used for velocity measurements in the surface plane, thus providing needed additional data to complement refraction measurements.

Determination of the ultrasonic velocity of a solid by critical angle measurements on a sample immersed in liquid was first suggested by Mayer in the 1960s [18]. This method has the advantages over transmission-type measurements that (1) there is no need for the sample to have two parallel surfaces and (2) access to only one surface is required. It was shown by Mayer that a sharp reflection maximum corresponds to the first critical angle for materials of not very high acoustic impedance (most composites qualify as such); therefore, velocity measurements may be made with sufficient precision if the temperature of the coupling liquid is stabilized. A detailed description of the critical angle method was given later by Fountain [19] who, in addition, observed in his experiments a dip corresponding to the Rayleigh angle.

While generalization of the critical angle method to anisotropic materials seems natural, it cannot be done in general as has been discussed by Henneke and Jones [20]. The critical angle condition for anisotropic materials is that the propagation of the energy flow of the elastic wave in the solid be parallel to the surface, *i.e.*, the refraction angle for the energy flow $\Theta_r = 90°$, which does not necessarily mean that the refraction angle for the phase velocity $\theta_r = 90°$ [20]. This means that, in general, the phase velocity cannot be measured using the critical angle in anisotropic materials. Possibly for this reason, further work for anisotropic materials has been performed by measuring the Rayleigh wave "critical" angle, or simply Rayleigh angle [21–23]. As long as the Rayleigh wave exists on the free surface without attenuation, both the energy flow and the phase velocity for this wave lie in the plane of the sample surface (this condition is also satisfied for pseudo-Rayleigh waves). Here, the phase velocity is related to the common real part of the complex wave numbers for the three evanescent partial waves.

In this section, the applicability of the critical angle method is demonstrated for composite materials following references [6, 24]. Besides the above-mentioned geometrical considerations, the method of bulk-wave critical angles offers the advantage of simple and unique calculation of the elastic constants of the material from the experimental data. We will show that for 2-D (transversely isotropic) composite materials, all three critical angles may be observed, and at the critical angle the phase velocity always lies in the surface plane and therefore may be calculated from Snell's law. All three in-plane velocity surfaces can be measured and the in-plane elastic constants of the composite material can be reconstructed from the experimental data. Using this technique in combination with the wave refraction method [8], the full matrix of elastic constants for an orthotropic material (9 independent parameters) can be reconstructed.

3.3.2 Concept of the method

In an orthotropic material, three elastic waves may propagate in general with different velocities: one quasilongitudinal and two quasitransverse. If the plane of incidence is oriented arbitrarily to the planes of symmetry of the specimen (as, *e.g.*, in Fig. 3.3), then all three elastic waves will be excited in the sample since all will have a normal component of displacement on the surface of the sample and will be coupled to the liquid. If the incident plane coincides with a plane of symmetry ((x_1, x_3) or (x_2, x_3) in Fig. 3.3), only two elastic waves will be excited in the solid since one of the transverse waves in this plane will be a pure SH wave with displacement parallel to the surface of the sample and will not be coupled to the liquid. Therefore, one may expect two critical angles for planes of symmetry and three critical angles for an arbitrary plane of incidence.

As discussed in Chapter 2, the definition of a critical angle for an anisotropic material must be more specific than that for an isotropic material, since the direction of the wave vector and the energy flow do not necessarily coincide in the anisotropic case. The critical angle is defined as the angle at which the energy flow (group velocity) in the refracted wave is parallel to the interface separating the two media, *i.e.*, $\Theta_r = 90°$ (see Fig. 3.19). At this angle of incidence, the wave vector of the refracted wave in general deviates from the interface, *i.e.*, $\theta_r \neq 90°$. Therefore, in an anisotropic material, at the critical angle both the refraction angle θ_r and the wave velocity V are unknown and cannot be easily calculated from Snell's law by assuming that $\theta_r = 90°$. This limitation does not apply, as we will see in a moment, for 2-D composite materials (*i.e.*, a composite with unidirectional fiber reinforcement showing either orthotropic or transversely isotropic symmetry), when the material surface is a plane of symmetry, as shown in Fig. 3.19.

Let us consider a unidirectional composite material with fibers in the direction of axis x_1 as previously shown in Fig. 3.3. Plane (x_1, x_2) will be parallel to the sample surface and plane (x_2, x_3) will be orthogonal to the

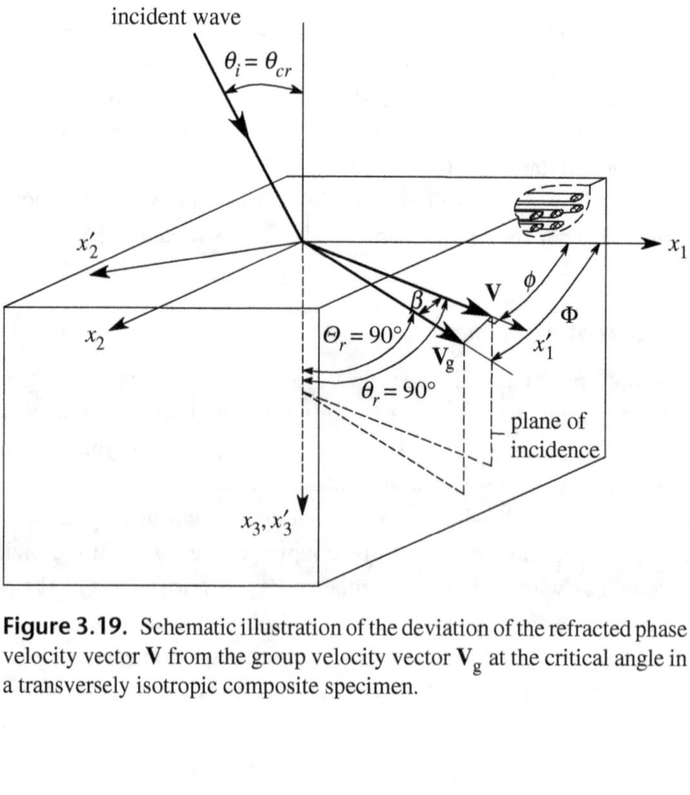

Figure 3.19. Schematic illustration of the deviation of the refracted phase velocity vector **V** from the group velocity vector $\mathbf{V}_g$ at the critical angle in a transversely isotropic composite specimen.

surface as well as to the fiber direction. Let us select a plane of incidence having arbitrary angle ϕ with the fiber direction. Let the incident angle in the liquid be θ_i and select one of the refracted waves with wave vector $\mathbf{k}$ lying in the incident plane at the angle of refraction θ_r. The ray vector, which coincides with the energy flow and group velocity directions, will deviate from the wave vector direction. This deviation may be represented by an in-incident-plane deviation angle α and an out-of-incident-plane deviation angle β. As discussed previously, the critical angle condition $\theta_i = \theta_{cr}$ is satisfied when the group velocity vector is parallel to the interface plane (x_1, x_2).

For an arbitrary direction of propagation in a 2-D composite, the angle of deviation between the phase and group velocity vectors has both in-plane and out-of-plane components. However, when the surface includes the fiber direction, at the critical angle both the group velocity (ray) vector and the wave vector will lie in the same plane (x_1, x_2) of symmetry as shown in Fig. 3.19. This means that at the critical angle of the refracted wave vector, the phase velocity vector will be parallel to the plate surface and, as always, will also lie in the plane of incidence. The group velocity vector will also be parallel to the plate surface but may not lie in the plane of incidence and will deviate from this plane by an angle β. This conclusion is further illustrated for a unidirectional graphite/epoxy composite material by calculations of the deviation angle versus angle of refraction in Fig. 3.20. The incident plane is rotated at an angle $\phi = 45°$ to the fiber direction. It can be seen

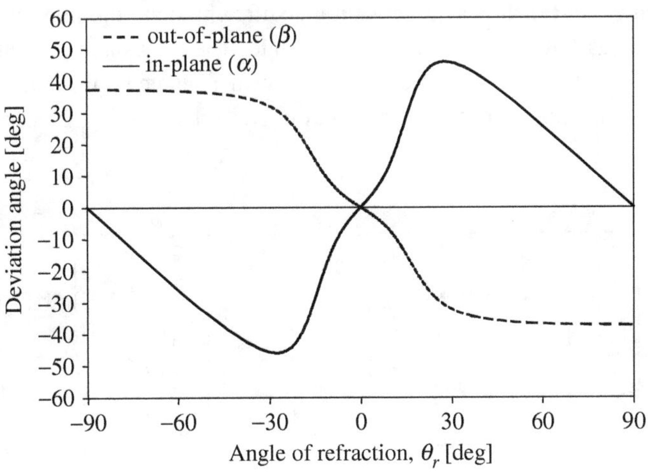

Figure 3.20. In-plane and out-of-plane angles of deviation between the group velocity and the wave normal as functions of the refraction angle for the quasilongitudinal wave in a transversely isotropic composite specimen at $\phi = 45°$.

that the in-plane deviation angle (solid curve) equals zero at a refraction angle of $\theta_r = 90°$, though the out-of-plane deviation angle (dashed curve) is more than 40°. The most important practical conclusion we can draw from this observation is that Snell's law may be used at the critical angle for the calculation of the in-plane phase velocity of a transversely isotropic composite plate, if the surface is normal to the plane of isotropy, *i.e.*, it includes the fiber direction. (Here, "in-plane velocity" refers to the plane of the composite plate's surface.)

This conclusion will hold for orthotropic composite materials if the plane of the surface is a plane of symmetry, but may not hold otherwise. When the surface of the specimen is not a plane of symmetry, the deviation angle between the phase and group velocity may have a component that is out of the plane of the surface (in the plane of incidence); therefore, the wave vector will not lie in the plane of the surface at the critical angle. The general algorithm for solving the reflection/transmission problem at an interface between two anisotropic media [25], which has also been adapted to composite materials [26], is discussed in the next chapter. This algorithm can be easily specialized to calculate the reflection coefficient of an immersed composite plate. The calculated energy reflection coefficient from a water–composite interface is shown in Fig. 3.21 for three different orientations of the incident plane relative to the fiber direction. The material used in these calculations was the same transversely isotropic unidirectional graphite fiber-reinforced epoxy matrix composite we used in Chapter 2 as an example. The elastic constants of this material were listed in Table 2.1.

Figure 3.21(a) shows the reflection coefficient along the fiber direction ($\phi = 0°$) as a function of the incident angle. The very sharp maximum at approximately $\theta_i = 10°$ corresponds to the critical angle for the longitudinal wave propagating along the fiber direction. The reflection coefficient reaches

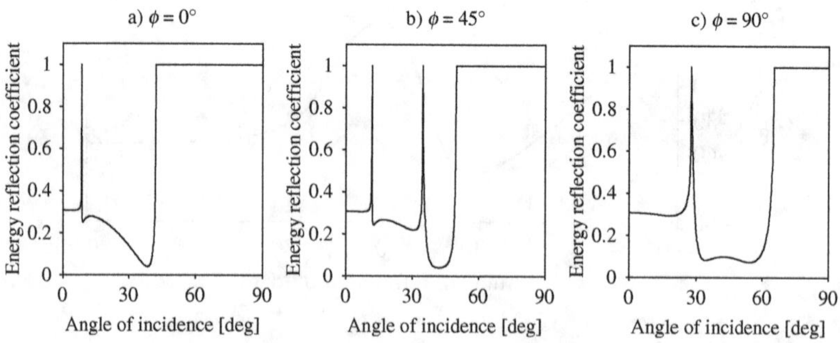

Figure 3.21. Calculated energy reflection coefficients from a water–composite interface for three different orientations of the incident plane relative to the fiber direction for a transversely isotropic graphite/epoxy composite.

unity again at the second critical angle and remains constant above it. The second critical angle at approximately $\theta_i = 40°$ corresponds to the transverse SV wave propagating along the fiber direction and polarized perpendicular to the interface. The transverse SH wave with polarization parallel to the surface is not coupled to the liquid and is not excited in the solid at this fiber orientation. Figure 3.21(b) shows the reflection coefficient when the incident plane is rotated by $\phi = 45°$ from fiber direction. At oblique incidence, three elastic waves are excited in the solid in this case, which produces three critical angles. The first very sharp maximum corresponds to the critical angle for the quasilongitudinal wave, the second is for the fast quasitransverse wave, and the reflection coefficient becomes unity at the third critical angle which corresponds to the slow quasitransverse wave. At other oblique angles of ϕ between the fiber direction (which is an axis of symmetry) and the plane of incidence, the dependence of the reflection coefficient on the angle of incidence resembles that shown for $\phi = 45°$, except that the positions of the critical angles will be shifted according to the angular dependence of the ultrasonic velocities in the plate.

The case of $\phi = 90°$, when the incident wave is perpendicular to the fibers, is shown in Fig. 3.21(c). There, the only peak corresponds again to the critical angle for a longitudinal wave that propagates perpendicular to the fibers. After the first critical angle, only the slow transverse wave of SV polarization propagates. It has displacements in the plane perpendicular to the fiber direction and has a velocity only slightly above the sound velocity in water. In many cases (not shown in the figure), the slow transverse wave velocity normal to the fiber direction is so low that it remains below the sound velocity in water. Therefore, a second critical angle does not exist for incident planes close to the plane of isotropy, and the reflection coefficient reaches unity only at grazing incidence. The fast transverse wave exhibits SH character for reasons of symmetry and is not coupled to the liquid. Moreover, this fast transverse wave at $\phi = 90°$, which is polarized in the fiber direction, has the same velocity as the transverse SV wave that propagates along the fiber direction and is polarized perpendicular to the fibers and the plate surface. The latter mode corresponds to the second critical angle in Fig. 3.21a) and can be readily detected at $\phi = 0°$.

3.3.3 Critical angle measurements

In the previous section we introduced the concept of critical angle measurements for characterization of the elastic properties of composites. These measurements can be made by the same experimental arrangement described earlier in Section 3.2.1 (Fig. 3.15) that is suitable for both doubly transmitted (through the sample) and doubly reflected (from the sample) ultrasonic measurements using a single ultrasonic transducer. In reflection mode, the incident and specularly reflected ultrasonic beams both propagate radially inside the cylinder. Therefore, after being reflected from the inside wall of the cylinder, the ultrasonic signal returns to the sample along the same path, reflects a second time from its surface, and finally retraces its path back to the receiving transducer. In critical angle measurements, only the first arriving signal from the sample front surface is considered, while the second and subsequent arrivals that reflect from the back wall of the specimen are neglected. The front wall reflection can become laterally shifted at the critical angles much like the previously described shift of the back wall signals, although to a lesser degree. This lateral shift is caused by the sharply changing phase of the reflection coefficient around the critical angles that causes non-specular reflection for finite-diameter ultrasonic beams. In the arrangement shown in Fig. 3.15, laterally shifted reflections reach the surface of the cylindrical reflector at slightly off-normal incidence, and therefore they cannot exactly retrace their paths back to the transducer. This minor misalignment at the critical angles slightly reduces the accuracy of the measurement, but will not result in significant loss of accuracy.

It should be mentioned that the well-known non-specular nature of the reflection of finite-diameter ultrasonic beams at the Rayleigh angle [27] is much more pronounced than at the longitudinal and shear critical angles, and can be easily exploited for ultrasonic characterization of solids even of very high acoustic impedance. We show later that the leaky Rayleigh wave technique is rendered almost useless for the characterization of composite materials by their low velocity and density. However, a similar technique based on Lamb wave propagation can be readily used for the characterization of thin composite plates. The Lamb wave method is based on the overall reflection from the plate, which includes the front reflection as well as all the multiple reflected signals from the back wall of the plate. This method is discussed in detail in Chapter 8.

The double-reflection coefficient measured in the arrangement shown in Fig. 3.15 corresponds to the single energy reflection coefficient that is presented in Fig. 3.21. One advantage of this method over the conventional single-reflection pitch–catch measurement is that all the features are enhanced by the second interaction and can be more easily and accurately identified. The ultrasonic phase velocity in the plane of the sample may be calculated at each critical angle from

$$V_\phi = V_f / \sin\theta_i, \tag{3.28}$$

where V_f is the sound velocity in water. The measurements presented in this Section were taken at a stabilized temperature of 29.8°C to increase the measurement accuracy.

3.3.4 Experimental results for graphite/epoxy composites

Typical recordings of the amplitude of the doubly reflected signal versus incident angle θ_i are shown in Fig. 3.22 [24]. The data were taken with an incident angle step size of 0.05°. The doubly reflected signal disappears for incident angles below 4° and above 75° due to the openings in the cylinder wall made for the ultrasonic transducer and back reflector (see Fig. 3.15). The first recording was taken in the plane coinciding with the fiber direction ($\phi = 0°$). The experiments were repeated by rotating the sample in 5°-steps from parallel to the fiber direction to perpendicular to the fiber direction.

As can be seen from the data, the first maximum is usually not very pronounced and cannot be identified directly. Its position may be located with the help of the minimum that follows it (see also Fig. 3.21). Polishing of the sample surface, which is originally rough, changes the results very little. The maximum is weak because it is very narrow, so its appearance depends critically upon the ultrasonic attenuation and the

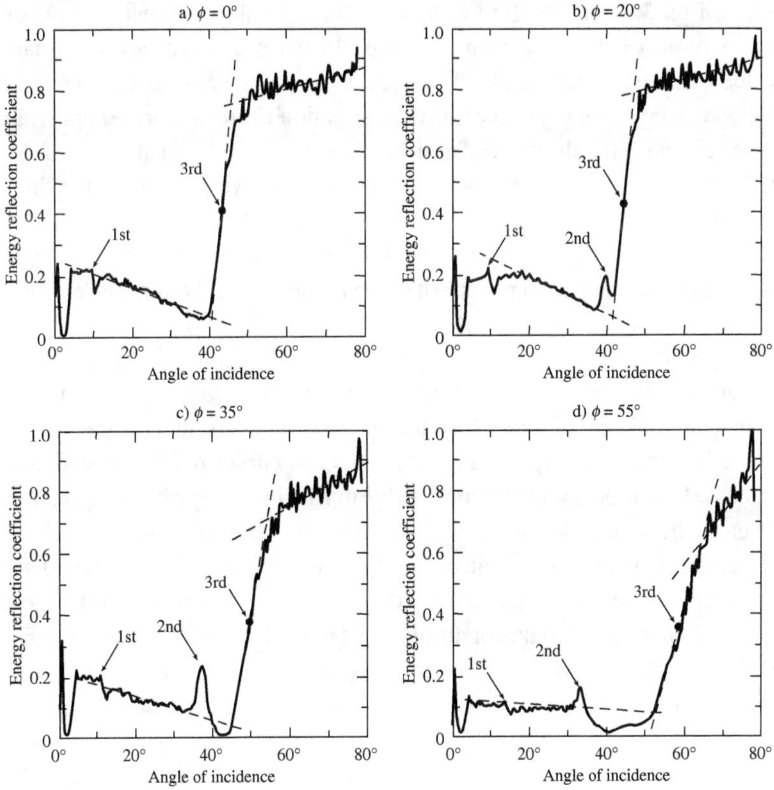

Figure 3.22. Typical recordings of the amplitude of the doubly reflected signal versus incident angle θ_i at four different fiber orientations with respect to the plane of incidence.

effect of finite beam width. These effects have been discussed theoretically in reference [28] for isotropic materials. To test the assumption that the experimentally observed maximum is very weak because it is very narrow, the measurements were repeated for the same experimental conditions with Plexiglas (polymethylmethacrylate). Plexiglas has acoustic characteristics similar to those of the composite matrix and theoretical calculations show a very strong, wide maximum at the first critical angle. The experimental maximum at the first critical angle in this case was very strong, supporting the above assumption, and the results confirm the measurements on Plexiglas done by Mayer [18].

In spite of the weakness of the maximum at the first critical angle, it can be located fairly easily in the experimental data, as mentioned above, with the help of the minimum following it. In comparison, the second maximum that corresponds to the critical angle for the fast transverse wave is usually very clear and sharp and its position can be easily measured. The third critical angle, which corresponds to the slow transverse wave, is also problematic, though in a different way from the first one. As one can see in Fig. 3.21, at the third critical angle, the reflection coefficient does not exhibit a local maximum, *i.e.*, a peak, but rather reaches a constant value of unity. Theoretically, this feature can be detected just as easily as a slope change, but experimentally the situation is much more complex. In practice, the difficulty arises from the facts that (1) the final slope of the reflection coefficient versus angle of incidence curve just before the third critical angle is finite rather than infinite and (2) the reflection coefficient above the third critical angle slowly increases with the angle of incidence rather than reaching a flat plateau of unity. In this way, although the detection of the third critical angle is quite easy, its accurate location is rather difficult to pinpoint.

One simple approach, which has been successfully tested on numerous composite specimens, is matching the three more-or-less linear regions of the reflection coefficient versus angle of incidence curves before, around, and after the transition that corresponds to the third critical angle by straight lines. Then we find the mid-point between the two intersections of these tangential lines. These points are indicated by solid circles in Fig. 3.22(a)–(d) and are assumed to mark the experimentally determined location of the third critical angle. Experimental data for in-plane velocities of the quasilongitudinal and two quasitransverse waves are shown in Fig. 3.23 as functions of the rotation angle ϕ of the plane of incidence from the fiber direction. These data represent the experimentally determined dependence of the phase velocity in the plane of the surface on the direction of wave propagation. The solid lines also shown in Fig. 3.23 represent the theoretical predictions for a transversely isotropic unidirectional graphite fiber-reinforced epoxy matrix composite material, the properties of which were previously listed

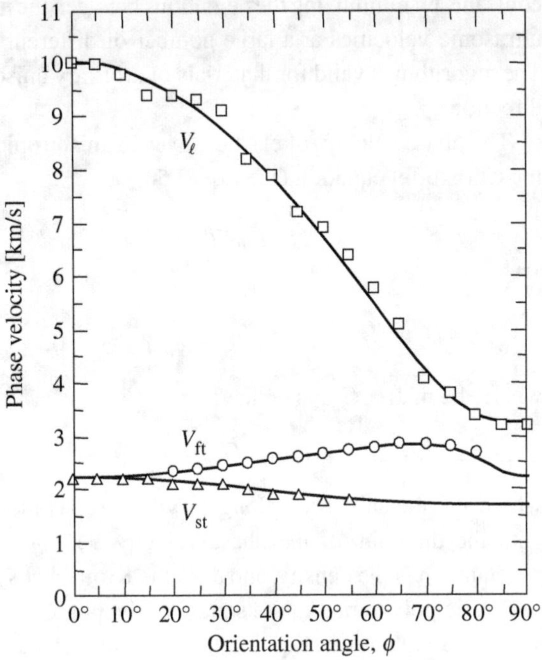

Figure 3.23. The phase velocity of in-plane ultrasonic waves in a transversely isotropic graphite/epoxy composite as function of the orientation angle between the direction of wave propagation and the fiber direction. Experimental velocity values were extracted from critical angle measurements and are indicated by symbols. Solid lines represent calculated values based on elastic constants listed in Table 2.1.

in Table 2.1. The five independent elastic parameters were determined by best fitting the appropriate theoretical predictions to the experimental data obtained by both double transmission and critical angle measurements (while the density was independently measured). At ϕ close to 90°, the fast transverse wave approaches SH character and therefore cannot be excited from the liquid. Data for the slow transverse wave at angles above $\phi = 55°$ also could not be obtained since its velocity drops so low that, although it is still slightly above the sound velocity in water, for all practical purposes the critical angle disappears.

3.4 Determination of Elastic Constants from Phase Velocity Data

3.4.1 Reconstruction of elastic constants

The elastic constants may be calculated from phase velocity measurements in a limited number of specific directions of the anisotropic material. While this approach is possible, it is somewhat sensitive to errors of measurement when reconstruction of all the elements of a full matrix of elastic constants is desired. For this reason, we use another approach—a nonlinear least squares optimization technique. This technique determines the set of elastic constants by minimizing the deviations between the measured and calculated ultrasonic velocities at a large number of different angles of propagation. The algorithm is valid for materials of arbitrary anisotropy and for arbitrary directions.

The phase velocity of elastic waves in anisotropic materials must satisfy the Christoffel equation (see Eq. (2.50)):

$$|c_{ij\ell m}n_j n_\ell - \delta_{im}\rho V^2| = 0 \tag{3.29}$$

or

$$|G_{im} - \delta_{im}\rho V^2| = 0, \tag{3.30}$$

where the matrix G_{im} is defined as

$$G_{im} = c_{ij\ell m}n_j n_\ell, \tag{3.31}$$

where n_i is the unit vector along the wave propagation direction (**n** coincides with the direction of the phase velocity **V**), $c_{ijk\ell}$ is the tensor of elastic constants, ρ is the density, and δ_{ij} is the Kronecker symbol. Equation (3.30) is a cubic polynomial in the square of the phase velocity V. If the material properties are known, Eq. (3.30) defines three independent wave velocities for a given direction of the wave normal (direction of wave propagation). On the other hand, if the phase velocities of waves for a given propagation direction are known, we can obtain the material properties from Eq. (3.30) by inversion.

To extract elastic constants of composites from the experimental angle-dependent phase velocity data by the nonlinear least squares optimization technique, one can use analytical solutions of the Christoffel equation that relate the propagation directions, the phase velocities and the elastic constants. This is done using Cardan's solution of the cubic equation that was discussed in Chapter 2 (see Eqs. (2.57–2.63)). It should be emphasized that these solutions are applicable for arbitrary anisotropy and arbitrary direction of propagation in the material. Moreover, a numerical solution of Eq. (3.30) can be used with the same success.

The unknown material properties can be found by minimizing the sum of the squares of the deviations between the experimental and calculated velocities via changing the elastic constants (and, if necessary, the density) as variables in a multi-dimensional space $\Re^n$ of dimension n:

$$\min_{c_{ij\ell m} \in \Re^n} \sum_{q=1}^{Q} (V_q^{\text{exp}} - V_q^{\text{calc}})^2, \tag{3.32}$$

where n is the number of independent parameters to be extracted and Q is the number of velocity measurements in different directions, while V^{exp} and V^{calc} are the experimental and calculated phase velocities, respectively.

For orthotropic materials, the number of independent elastic constants n is nine. This method minimizes the effect of random deviations of experimental points on the results of reconstruction. In typical cases, even if the measured phase velocities have as much as 5% inherent random scattering, a complete set of elastic constants can be obtained. For details of the reconstruction algorithm, see references [8,29]. The full set of elastic constants reconstructed from the previously presented double transmission and critical angle phase velocity data was given in Table 2.1. Since, as discussed previously, the sample material is transversely isotropic (in planes perpendicular to the fiber direction), only five of the nine elastic constants are independent.

3.4.2 Stability of the reconstruction algorithm

The stability of the nonlinear least squares reconstruction algorithm can be easily demonstrated by computer simulations on a set of synthetic velocity data with random noise [6, 29]. Let us start with a selected set of elastic constants (the "original" set) and calculate the ultrasonic phase velocity. The calculations can be done at selected orientations of the incident plane for a discrete set of refraction angles. Next, we add random noise of varying magnitude (*e.g.*, relative levels of 0.1, 1, or 5%) to the calculated velocity data and consider the perturbed values as simulated experimental data. Finally, the elastic constants can be reconstructed from the noisy simulated experimental data and compared to the original set.

As an example, Fig. 3.24 shows the simulated velocity data for the plane parallel to the surface of the sample with relative noise of 5% along with the best-fitting theoretical curves in solid lines; scatter of the synthetic velocity data (plotted as points) is clearly visible. It should be mentioned that the full set of elastic constants was reconstructed from all the simulated velocity data in the plane of the surface $\theta_r = 90°$ (x_1, x_2) as well as in two other planes of symmetry, namely $\phi = 0°$ (x_1, x_3) and $\phi = 90°$ (x_2, x_3). In a similar way, Fig. 3.25 shows the simulated velocity data of 5% relative noise for varying angles of refraction in the the incident plane rotated by an angle $\phi = 30°$ relative to the fiber direction. This time, the full set of elastic constants was reconstructed from the simulated velocity data in three planes of incidence, namely $\phi = 0°$, $\phi = 30°$, and $\phi = 60°$.

The stability of these two realizations of the reconstruction procedure are compared in Tables 3.1–3.4 (all stiffness constants are in GPa). Table 3.1 shows the results of the reconstruction from synthetic velocity data in three planes of symmetry $\theta_r = 90°$, $\phi = 0°$, and $\phi = 90°$ (the case illustrated in Fig. 3.24). In comparison, Table 3.2 shows the results of the reconstruction from synthetic velocity data in three non-symmetry planes $\phi = 30°$, $\phi = 45°$, and $\phi = 60°$. In both cases, the second and third rows

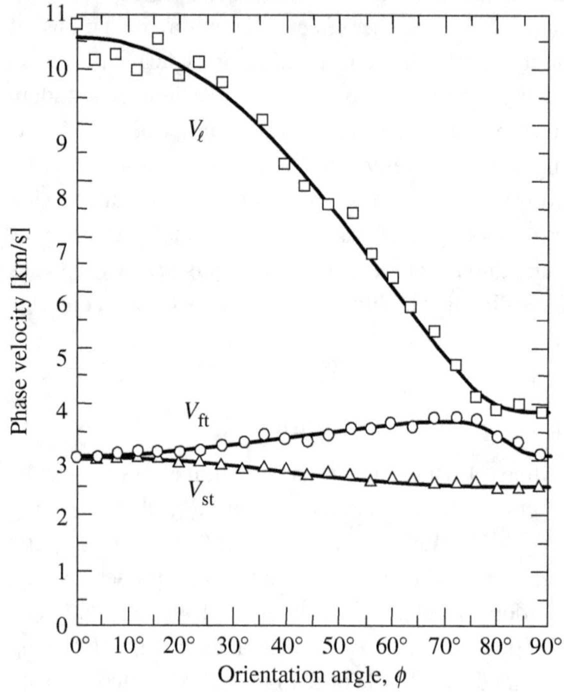

Figure 3.24. Simulated velocity data for the plane parallel to the surface of the sample with relative noise of 5% along with the best-fitting theoretical curves in solid lines.

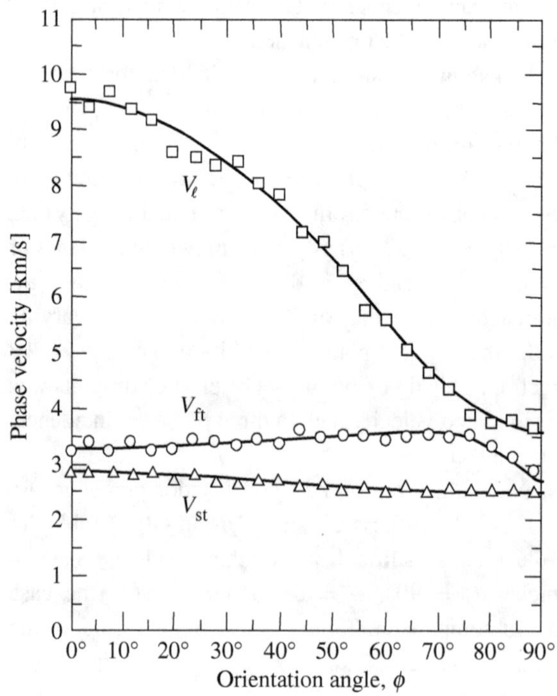

Figure 3.25. Points are synthetic velocity data with random noise of 5%. The incident plane deviates $\theta = 30°$ from the fiber direction. Solid lines are calculations based on reconstructed set of data in incident planes $\theta = 30°$, $45°$, and $60°$.

Table 3.1 The results of the reconstruction from synthetic phase velocity data in three planes of symmetry $\theta_r = 90°$, $\phi = 0°$, and $\phi = 90°$.

Elastic constant	C_{11}	C_{22}	C_{33}	C_{12}	C_{13}	C_{23}	C_{44}	C_{55}	C_{66}
Original value	144.0	13.6	12.0	5.47	5.0	7.0	3.7	6.0	6.5
Initial value	130.0	12.0	11.0	5.0	4.0	6.0	3.0	5.0	5.0
Noise									
0.00	144.0	13.6	12.0	5.47	4.93	7.0	3.7	6.0	6.5
0.001	143.99	13.59	11.99	5.46	4.81	7.0	3.7	6.0	6.5
0.01	144.54	13.62	11.95	5.46	4.56	7.02	3.7	5.99	6.49
0.05	143.89	13.27	12.1	4.95	4.54	6.96	3.65	6.0	6.54

Table 3.2 The results of the reconstruction from synthetic phase velocity data in three non-symmetry planes $\phi = 30°$, $\phi = 45°$, and $\phi = 60°$.

Elastic constant	C_{11}	C_{22}	C_{33}	C_{12}	C_{13}	C_{23}	C_{44}	C_{55}	C_{66}
Original value	144.0	13.6	12.0	5.47	5.0	7.0	3.7	6.0	6.5
Initial value	130.0	12.0	11.0	5.0	4.0	6.0	3.0	5.0	5.0
Noise									
0.00	144.0	13.6	12.0	5.47	5.0	7.0	3.7	6.0	6.5
0.001	143.97	13.61	12.0	5.52	4.98	7.0	3.7	6.0	6.49
0.01	143.93	13.48	11.95	5.19	4.97	6.94	3.71	5.98	6.52
0.05	146.34	12.87	12.03	−2.00	5.07	6.36	3.72	5.94	6.21

Table 3.3 The results of reconstruction from phase velocity data in one plane of symmetry ($\phi = 0°$).

Elastic constant	C_{11}	C_{22}	C_{33}	C_{12}	C_{13}	C_{23}	C_{44}	C_{55}	C_{66}
Original value	144.0	13.6	12.0	5.47	5.0	7.0	3.7	6.0	6.5
Initial value	130.0	12.0	11.0	5.0	4.0	6.0	3.0	5.0	5.0
Noise									
0.00	144.0	—	12.0	—	5.0	—	3.7	6.0	6.5
0.001	143.92	—	12.0	—	5.04	—	3.7	6.0	6.5
0.01	143.23	—	11.99	—	5.47	—	3.68	6.0	6.47
0.05	144.85	—	12.17	—	5.65	—	3.61	5.92	6.65

in the tables list the "original" elastic constants selected for the material and the set of initial data for reconstruction, respectively. The next rows show the elastic constants obtained after application of the least squares minimization procedure, where the first column in these rows gives the values of the random noise (starting with zero noise, going up to 5% noise). All nine elastic constants can be reconstructed in this case. It is clear that reconstruction based on the three orthogonal planes of symmetry is much more stable, although at the highest 5% noise level C_{12} and C_{13} are underestimated by as much as 10%. Not surprisingly, the reconstruction from the three non-symmetry planes is less stable, especially where C_{12} is concerned.

Table 3.4 The results of reconstruction from velocity data in one non-symmetry plane $\phi = 30°$.

Elastic constant	C_{11}	C_{22}	C_{33}	C_{12}	C_{13}	C_{23}	C_{44}	C_{55}	C_{66}
Original value	144.0	13.6	12.0	5.47	5.0	7.0	3.7	6.0	6.5
Initial value	130.0	12.0	11.0	5.0	4.0	6.0	3.0	5.0	5.0
Noise									
0.00	144.13	13.88	12.0	4.32	5.03	6.89	3.7	6.0	6.33
0.001	143.82	11.08	12.0	3.06	5.06	6.87	3.7	6.0	7.25
0.01	144.15	14.41	11.96	6.93	4.81	7.02	3.72	5.97	6.33
0.05	140.69	16.68	12.23	13.05	5.17	7.45	3.75	6.25	6.2

Generally speaking, reconstruction from phase velocity data in a single plane is less stable and cannot necessarily provide all the elastic constants even when the measurement accuracy is very high. Table 3.3 shows results of the reconstruction from one plane of symmetry, namely, $\phi = 0°$. One can see that not all elastic constants can be recovered (C_{12}, C_{22}, and C_{23} are not accessible by measurement in this plane). In addition, although the constants C_{44} and C_{66} are apparently recovered since data for SH wave velocity have been assumed, in the plane of symmetry this wave cannot be excited by an incident longitudinal wave (see Section 3.22).

Finally, Table 3.4 shows results of the reconstruction from one non-symmetry plane, namely, $\phi = 30°$. Here, all three velocities can be measured and all elastic constants can be reconstructed for small noise. The reconstruction is sensitive to experimental errors and noise even as low as 0.1% can change the values of C_{12} by as much as 30%. The constant C_{22} is also not very stable in this case. One may conclude that, while in planes of symmetry it is possible to reconstruct fewer elastic constants than off the plane of symmetry, these results are less susceptible to noise. One advantage of using data from off-symmetry planes is that all elastic constants can be obtained by the same measurement technique. Recall that the data in Table 3.1 for one of the planes of symmetry had to be obtained by a different technique, namely, the critical angle method. Another advantage of measurements in off-symmetry planes is that all three velocities can be measured for each plane. What can be recommended is a combination of results from measurements in off-symmetry and symmetry planes. This action will give the best reconstruction results for all elastic constants of orthotropic materials (see also the discussion in Section 3.5.4).

3.5 Group Velocity Measurements

The majority of nondestructive techniques for elastic constant measurement utilize phase velocity data, as discussed in the previous section.

This is partly because it is easier to determine the elastic constants from the phase velocity than from the group velocity (see Chapter 2) and partly because the phase velocity is often easier to measure, especially if we consider the necessity to measure the group velocity direction in addition to its magnitude. Group velocity data, however, can also be used to recover the elastic constants. The data can be obtained using, for example, a point source–point receiver technique [30] that allows one to measure group velocities in different directions including those in a cuspidal region [31]. The advantage of this technique is that it does not require rotating the sample for velocity measurements in different directions. Moreover, techniques with one-sided access to the sample are available for group velocity measurements [32]. These techniques are discussed next. We should note that in composite materials these group velocity measurement techniques require, in addition to accurate time delay measurements, accurate determination of the group velocity propagation direction.

3.5.1 Reconstruction of elastic constants from group velocity data

Several methods have been proposed for determining elastic constants from group velocity data. Every and Sachse [33] used nonlinear least squares fitting to determine three independent elastic constants of a cubic material from group velocity data outside the cuspidal regions. Kim [34] showed that mixed-index elastic constants, namely C_{12}, C_{13}, and C_{23}, in orthotropic materials can be determined by measuring the group velocity of a quasilongitudinal or quasishear wave propagating obliquely in the symmetry plane assuming that pure-index elastic constants are known from measurements along symmetry axes. Kim *et al.* [35] applied this approach to determine all nine elastic constants of an orthotropic plate. Minachi *et al.* [32] used a perturbation method and a nonlinear least squares procedure for a transversely isotropic material to determine C_{11} and C_{13} from measurements at oblique incidence in the symmetry plane, while the other elastic constants (C_{33}, C_{44}, and C_{55}) were assumed to be known from normal incidence measurements.

In these techniques, some or all pure-index elastic constants are determined from measurements along symmetry axes. As mentioned above, such measurements may not always be possible. For shear modulus measurement in the direction normal to the plate surface, one needs to use a shear wave. Although such normal incidence shear waves can be excited using noncontacting electromagnetic transducers in metals, shear wave measurements using contact transducers are difficult to implement for scanning when nondestructive mapping of elastic constants is required. Therefore, oblique excitation of transverse waves is preferred for shear elastic constant measurement.

In this chapter, we describe an inversion method [36] to determine the elastic constants of an orthotropic material from group velocity data of longitudinal and shear waves in different directions of symmetry and non-symmetry planes using a nonlinear least squares algorithm that minimizes the overall difference between experimental and calculated group velocity data. No elastic constants except one (determined from the longitudinal wave at normal incidence) are assumed to be known from measurements along symmetry axes. It will be shown that experimental data in a cuspidal region can be utilized for the inversion. This eliminates the necessity of obtaining quasishear wave data away from folded regions. The reconstruction is illustrated on synthetic sets of experimental data corresponding to materials with different degrees of anisotropy. We will compare the reconstruction results to those obtained from phase velocity measurements, also using nonlinear least squares minimization.

We will show that seven out of nine elastic constants for an orthotropic medium can be determined from group or phase velocity data in two symmetry planes perpendicular to the specimen plane that is assumed to be a plane of symmetry. To determine the remaining two elastic constants, velocity data in a non-symmetry plane must be used. Such determination from phase velocity data was discussed briefly in the previous section and in more detail in [8, 37]. Here, following [36], we simply demonstrate that group velocity data can be used for this purpose. In addition, it is shown that the five independent elastic constants of a hexagonal (transversely isotropic) material can be determined from group velocity data obtained from a single plane of incidence. Simple analytical formulas discussed in Section 2.9.4 will be used for this case. As we discussed in that section, this technique provides good approximations for the wave velocities in non-symmetry planes of weakly orthotropic (with respect to transversely isotropic) materials.

3.5.2 Determination of elastic constants from group velocity data in a symmetry plane

To understand the reconstruction method better, let us consider the generic through-transmission arrangement for group velocity measurement shown in Fig. 3.26. The ray (group velocity) direction can be determined from the receiver position, for example, point A on the back surface of the sample (Fig. 3.26). In addition, the group velocity can be measured from time of flight of the ultrasonic signal (the measurements described by Kim and Sachse [31] and by Minachi *et al.* [32], which were discussed in the previous section, were performed by slightly different techniques). The ray

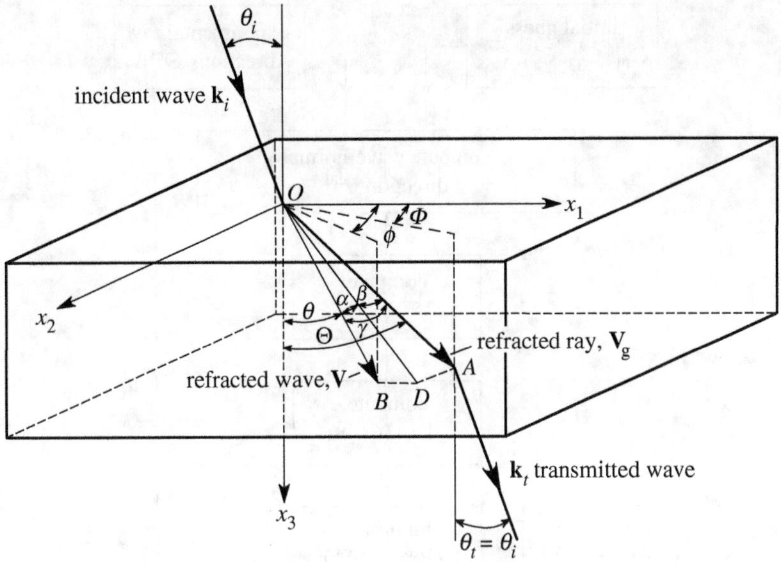

Figure 3.26. Through-transmission arrangement for group velocity measurement showing the different wave and ray directions for an arbitrary plane of incidence.

direction is defined by a pair of azimuthal and polar refraction angles (Φ, Θ). Generally, the ray direction is skewed with respect to the wave direction (ϕ, θ) in both dimensions (for simplicity, we dropped the subscripts r). However, in a plane of symmetry (*e.g.*, in the (x_1, x_3) plane), the out-of-plane component of the ray deflection vanishes by virtue of symmetry $(\Phi = \phi)$ and the localization of the refracted ray becomes a much easier 1-D problem.

From a set $\{V_g^{\exp}(\Theta^{\exp})\}$ of experimental quasilongitudinal (QL) and quasitransverse (QT) group velocity data in the (x_1, x_3) plane, four unknown elastic constants can be determined. Here, a pair of braces { } denotes a finite set of data. Figure 3.27 shows a flow chart of the inversion procedure that can be used for elastic constant determination from group velocities [36]. It is based on the nonlinear least squares optimization algorithm. Initial guesses of the elastic constants are used to determine the wave directions $\{\theta^{\text{calc}}\}$ corresponding to the measured ray directions $\{\Theta^{\exp}\}$ by searching for solutions of Eq. (3.33):

$$\frac{V_{g1}(\theta^{\text{calc}})}{V_{g3}(\theta^{\text{calc}})} = \tan(\Theta^{\exp}) \tag{3.33}$$

in the interval $\theta^{\text{calc}} \in [0, \pi]$. The phase velocity V^{calc} can now be determined for this particular propagation direction using the initial guess of the elastic constants. For a QT wave, the solution of Eq. (3.33) is not unique in

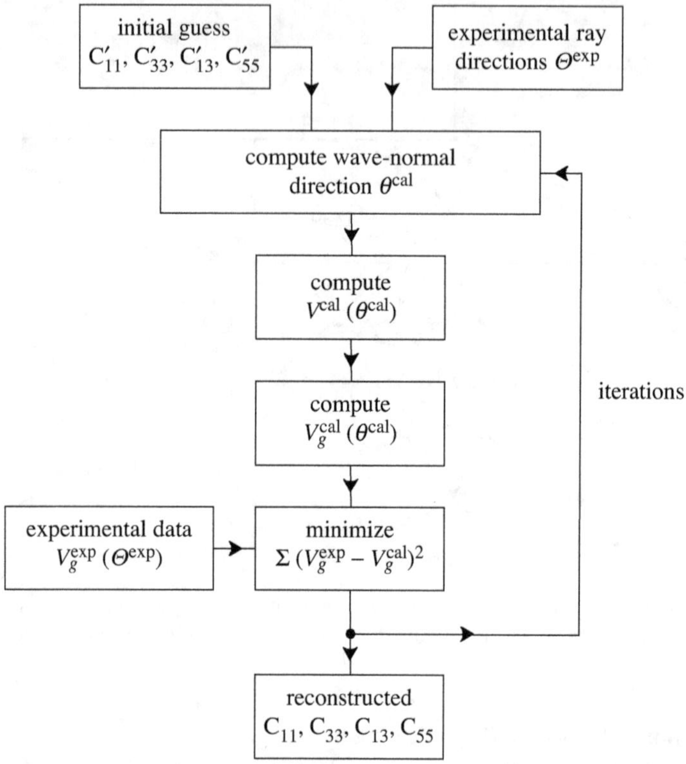

Figure 3.27. Flow chart of the inversion procedure that can be used for elastic constant determination from group velocities.

the cuspidal regions, which means that there are several quasitransverse waves propagating in the same ray direction with different group velocities each corresponding to the different phase velocities and wave normals. When Eq. (3.33) yields multiple solutions $(\theta_1, \theta_2, \theta_3)$ for the wave-normal direction in a cuspidal region, we choose the one that gives the smallest deviation between the possible phase velocity $V(\theta_\ell)$ $(\ell = 1, 2, 3)$ and the projection of the experimental group velocity on the calculated wave direction $V_g^{\exp}\cos(\Theta^{\exp} - \theta_\ell)$. Of course, for the exact set of elastic constants these values would be equal since the projection of the group velocity on the wave-normal direction is equal to the phase velocity. We take this choice as the calculated wave normal angle θ^{calc}, *i.e.*,

$$\left\{\min_{\ell=1,2,3}\left[V_g^{\exp}\cos(\Theta^{\exp} - \theta_\ell) - V(\theta_\ell)\right]\right\} \to \left\{V_g^{\exp}(\Theta^{\exp} - \theta^{\text{calc}})\right\}. \tag{3.34}$$

Knowing the wave-normal direction, we can determine the group velocity by following the procedure described in Chapter 2 or using explicit analytical formulas available for the group velocity in planes of symmetry for materials of orthotropic or higher symmetry [34]. Repeating this procedure for each experimental group velocity direction, we obtain a calculated set of group velocity data $\{V_g^{\mathrm{calc}}(\Theta^{\mathrm{exp}})\}$ for a given set of initial guesses of the elastic constants. Now, considering the elastic constants as variables in a 4-D space, a nonlinear least square minimization of deviations between calculated and experimental group velocities can be performed to determine the desired set of elastic constants

$$\min_{C_{11}, C_{33}, C_{13}, C_{55}} \sum_{q=1}^{Q} [V_g^{\mathrm{exp}}(q) - V_g^{\mathrm{calc}}(q)]^2 , \tag{3.35}$$

where Q is the number of velocity measurements for different directions, and V_g^{exp} and V_g^{cal} are the experimental and calculated group velocities, respectively.

3.5.3 Comparison of the reconstruction results from group and phase velocity data in symmetry planes

The efficiency of the above-described reconstruction algorithm can be demonstrated through the examples of hexagonal crystalline zinc and graphite/epoxy composite (also transversely isotropic). Like before, we can simulate both QL and QT waves to obtain a synthetic set of group $V_g^{\mathrm{exp}}(\Theta^{\mathrm{exp}})$ and phase $V^{\mathrm{exp}}(\theta^{\mathrm{exp}})$ velocity data for the (x_1, x_3) plane of symmetry assuming a certain set of elastic constants that will be referred to as "original" values. Then, different levels of "noise" are introduced in the synthetic group and phase velocities, as well as in the group velocity directions, to simulate the effect of random scattering in the measurements. In phase velocity measurements, the direction of the phase normal typically is known or can be estimated accurately from the measured velocity itself. Therefore, it was assumed that the wave-normal direction can be determined with high precision. Four elastic constants (C_{11}, C_{33}, C_{13}, and C_{55}) can be determined from velocity data in the (x_1, x_3) plane.

The constant C_{33} can be directly determined from longitudinal wave velocity measurement at normal incidence and then fixed for further reconstruction of the remaining three elastic constants. Simulation results, however, indicate that this does not improve the accuracy of C_{33} itself or that of the other elastic constants. Thus, we consider simultaneous determination of all four elastic constants and since C_{33} is effectively decoupled in the equations, it is found most precisely. For the zinc crystal, the angular range and the number of points for the group velocity data as well as original values for the elastic constants are taken from [31]. Synthetic sets of "experimental" group velocity data with 1% scattering in both velocity values and directions of propagation were generated together with corresponding phase velocity data with 1% scattering.

Here, for simplicity, we assume that the ray angle and group velocity data have the same relative error. In an actual experiment, however, the ray angle is very difficult to measure accurately. In contrast to the phase velocity measurement method, the wave normal angle is accurately determined from the incident angle, which is easier to measure accurately. Synthetic data are presented in Fig. 3.28 and Fig. 3.29 for the group (a) and phase (b) velocities in transversely isotropic crystalline zinc and graphite/epoxy composite with 1% scattering, respectively. The angular range and the number of

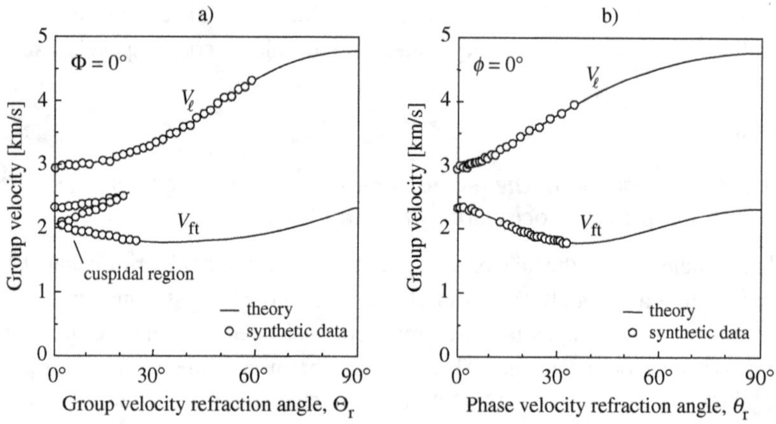

Figure 3.28. Synthetic data for the group (a) and phase (b) velocities in transversely isotropic crystalline zinc with 1% scattering.

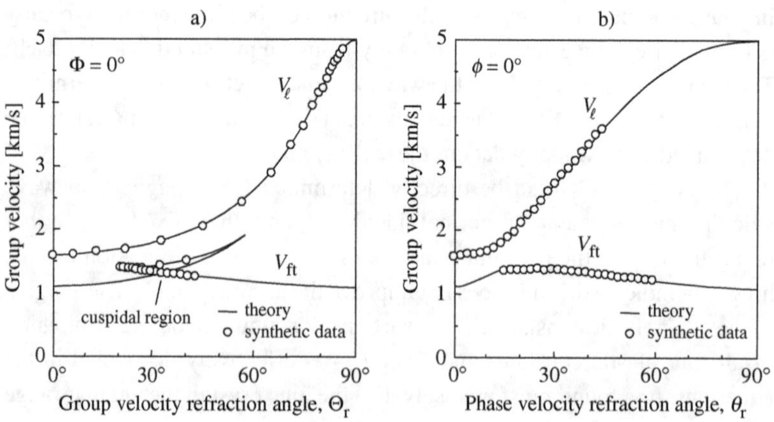

Figure 3.29. Synthetic data for the group (a) and phase (b) velocities in transversely isotropic graphite/epoxy composite (see Table 2.1) with 1% scattering.

"experimental" points for the graphite/epoxy composite were chosen such that the corresponding phase velocity data range is experimentally available using the double through-transmission technique described in [13]. Note that for both materials, the group velocity data include a cuspidal region. For both sets of "experimental" group velocity data, the reconstruction was performed using the procedure described above. The initial guess was taken to be 20% from the original elastic constants. One hundred inversions (runs) were made for randomly synthesized velocity data sets to compute the standard deviation of the reconstructed moduli from the original values. A similar reconstruction was performed from the phase velocity data as well. Several sets of the synthesized velocity data with scattering levels of 0.5, 1, 2, and 5% were also used. The standard deviation of the recovered elastic constants from the original ones is shown in Fig. 3.30 for crystalline zinc in frame (a) and graphite/epoxy composite in frame (b). We see that the accuracy of determination of elastic constants from phase and group velocities is approximately the same for both the zinc crystal and the graphite/epoxy composite.

3.5.4 Reconstruction from group velocity data in non-symmetry planes

In non-symmetry planes of incidence, the ray vector deviates from the incident plane as shown in Fig. 3.26. Group velocity measurements can be performed by scanning the receiver in two dimensions to find the energy propagation direction and measure the time of flight. In this section, we demonstrate that for transversely isotropic materials, all five elastic constants can be determined from experimental quasilongitudinal and fast and slow

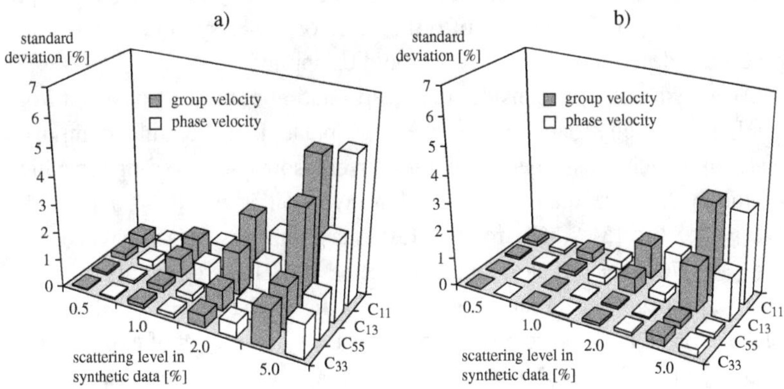

Figure 3.30. Standard deviation of the recovered elastic constants from the original ones for crystalline zinc (a) and graphite/epoxy composite (b) based on 100 separate runs.

quasitransverse group velocity data $\{V_g^{\exp}(\Theta^{\exp}, \Phi^{\exp})\}$ in a single non-symmetry incident plane of ϕ with respect to the principal (*e.g.*, fiber) direction. Reconstruction of nine elastic constants of an orthotropic material from phase or group velocity data obtained from a single non-symmetry plane, however, is not stable against scatter and initial selection of guesses. Nonetheless, one can determine the remaining two elastic constants (C_{12} and C_{66}) assuming that seven elastic constants (C_{11}, C_{22}, C_{33}, C_{13}, C_{23}, C_{44}, and C_{55}) were found from measurements in the (x_1, x_3) and (x_2, x_3) symmetry planes.

The inversion procedure is similar to the one described in Section 3.5.2 for determination of elastic constants from group velocity data in symmetry planes. First, given the experimentally measured energy propagation direction $(\Theta^{\exp}, \Phi^{\exp})$ and an initial guess for the elastic constants, we determine the wave-normal direction $(\theta^{\text{calc}}, \phi^{\text{calc}})$ by solving the following system of equations

$$\left\{\frac{V_{g3}(\theta^{\text{calc}}, \phi^{\text{calc}})}{V_g(\theta^{\text{cal}}, \phi^{\text{cal}})}\right\} = \{\cos\Theta^{\exp}\} \tag{3.36}$$

$$\left\{\frac{V_{g2}(\theta^{\text{calc}}, \phi^{\text{calc}})}{V_{g1}(\theta^{\text{calc}}, \phi^{\text{calc}})}\right\} = \{\tan\Phi^{\exp}\} \tag{3.37}$$

in the interval $\theta^{\text{calc}} \in [0, \pi]$, $\phi^{\text{calc}} \in [-\pi/2, \pi/2]$. If the solution is not unique, we follow the above-described procedure (see Eq. (3.34)) and choose the one for which the projection of the measured group velocity is the closest to the calculated phase velocity. After the wave normal direction is found, we employ the nonlinear least square minimization procedure described in Section 3.5.2. To calculate the group velocities we can use the explicit analytical formulas available for the group velocity in transversely isotropic materials [36]. For materials of orthotropic or lower symmetry, the exact procedure described in Chapter 2 should be applied.

As an example, we consider the determination of elastic constants from synthetic velocity data in the $\phi = 45°$ plane in a ceramic composite material, which is assumed to be transversely isotropic. The range in which synthetic data are made available for inversion is $\theta \in [0°–50°]$ for the QL wave, $\theta \in [35°–75°]$ for the fast quasitransverse (QT1) wave, and $\theta \in [30°–70°]$ for the slow quasitransverse (QT2) wave. In this case, the energy propagation direction $\mathbf{V}_g$ does not lie in the $\Phi = 45°$ plane (see Fig. 3.26) and must be computed to produce a synthetic set of experimental data $\{V_g^{\exp}(\Theta^{\exp}, \Phi^{\exp})\}$. One particular set of synthetic group velocity data used in this example for inversion is shown in Fig. 3.31. Open circles represent the data with ray direction angle $\Phi \le 45°$ and closed circles correspond to data with $\Phi > 45°$. The total number of experimental points

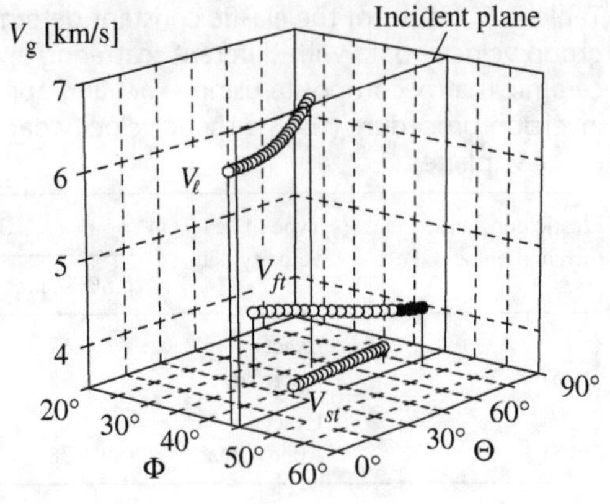

Figure 3.31. The set of synthetic group velocity data in the $\phi = 45°$ plane of a transversely isotropic ceramic matrix composite that was used in this example.

in one set is 71. Scattering of 0.5, 1, 2, and 5% is introduced into the group velocities using a random function generator. It is assumed that the energy propagation direction ($\Theta^{\exp}$, $\Phi^{\exp}$) is known precisely. The inversion procedure described in the previous section is applied. Table 3.5 presents the results of the elastic constant determination from synthetic group velocity data with different scattering levels for a transversely isotropic ceramic matrix composite along with the results of reconstruction from phase velocity data, which are described in [37]. Standard deviations of the reconstructed elastic constants from the "original" value were computed using 100 runs for each level of velocity data scatter. Again, we see that the

Table 3.5 Results of the elastic constant determination from synthetic group velocity data with different scattering levels for a transversely isotropic ceramic matrix composite along with the results of reconstruction from phase velocity data (the synthetic velocity data were assumed in the $\phi = 45°$ plane and are shown in Fig. 3.31).

Elastic constant (original value, GPa)	Type of velocity data	Standard deviation σ, GPa		
		0.5% scatter	1.0% scatter	2.0% scatter
C_{11} (150)	Phase	1.28	2.51	4.20
	Group	1.38	2.66	5.62
C_{33} (100)	Phase	0.12	0.31	0.56
	Group	0.15	0.33	0.54
C_{13} (32)	Phase	0.39	0.78	1.53
	Group	0.45	0.86	1.83
C_{55} (36)	Phase	0.31	0.69	1.00
	Group	0.13	0.25	0.50
C_{23} (36)	Phase	0.29	0.59	0.91
	Group	0.19	0.39	0.75

Table 3.6 Results of the elastic constant determination from synthetic group velocity data with different scattering levels for an orthotropic ceramic matrix composite using exact and approximate equations in the inversion procedure (the synthetic velocity data were assumed in the $\phi = 45°$ plane).

Elastic constant (original value, GPa)	Type of velocity data	Standard deviation, GPa		
		0.5% scatter	1.0% scatter	2.0% scatter
C_{12} (25.6)	Exact	1.67	2.89	3.45
	Approximate	2.38	3.32	5.86
C_{66} (28.8)	Exact	1.23	1.68	2.56
	Approximate	1.78	2.37	3.87

accuracy of elastic constant determination from group and phase velocities is approximately the same.

In a final example, let us determine two elastic constants (C_{12} and C_{66}) from synthetic velocity data in the $\theta = 45°$ plane of an orthotropic ceramic composite material. The orthotropy is introduced by decreasing C_{22}, C_{12}, and C_{66} by 10% and increasing C_{33}, C_{13}, and C_{55} by 10%. It is assumed that the other elastic constants are known from measurements in the (x_1, x_3) and (x_2, x_3) symmetry planes. The angular ranges and number of experimental points are the same as in the previous example. The inversion is performed from synthetic group velocity data using exact (see Chapter 2) and approximate analytical equations [36] for the group velocities (the approximate equations are exact for transversely isotropic media). The reconstruction results are summarized in Table 3.6. The inversion results using the approximate equations are slightly less accurate, but still comparable with those using the exact formulas.

Bibliography

1. M. F. Markham, "Measurement of the elastic constants of fibre composites by ultrasonics," Composites **1**, 145–149 (1970).
2. R. E. Smith, "Ultrasonic elastic constants of carbon fibers and their composites," J. Appl. Phys. **43**, 2555–2562 (1972).
3. J. M. Gieske and R. E. Allred, "Elastic constants of B–Al composites by ultrasonic velocity measurements," Exp. Mech. **14**, 158–165 (1974).
4. L. H. Pearson and W. J. Murri, "Measurement of ultrasonic wavespeeds in off-axis directions of composite materials," in *Review of Progress in Quantitative Nondestructive Evaluation*, eds. D. O. Thompson and D. E. Chimenti, vol. **6B** (Plenum Press, New York, 1987), pp. 1093–1101.
5. B. Hosten, M. Deschamps, and B. R. Tittman, "Inhomogeneous wave generation and propagation in lossy anisotropic solids. Application to the characterization of viscoelastic composite materials," J. Acoust. Soc. Am. **82**, 1769–1770 (1987).

6. S. I. Rokhlin and W. Wang, "Ultrasonic evaluation of in-plane and out-of-plane elastic properties of composite materials," in *Review of Progress in Quantitative Nondestructive Evaluation*, eds. D. O. Thompson and D. E. Chimenti, vol. **8B** (Plenum Press, New York, 1989), pp. 1489–1497.
7. L. Wang, A. I. Lavrentyev, and S. I. Rokhlin, "Beam and phase effects in angle-beam through-transmission method of ultrasonic velocity measurement," J. Acoust. Soc. Am. **113**, 1551–1559 (2003).
8. S. I. Rokhlin and W. Wang, "Double through-transmission bulk wave method for ultrasonic phase velocity measurement and determination of elastic constants of composite materials," J. Acoust. Soc. Am. **91**, 3303–3312 (1992).
9. R. A. Kline and Z. T. Chen, "Ultrasonic technique for global anisotropic property measurement in composite materials," Mater. Eval. **46**, 986–992 (1988).
10. P. B. Nagy, W. R. Rose, and L. Adler, "A single-transducer broadband technique for leaky Lamb wave detection," in *Review of Progress in Quantitative Nondestructive Evaluation*, eds. D. O. Thompson and D. E. Chimenti, vol. **6** (Plenum, New York, 1987), pp. 483–490.
11. M. Hefetz and S. I. Rokhlin, "Thermal shock damage assessment in ceramics using ultrasonic waves," J. Am. Ceram. Soc. **75**, 1839–1845 (1992).
12. Y. C. Chu and S. I. Rokhlin, "Determination of macro- and micromechanical and interfacial elastic properties of composites from ultrasonic data," J. Acoust. Soc. Am. **92**, 920–931 (1992).
13. Y. C. Chu and S. I. Rokhlin, "Comparative analysis of through-transmission ultrasonic bulk wave methods for phase velocity measurements in anisotropic materials," J. Acoust. Soc. Am. **95**, 3204–3212 (1994).
14. A. I. Lavrentyev, A. D. Degtyar, and S. I. Rokhlin, "New method for determination of applied and residual stresses in anisotropic material from ultrasonic velocity measurement," Mater. Eval. **55**, 1162–1168 (1997).
15. K. W. Hallman and C. M. Fortunko, "An accurate method for measurement of transverse elastic-wave velocities," Meas. Sci. Technol. **9**, 1721–1727 (1998).
16. A. I. Lavrentyev and S. I. Rokhlin, "Phase correction for ultrasonic bulk wave measurements of elastic constants in anisotropic materials," in *Review of Progress in Quantitative Nondestructive Evaluation*, eds. D. O. Thompson and D. E. Chimenti, vol. **16** (Plenum, New York, 1997), pp. 1367–1374.
17. L. Wang and S. I. Rokhlin, "Floquet wave ultrasonic method for determination of single ply moduli in multi-directional composites," J. Acoust. Soc. Am. **112**, 916–924 (2002).
18. W. G. Mayer, "Determination of ultrasonic velocities by measuring of angles of total reflection," J. Acoust. Soc. Am. **32**, 1213–1215 (1960).
19. L. S. Fountain, "Experimental evaluation of the total-reflection method of determining ultrasonic velocity," J. Acoust. Soc. Am. **42**, 242–247 (1966).
20. E. G. Henneke II and G. L. Jones, "Critical angle for reflection at a liquid–solid interface in single crystals," J. Acoust. Soc. Am. **59**, 204–205 (1976).
21. H. Enfan, K. A. Ingerbrigtsen, and A. Tonning, "Elastic surface wave in α-quartz: Observation of leaky surface waves," Phys. Lett. **10**, 311–313 (1967).
22. F. R. Rollins, Jr., T. C. Lim, and G. W. Farnell, "Ultrasonic reflectivity and surface wave phenomenon on surfaces of copper single crystals," Appl. Phys. Lett. **12**, 236–238 (1968).
23. O. I. Diachok, R. J. Hallermier, and W. G. Mayer, "Measurement of ultrasonic surface wave velocity and absorptivity on single-crystal copper," Appl. Phys. Lett. **17**, 288–289 (1970).

24. S. I. Rokhlin and W. Wang, "Critical angle measurement of elastic constants in composite material," J. Acoust. Soc. Am. **86**, 1876–1822 (1989).
25. S. I. Rokhlin, T. K. Bolland, and L. Adler, "Reflection and refraction of elastic waves on a plane interface between two generally anisotropic media," J. Acoust. Soc. Am. **79**, 906–918 (1986).
26. S. I. Rokhlin, T. K. Bolland, and L. Adler, "Effects of reflection and refraction of ultrasonic waves on the angle beam inspection of anisotropic composite materials," in *Review of Progress in Quantitative Nondestructive Evaluation*, eds. D. O. Thompson and D. E. Chimenti, vol. **6** (Plenum, New York, 1987), pp. 1103–1110.
27. H. L. Bertoni and T. Tamir, "Unified theory of Rayleigh-angle phenomena for acoustic beams at liquid–solid interfaces," Appl. Phys. **2**, 157–172 (1973).
28. T. D. K. Ngoc and W. G. Mayer, "Ultrasonic nonspecular reflectivity near longitudinal critical angle," J. Appl. Phys. **50**, 7948–7951 (1979).
29. Y. C. Chu and S. I. Rokhlin, "Stability of determination of composite moduli from velocity data in planes of symmetry for weak and strong anisotropies," J. Acoust. Soc. Am. **95**, 213–225 (1994).
30. W. Sachse and K. Y. Kim, "Novel approaches for the ultrasonic NDE of thick and other composites," in *Review of Progress in Quantitative Nondestructive Evaluation*, eds. D. O. Thompson and D. E. Chimenti, vol. **8** (Plenum, New York, 1989), pp. 1473–1480.
31. K. Y. Kim and W. Sachse, "Direct determination of group velocity surfaces in a cuspidal region in zinc," J. Appl. Phys. **75**, 1435–1441 (1994).
32. A. Minachi, D. K. Hsu, and R. B. Thompson, "Single-sided determination of elastic constants of thick composites using acoustoultrasonic technique," J. Acoust. Soc. Am. **96**, 353–362 (1994).
33. A. G. Every and W. Sachse, "Determination of all the elastic constants of anisotropic solids from acoustic-wave group-velocity measurements," Phys. Rev. **B42**, 8196–8205 (1990).
34. K. Y. Kim, "Analytic relations between the elastic constants and group velocity in an arbitrary direction of symmetry planes of media with orthorhombic or higher symmetry," Phys. Rev. **B49**, 3713–3724 (1994).
35. K. Y. Kim, T. Ohtani, A. R. Baker, and W. Sachse, "Determination of all elastic constants of orthotropic plate specimens from group velocity data," Res. Nondestr. Eval. **7**, 13–29 (1995).
36. A. D. Degtyar and S. I. Rokhlin, "Comparison of elastic constant determination in anisotropic materials from ultrasonic group and phase velocity data," J. Acoust. Soc. Am. **97**, 3458–3466 (1997).
37. Y. C. Chu, A. D. Degtyar, and S. I. Rokhlin, "On determination of orthotropic material moduli from ultrasonic velocity data in non-symmetry planes," J. Acoust. Soc. Am. **95**, 3191–3203 (1994).

4

Reflection and Refraction of Waves at a Planar Composite Interface

4.1 Introduction

Nondestructive ultrasonic testing of composite materials is affected by several special features of wave propagation that arise from the strong anisotropy and inhomogeneity of these materials. The resulting complexity requires re-examination of old testing methodologies and development of new ones. One of the most fundamental phenomena in ultrasonic nondestructive evaluation is the reflection–refraction of ultrasonic waves at a plane interface. Even the simplest test procedure requires understanding of mode conversion and knowledge of elastic wave reflection and transmission coefficients and refraction angles [1]. Reflection–refraction phenomena, while straightforward and well documented for isotropic materials, are much more complicated for anisotropic materials. When analyzing the oblique incidence inspection method for composite materials, one first has to address the problem of wave propagation through the interface between the coupling medium and the composite material. For example, there is an inherent fluid/composite interface in the immersion technique and a perspex/composite interface in the contact method. In the latter case, assuming that a thin fluid layer is applied to facilitate coupling through the interface, slip rather than welded boundary conditions prevail. Another example of great practical importance is the case of multidirectional fiber plies in a composite laminate, when the reflection and transmission of ultrasonic waves from one ply to another with a different orientation must be considered.

4.2 Background

4.2.1 Snell's law

Before discussing the general problem of wave refraction in anisotropic composite materials, let us review the simple isotropic case. Consider a plane interface between two isotropic elastic media in "welded" (perfectly bonded) contact, implying continuity of tractions and displacements across the interface, although the boundary conditions are not important at this point. Figure 4.1 shows a schematic diagram of a plane wave with wavenumber k_i incident on the interface at angle θ_i. The parallel lines with spacing equal to the incident wavelength λ_i correspond to equal-phase planes orthogonal to the incident plane. By definition, the wavenumber $k_i = 2\pi/\lambda_i$ is the magnitude of the wave vector $\mathbf{k}_i$. The incident wave is converted at the interface into reflected and transmitted waves. The refraction angle of the transmitted wave is θ_r and its wavenumber is k_r. The wave fronts of the transmitted and reflected (not shown in Fig. 4.1) waves should match that of the incident wave along the entire interface. This follows from the requirement that the interface continuity conditions be satisfied not only at one particular point (*e.g.*, at the origin), but everywhere along the interface. Any mismatch among the phases of the incident, reflected, and transmitted waves, which together must satisfy the boundary conditions, would lead to an inevitable variation of their relative contributions along the interface and thereby would prevent them from perfectly balancing each other over the whole boundary.

From the universal phase-matching requirement along the interface we get

$$\Lambda \sin\theta_i = \lambda_i \quad \text{and} \quad \Lambda \sin\theta_r = \lambda_r, \tag{4.1}$$

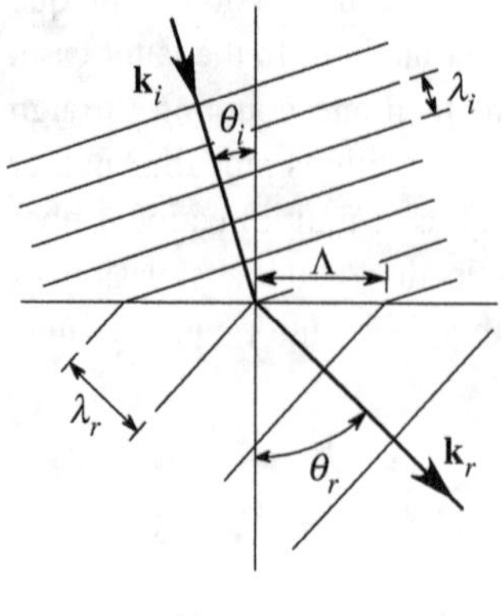

Figure 4.1. A schematic diagram of a plane wave with wavenumber k_i incident on the interface at angle θ_i producing a single refracted wave with wavenumber k_r at angle θ_r. The wavelengths of the incident and refracted waves are shown as λ_i and λ_r, while the *trace* wavelength is illustrated as Λ.

where λ_i and λ_r are the wavelengths for the incident and transmitted waves and Λ, as shown in Fig. 4.1, is the spatial period of the field distribution along the interface. Equation (4.1) is Snell's law that can be also written in terms of the wavelengths

$$\Lambda = \frac{\lambda_i}{\sin\theta_i} = \frac{\lambda_r}{\sin\theta_r}, \tag{4.2}$$

in terms of the velocities

$$V = \frac{V_i}{\sin\theta_i} = \frac{V_r}{\sin\theta_r}, \tag{4.3}$$

or in terms of the wavenumbers

$$K = k_i \sin\theta_i = k_r \sin\theta_r. \tag{4.4}$$

Here, V_i and V_r are the velocities of the incident and refracted waves, respectively, and V and K are the common phase velocity and wavenumber of the incident and refracted (including both reflected and transmitted) waves along the interface. Figure 4.2 shows the geometrical interpretation of Snell's law in terms of velocity vectors (a) and wave vectors (b), which can be summarized in the following way:

1. The *traces*, or lengths projected onto the interface, of the wavelength (intersection of wavefronts with the interface) in Eq. (4.2) and velocities (speeds projected along wavefronts to the interface) in Eq. (4.3) of the incident, reflected, and refracted waves onto the interface must be equal. The trace of the wavelength Λ is shown in Fig. 4.1 and that of the phase velocity **V** is shown in Fig. 4.2(a).
2. The horizontal x-projections of the wave vectors of the incident, reflected, and refracted waves onto the interface must be equal. The wave vector projection **K** is shown in Fig. 4.2(b), where $K = \omega/V$.

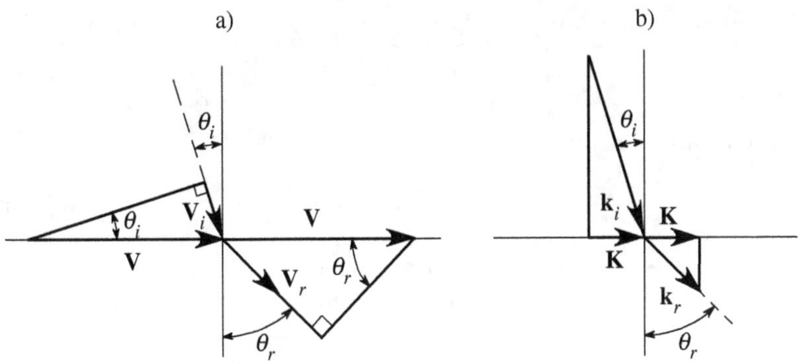

Figure 4.2. Geometrical interpretation of Snell's law in terms of velocity vectors (a) and wave vectors (b).

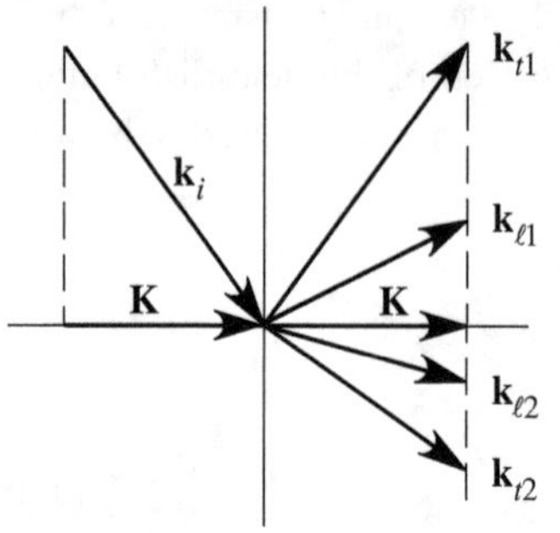

Figure 4.3. Reflected and transmitted wave vectors produced at an isotropic solid/solid interface by an obliquely incident shear wave.

The same conclusion holds for multilayered media. The projections on the interfaces of the transmitted and reflected wave vectors always equal that of the incident wave. As we will see later, for wave refraction on an interface between elastic media, in general, the incident wave is converted into longitudinal and transverse waves in such a way that Snell's law will hold for both types of waves, as shown in Fig. 4.3. Here, the incident (assumed to be shear) wave $\mathbf{k}_i$ produces four waves, namely a reflected longitudinal wave $\mathbf{k}_{\ell 1}$, a reflected transverse wave $\mathbf{k}_{t1}$, a transmitted longitudinal wave $\mathbf{k}_{\ell 2}$, and a transmitted shear wave $\mathbf{k}_{t2}$. According to Snell's law, all five (incident, two reflected, and two transmitted) waves have the same wave vector projection $\mathbf{K}$ on the interface. We will see later that in the anisotropic case, the incident wave is generally converted at the interface into three possible modes of elastic waves and Snell's law is similarly satisfied.

4.2.2 Isotropic media analysis using slowness surfaces

The concept of the slowness surface has been introduced in Section 2.9. In any particular direction defined by the unit vector $\mathbf{n}$, the wave vector $\mathbf{k}$ can be expressed by the ratio of the angular frequency ω and the phase velocity V as follows

$$k_\ell = \frac{\omega}{V} n_\ell. \tag{4.5}$$

The quantity

$$m_\ell = \frac{n_\ell}{V} \tag{4.6}$$

was defined as the slowness vector $\mathbf{m}$. The loci of the slowness vectors in different propagation directions form the slowness surfaces (while the loci of the phase velocity vectors form the phase velocity surfaces). Generally, the radial coordinate of the slowness surface in a polar coordinate system can be written as $S(\theta, \phi) = 1/V(\theta, \phi)$ where $V(\theta, \phi)$ is the phase velocity plotted as a function of the polar θ and azimuthal ϕ propagation angles.

The usefulness of the slowness concept lies in its proportionality to the wavenumber $k(\theta, \phi) = \omega/V(\theta, \phi)$. Thus, slowness can be regarded as a wavenumber at the angular frequency of unity. Later in this chapter, we will make extensive use of slowness surfaces to assist our understanding of wave scattering from interfaces.

Here, we illustrate the basic features of scattering from an interface between two isotropic fluids. Since, in isotropic materials, the wave velocity is independent of the propagation direction, the slowness surface is a sphere with radius $S = 1/V$. As an example, Fig. 4.4(a) illustrates the case of a plane wave incident from a faster to a slower fluid (mode conversion is avoided for the sake of simplicity), *i.e.*, $V_1 > V_2$. The wave vectors for the incident, reflected, and transmitted waves are $\mathbf{k}_i$, $\mathbf{k}_1$, and $\mathbf{k}_2$, respectively, and θ_i is the incident angle. If we take $\omega = 1$, the wave and slowness vectors will be identical. The wave vector of the incident wave originates on the slowness surface and is directed at angle θ_i toward the origin as shown in the figure. To find the angles of the reflected θ_1 and transmitted θ_2 waves, we use Eq. (4.4), *i.e.*, we equate the projections on the interface of all wave vectors involved. Of course, the reflection angle equals the incident angle. The refraction angle of the transmitted wave is uniquely defined by Snell's law ($\theta_2 < \theta_i$).

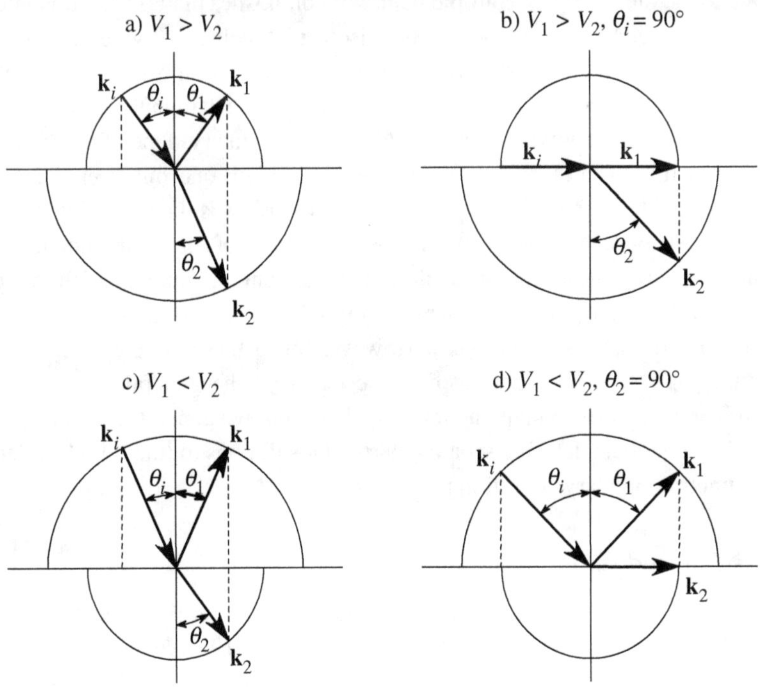

Figure 4.4. Reflected and transmitted wave vectors produced at a fluid/fluid interface under different conditions.

Figure 4.4(b) illustrates the case of incidence at grazing angle $\theta_i = 90°$. The refraction angle is clearly determined from the figure as $\sin\theta_2 = V_2/V_1$.

The case of a wave arriving from a slow to a fast medium at oblique incidence is illustrated in Fig. 4.4(c). Here the refraction angle is larger than the incident one, $\theta_2 > \theta_i$. An important phenomenon occurs when θ_i reaches the critical angle θ_{cr} as shown in Fig. 4.4(d). At the critical angle, the refraction angle $\theta_2 = 90°$ and above the critical angle, the refracted wave becomes evanescent. In the absence of attenuation, this wave has only reactive power and does not carry energy; however, it may affect the phase of the reflected signal. If both media support only one type of wave (like fluids), the reflection coefficient beyond the critical angle becomes unity.

In the case of isotropic solids, the situation is somewhat more complicated due to the existence of two waves with different velocities. If a wave is incident from the fluid onto a "fast" solid surface, two critical angles will exist, one for the longitudinal and one for the transverse wave. This situation will be discussed in much more detail next.

4.3 Scattering at an Interface between Two Generally Anisotropic Media

Due to the inherently anisotropic nature of composites materials, ultrasonic wave interaction between generally anisotropic media is of great practical importance in their nondestructive testing. Many composites consist of orthotropic or transversely isotropic plies, but the laminate as a whole can exhibit either lower or higher degree of symmetry depending on the ply structure. Therefore, the most general problem one might encounter in ultrasonic material characterization of composites is the scattering at an interface between two generally anisotropic media. Of course, in immersion and contact angle-beam inspection, the incident medium is a fluid or isotropic solid, respectively, which could be exploited to simplify the reflection–transmission problem. However, even these simpler cases are conceptually easier to approach by specializing the general solution of ultrasonic scattering at a plane interface between two generally anisotropic media rather than deriving separate particular solutions for these high-order symmetry boundary condition problems.

4.3.1 Introduction

Musgrave [2] and Fedorov [3] summarized the general theoretical results for the reflection–refraction problem in their comprehensive books. Several examples for planes of symmetry were discussed by Auld [4]. Henneke [5] reviewed Fedorov's method and discussed critical-angle phenomena for the anisotropic case. Rokhlin *et al.* have described a unified approach to the numerical solution of the reflection–refraction problem for generally anisotropic media [6, 7]. This method was also applied to the analysis of ultrasonic wave scattering from interfaces in composites or from interfaces between fluids and composites [8]. Analysis of the problem, as well as application to composites, has also been described by Nayfeh [9].

In this chapter, we will first review the general theory of elastic wave scattering from interfaces between anisotropic media following [6, 7], and then consider numerous examples relevant to composites. We consider the Christoffel equations and interface boundary conditions in "physical" coordinate systems formed by the incident and interface planes rather than in principal material coordinates. This makes it possible to write simple unified algorithms to find wave vectors, group velocity vectors, slowness vectors, and polarization (unit displacement) vectors. The goal here is to find the amplitudes of the displacement vectors. To do this, we will follow the approach of Fedorov [3]. Special care will be taken to include in the general algorithm, the special cases of propagation along so-called acoustic axes. Next, as an example, we consider a unidirectional graphite/epoxy composite. We present energy conversion coefficients for reflected and transmitted waves and discuss reciprocity relationships. Finally, using slowness surfaces, we consider the definitions of "grazing" and "critical angles" for incident waves. We demonstrate that a grazing angle exists when the incident ray (and not the wave vector) is parallel to the interface. The incident wave angle for grazing incidence may be greater or less than 90°. In fact, for certain lamina orientations, the domain of permissible incident wave vector angles can split into disjoint segments.

4.3.2 Snell's law (anisotropic case)

Let a monochromatic plane wave be incident on the interface between two anisotropic solids as shown in Fig. 4.5. The displacement vector u_j of an elastic wave can be given by either of the following two forms

$$u_j = Ap_j \exp[i(k_\ell x_\ell - \omega t)] = Ap_j \exp[ik(n_\ell x_\ell - Vt)], \tag{4.7}$$

where A is the amplitude, p_j is the polarization unit vector, and k_ℓ is the wave vector previously defined by Eq. (4.5), k is the wave number, *i.e.*, the magnitude of the wave vector, n_ℓ is the wave unit vector, and V is the phase velocity along the n_ℓ direction. With the previously defined slowness vector m_ℓ of Eq. (4.6), the displacement vector can also be written in the following alternative form

$$u_j = Ap_j \exp[i\omega(m_\ell x_\ell - t)]. \tag{4.8}$$

In a real experiment, instead of an infinite harmonic plane wave, pulsed ultrasonic beams are commonly used, which are bounded in both the time

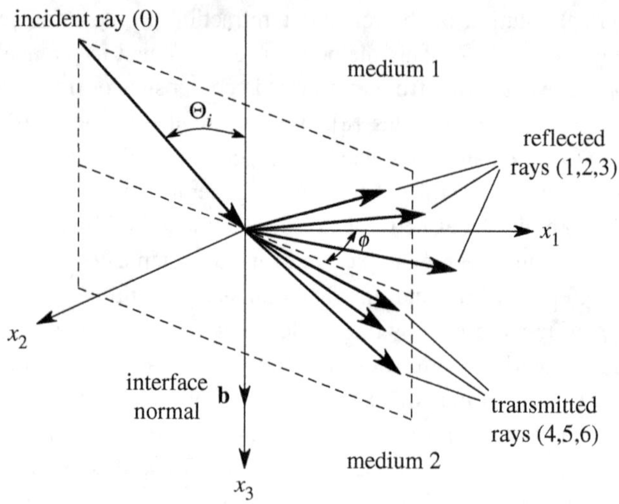

Figure 4.5. Acoustic wave incident on an interface between two different anisotropic media. **b** is the normal unit vector of the interface plane. The interface coincides with the (x_1, x_2) plane of the coordinate system therefore **b** is parallel to the x_3 axis. In this case the incident ray lies in the incident plane, which is a plane of symmetry for medium 1. Generally, the incident, reflected, and transmitted rays (or energy velocity vectors) do not lie in the plane of incidence (they are numbered from 0 to 6 for identification purposes).

and space domains. The monochromatic plane wave representation (Eqs. 4.7 and 4.8) is only a single Fourier component of the real ultrasonic field. If both geometric and material dispersions are negligible, the monochromatic plane wave model is a good first-step approximation in the analysis of wave–material interaction. This is not always the case for guided waves, which are usually strongly dispersive and will be considered separately in subsequent chapters. One should also note, as we argue later in the book, that ultrasonic beam and pulse solutions are commonly obtained by Fourier transforms in time and space domains. Let us assume welded boundary conditions at the interface, which require continuity of both displacements and tractions across the boundary. Later, we will also consider the more complicated case of imperfect boundary conditions. (Imperfect or "spring" boundary conditions will not change the general conclusions obtained in this section [10, 11].)

Then, continuity of the displacements and tractions at the boundary $(x_3 = 0)$ is given by

$$u_i^0 + \sum_{\alpha=1}^{3} u_i^\alpha = \sum_{\alpha=4}^{6} u_i^\alpha \tag{4.9}$$

and

$$\sigma_{i3}^{0} + \sum_{\alpha=1}^{3} \sigma_{i3}^{\alpha} = \sum_{\alpha=4}^{6} \sigma_{i3}^{\alpha}. \tag{4.10}$$

Here, we assumed that the interface plane coincides with the (x_1, x_2) plane of the coordinate system so that the normal unit vector **b** is parallel to the x_3 axis. In Fig. 4.5, the incident wave is marked with superscript 0, and the three reflected and three transmitted waves are marked with superscripts $1, 2, 3$ and $4, 5, 6$, respectively. It might appear that the incident wave $\alpha = 0$ could be easily included in the summation on the left sides of both equations, but later it will be advantageous to treat the incident wave separately.

Using the relevant displacement–strain and constitutive relationships, the stress tensor can be expressed from Eq. (4.8) as follows

$$\sigma_{ik} = c_{ikjm} \frac{\partial u_j}{\partial x_m} = i\omega m_m c_{ikjm} A p_j \exp[i\omega(m_\ell x_\ell - t)]. \tag{4.11}$$

Substituting Eq. (4.8) into (4.9) we get for the displacement continuity

$$A^0 p_j^0 \exp(i\omega m_\ell^0 x_\ell) = -\sum_{\alpha=1}^{3} A^\alpha p_j^\alpha \exp(i\omega m_\ell^\alpha x_\ell) + \sum_{\alpha=4}^{6} A^\alpha p_j^\alpha \exp(i\omega m_\ell^\alpha x_\ell), \tag{4.12}$$

where we omitted the common $\exp(-i\omega t)$ term. In a similar way, substituting Eq. (4.11) into (4.10) we get for the stress continuity

$$m_m^0 c_{i3jm}^I A^0 p_j^0 \exp(i\omega m_\ell^0 x_\ell)$$
$$= -\sum_{\alpha=1}^{3} m_m^\alpha c_{i3jm}^I A^\alpha p_j^\alpha \exp(i\omega m_\ell^\alpha x_\ell) + \sum_{\alpha=4}^{6} m_m^\alpha c_{i3jm}^{II} A^\alpha p_j^\alpha \exp(i\omega m_\ell^\alpha x_\ell), \tag{4.13}$$

where again we omitted the common $\exp(-i\omega t)$ term and also divided both sides by $i\omega$. In order to distinguish between the elastic coefficients of the first and second media, they are denoted by c_{ikjm}^I and c_{ikjm}^{II}, respectively.

The boundary conditions must be satisfied not only for all times t, which is a direct result of the common $\exp(-i\omega t)$ term in all equations, but also at all points $(x_1, x_2, x_3 = 0)$ on the interface. Since the exponential functions $\exp(i\omega m_\ell^\alpha x_\ell)$ are linearly independent for $\alpha = 0, 1, \ldots, 6$, Eqs. (4.12) and (4.13) will be satisfied only if all seven exponential functions are identically equal. Snell's law follows immediately from this simple fact since this condition requires that

$$m_\ell^0 x_\ell = m_\ell^1 x_\ell = m_\ell^2 x_\ell = m_\ell^3 x_\ell = m_\ell^4 x_\ell = m_\ell^5 x_\ell = m_\ell^6 x_\ell, \tag{4.14}$$

where in Eqs. (4.12–4.14) $x_3 = 0$, and therefore the summation might be limited to $\ell = 1, 2$. In the expressions above, the coordinate vector $\mathbf{x}$ lies on the interface plane therefore

$$\mathbf{b} \cdot \mathbf{x} = 0, \tag{4.15}$$

where $\mathbf{b}$ is normal to the interface as shown in Fig. 4.5. Another way of looking at Eq. (4.14) is that for any two different β and γ values from the set $(0, 1, \ldots, 6)$

$$(\mathbf{m}^\beta - \mathbf{m}^\gamma) \cdot \mathbf{x} = 0. \tag{4.16}$$

Comparing Eqs. (4.15) and (4.16) shows that the difference $\mathbf{m}^\beta - \mathbf{m}^\gamma$ is always parallel to $\mathbf{b}$, *i.e.*, the plane defined by any two of the seven slowness vectors is always perpendicular to the interface. By definition, the slowness vector of the incident wave $\mathbf{m}^0$ lies in the plane of incidence, therefore all the other slowness vectors must also lie in the plane of incidence. Since the direction of the slowness vector coincides with that of the wave vector, we can conclude that the wave vector of the incident wave as well as all the other wave vectors of the reflected and refracted waves lie in the same plane which is normal to the interface, *i.e.*, the plane of incidence. Furthermore,

$$(\mathbf{m}^\beta - \mathbf{m}^\gamma) \times \mathbf{b} = \mathbf{0} \qquad \text{or}$$
$$\mathbf{m}^\beta \times \mathbf{b} = \mathbf{m}^\gamma \times \mathbf{b}, \tag{4.17}$$

i.e., all vector products $\mathbf{m}^\alpha \times \mathbf{b}$ $(\alpha = 0, \ldots, 6)$ are equal, which is an alternative form of Snell's law. This relationship can be converted into a more familiar form by introducing polar angles θ^α between the interface normal $\mathbf{b}$ and the different wave normals $\mathbf{n}^\alpha$ for the incident wave and all refracted and reflected waves so that $\mathbf{b} \cdot \mathbf{n}^\alpha = \cos\theta^\alpha$. Then, according to the definition of the vector product

$$|\mathbf{m}^\alpha \times \mathbf{b}| = m^\alpha \sin\theta^\alpha, \quad (\alpha = 0, 1, \ldots, 6), \tag{4.18}$$

where we exploited the fact that $b = |\mathbf{b}| = 1$ since $\mathbf{b}$ is a unit vector. Equation 4.18 expresses Snell's law generalized for arbitrarily anisotropic media. It is important to note that the slowness values m^α in general depend on the angles ϕ^α due to anisotropy. Figure 4.6 illustrates the fundamental principle that the slowness vector of the incident wave and all the other slowness vectors of the reflected and refracted waves lie in the same plane which is normal to the interface, *i.e.*, the plane of incidence, and that they all have the same projection on the plane of the interface. The same conclusion

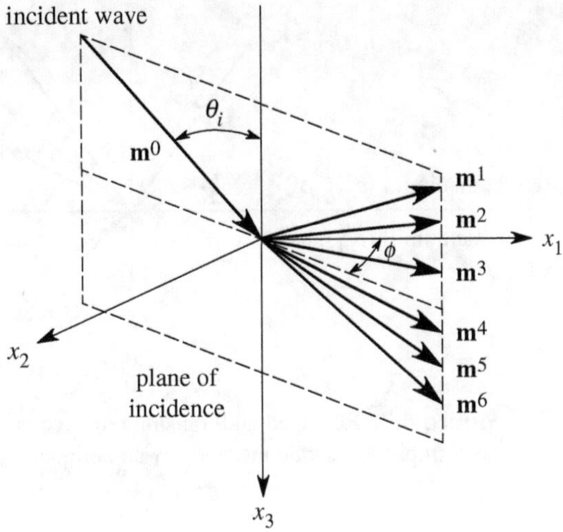

Figure 4.6. According to the generalized Snell's law, the slowness vector of the incident wave and all the other slowness vectors of the reflected and refracted waves lie in the same plane which is normal to the interface, *i.e.*, the plane of incidence, and they all have the same projection on the plane of the interface.

holds for the projections of the wave vectors since $\mathbf{k}^\alpha = \mathbf{m}^\alpha \omega$. In terms of the wave vectors, we can summarize Snell's law as follows:

- The wave vectors for the incident and reflected/refracted waves all lie in a single plane, which also contains the interface normal and is called the incident plane.
- All their projections on the interface are equal.

4.3.3 *Determination of the slowness vectors of reflected and refracted waves*

As the incident and all reflected and refracted waves lie in a single plane, *i.e.*, in the incident plane, it is advantageous to rotate the coordinate system so as to describe all phenomena in this plane. Let us select the incident plane to be the x_1, x_3 plane as shown in Fig. 4.7. In this "physical" coordinate system, the slowness vector $\mathbf{m}^\alpha$ has only two components; m_1^α and m_3^α, while the components $m_2^\alpha = 0$ identically vanish. From Snell's law (Eq. 4.14) the tangential components m_1^α in the plane of the interface are all equal to each other:

$$m_1^0 = m_1^1 = \ldots = m_1^6 = \frac{n_1^0}{V^0}. \tag{4.19}$$

Here, $n_1^0 = \sin\theta^0$, where $\mathbf{n}^0$ is wave normal for incident wave, $\theta^0 = \theta_i$ is the incident angle, and V^0 is the phase velocity of the incident wave. Since m_1^0 for the incident wave is known, the tangential components of the slowness vectors m_1^α (*i.e.*, the projections of the slowness vectors on the interface) are also known. Equation (4.19) can be re-written in terms of the corresponding

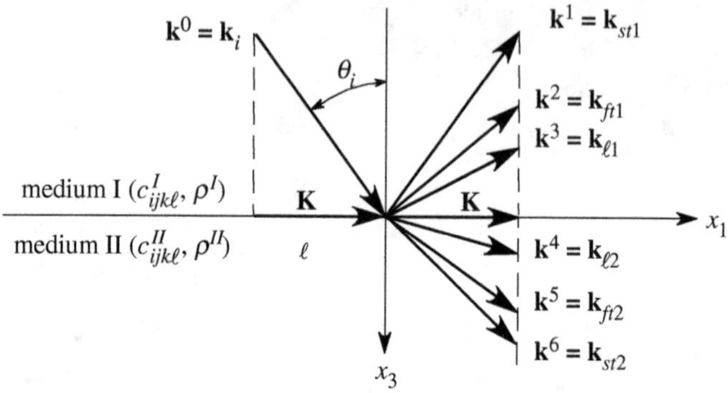

Figure 4.7. Reflected and transmitted wave vectors produced at an anisotropic solid/solid interface by an obliquely incident wave.

wave vector components as follows

$$k_1^0 = k_1^1 = \ldots = k_1^6 = k_i \sin\theta_i. \tag{4.20}$$

In the simpler isotropic case, the velocity is independent of the propagation direction and the slowness components m^α are easily calculated from

$$m^\alpha = \frac{1}{V^\alpha}, \tag{4.21}$$

where V^α are the velocities for longitudinal or transverse waves in the upper or lower halfspace. Therefore, the reflection and refraction angles can be found immediately by substituting Eq. (4.21) into Eq. (4.18). In the anisotropic case, the slowness vectors in Eq. (4.18) depend on the propagation direction, a fact that considerably complicates the situation. From Eq. (4.19), only the tangential components m_1^α of the slowness vectors $\mathbf{m}^\alpha$ are known and the normal slowness components m_3^α of the reflected and refracted waves are still unknown and must be determined. Therefore, there are two sets of unknowns in Eq. (4.18); the reflection/refraction angles θ^α and the slowness values m^α. Since the slowness is a known function of the propagation direction $m^\alpha = m^\alpha(\theta^\alpha)$ for each particular wave mode, Eq. (4.18) provides an implicit equation for the propagation directions θ^α rather than the simpler explicit equation found in the isotropic case.

To find the slowness vectors for the reflected and refracted waves, we need Snell's law and one additional equation that is derived next. As we have shown in Chapter 2, the acoustic wave must satisfy the so-called Christoffel equation (see Section 2.2)

$$[c_{ijk\ell} n_j n_k - \rho V^2 \delta_{i\ell}]\, p_\ell = 0. \tag{4.22}$$

In terms of the slowness vector $m_j = n_j/V$, this equation can be re-written as follows

$$[c_{ijk\ell}m_jm_k - \rho\delta_{i\ell}]\, p_\ell = 0. \tag{4.23}$$

In order to have a nontrivial solution for p_ℓ, the secular determinant of the matrix formed by the coefficients of the linear equation (Eq. 4.23) must vanish, leading to the following characteristic equation

$$| T_{i\ell} | = 0, \tag{4.24}$$

where we introduced the following matrix

$$T_{i\ell} = c_{ijk\ell}m_jm_k - \rho\delta_{i\ell}. \tag{4.25}$$

In the chosen coordinate system, the plane of incidence coincides with the (x_1, x_3) plane, therefore $m_2 = 0$ and Eq. (4.25) reduces to

$$T_{i\ell} = [c_{i11\ell}(m_1^0)^2 + (c_{i31\ell} + c_{i13\ell})m_3m_1^0 + c_{i33\ell}m_3^2] - \rho\delta_{i\ell}, \tag{4.26}$$

where m_1^0 is used for all modes instead of m_1 due to Snell's law. This is a second-order polynomial in the unknown normal component m_3 of the slowness vector and can be written as

$$T_{i\ell} = a_{i\ell} + b_{i\ell}m_3 + d_{i\ell}m_3^2, \tag{4.27}$$

where

$$a_{i\ell} = c_{i11\ell}(m_1^0)^2 - \rho\delta_{i\ell},$$

$$b_{i\ell} = (c_{i31\ell} + c_{i13\ell})m_1^0,$$

$$d_{i\ell} = c_{i33\ell},$$

$$a_{i\ell} = a_{\ell i}, \quad b_{i\ell} = b_{\ell i}, \quad d_{i\ell} = d_{\ell i}.$$

Note that the $T_{i\ell}$ matrix is symmetric, *i.e.*, $T_{i\ell} = T_{\ell i}$. Further details of the calculation are given in the next section.

4.3.4 Calculation of the wave and polarization vectors

In this section we consider the algorithm to calculate polarization and wave vectors for reflected and refracted waves. The directions of the wave vectors determine the reflection and refraction angles. First, arbitrary directions in anisotropic media will be considered. Unfortunately, the resulting equations cannot be used along so-called acoustic axes, *i.e.*, in propagation directions where the velocities of the two transverse waves are equal. Therefore, this case will be addressed separately in the second part of this section.

Propagation in an arbitrary direction

Let us assume that the incident and interface planes are specified. We can then select the physical coordinate system as it was shown in Fig. 4.7 so that the plane of the interface coincides with the (x_1, x_2) plane and the plane of incidence coincides with the (x_1, x_3) plane. We can transform the elastic constants of the upper and lower anisotropic half-spaces from their respective principal material (or crystallographic) coordinate systems to the physical coordinate system. To streamline and simplify our terminology, we will now change the way we designate the rotated and un-rotated coordinate systems. In this chapter, and later in the book, most of our equations are formulated in the incident plane (the rotated, physical coordinate system), so we depart here from our previous notation and omit the prime for the rotated system and the elastic constants in this system. Instead, we denote the elastic constants in principal material coordinate system by an *overbar* ($^-$) to distinguish them from those in the rotated physical coordinate system, without bar or prime. Now, we can either transform the fourth-order stiffness tensors according to the rules of tensor transformation

$$c_{ijk\ell} = a_{im}a_{jn}a_{kp}a_{\ell q}\bar{c}_{mnpq}, \tag{4.28}$$

where a_{im} is a 3×3 transformation matrix formed by the cosines of the direction angles between the mth axis of the original coordinate system and the ith axis of the rotated system $(i, j, k, \ell, m, n, p, q = 1, 2, 3)$. Alternatively, using Bond's method, which was described in Chapter 1 (see also Auld [4]), the transformation of the 6×6 stiffness matrices can be done as follows

$$C_{ik} = M_{i\ell}M_{kn}\overline{C}_{\ell n}, \tag{4.29}$$

where $\overline{C}_{\ell n}$ is the matrix of elastic constants before rotation (*i.e.*, in the principal material coordinates), C_{ik} is the matrix of elastic constants after rotation (*i.e.*, in the physical coordinates), and $M_{i\ell}$ is the 6×6 Bond transformation matrix given in the Appendix of Chapter 1. For a general rotation of coordinates, the subsequent application of two simple transformation matrices are required.

After the elastic constants in the physical coordinates are determined, the phase velocity of the incident wave V^0 can be found from Christoffel's equation for any given incident wave direction $\mathbf{n}^0$. Then, the tangential slowness component of the incident wave m_1^0 can be calculated from

$$m_1^0 = n_1^0/V^0. \tag{4.30}$$

Here, $n_1^0 = \sin\theta_i$, where θ_i is the angle of incidence. Due to the particular selection of the physical coordinate system, the direction cosines n_2^α are zero for all modes. The next step is to find the remaining unknown m_3^α normal components of the slowness vectors for the transmitted and reflected waves as solutions of the sixth-degree characteristic equation given in Eq. (4.24).

The characteristic equation of Eq. (4.24)

$$|T_{i\ell}| = \begin{vmatrix} T_{11} & T_{12} & T_{13} \\ T_{12} & T_{22} & T_{23} \\ T_{13} & T_{23} & T_{33} \end{vmatrix} = 0 \tag{4.31}$$

can be solved numerically after substituting the matrix elements $T_{i\ell}$ from Eq. (4.26), each of which are quadratic functions of the sought normal slowness vector component m_3^α thereby resulting in a sixth-order polynomial for the characteristic equation. This equation is formulated and solved separately for each of the two media in contact. One should distinguish the cubic polynomial resulting from Christoffel's equation Eq. (4.22) for the phase velocity squared V^2, which can have only three real positive solutions for any physically meaningful anisotropic material system without attenuation, from that of the sixth-order Eq. (4.24) or Eq. (4.31) relative to the slowness vector components m_3^α. The slowness vector component m_1^α is predetermined from Snell's law and $m_2^\alpha = 0,\ (\alpha = 1, \ldots, 6)$.

Six solutions of Eq. (4.31) will be found in each medium, of which only three correspond to physically meaningful solutions. In isotropic materials, the roots are either purely real or purely imaginary. A purely real root corresponds to a propagating wave while purely imaginary roots correspond to exponentially decaying evanescent waves. In anisotropic materials, complex conjugate roots appear just beyond critical angles, and these correspond to highly damped inhomogeneous waves of exponentially decaying amplitude. In some cases, two inhomogeneous waves with complex conjugate roots (m_3^α) combine to form an evanescent wave [12]. Purely imaginary and complex roots represent physical solutions only when the sign of their imaginary parts are such that they lead to diminishing amplitude with increasing distance from the interface, forming the basis of the root selection process. The inhomogeneous waves do not carry energy in a direction normal to the interface [2].

For a propagating reflected or transmitted wave to be physically realizable, its group velocity $\mathbf{V}_g$ must be oriented so that the energy flow will be upward for waves reflected into the upper halfspace and downward for waves transmitted into the lower halfspace. From Eq. (2.193), the group velocity can be written as follows

$$V_{gi} = \frac{c_{ij\ell m}}{\rho} m_\ell p_m p_j, \tag{4.32}$$

from which the crucial normal component can be calculated by simply substituting $i = 3$. Owing to the inherent symmetry of slowness surfaces with respect to the center of the coordinate system, for any physically allowable incident wave there must be at least two real roots in the upper medium, namely one for the incident and one for the reflected wave. The magnitude of the associated reflection coefficient, however, will be unity if all the other roots are complex, so that the remaining two reflected and all three transmitted waves are inhomogeneous (corresponding to total reflection). It is also possible that all roots of Eq. (4.31) are pure real so that all waves are propagating modes. It is essential that all allowable roots (real and complex, a total three for each medium) be substituted into the boundary conditions to assure correct amplitudes and phases of the reflected and transmitted waves. While inhomogeneous waves do not carry energy from the interface, they contribute to wave motion at the interface and to the balance of stresses and displacements at the interface.

To calculate the group velocity, the polarization unit vector p_i (eigenvector) must be first determined for each slowness vector m_i (eigenvalue) from Eq. (4.23) as described in Chapter 2. In each wave direction $\mathbf{n}^\alpha = \mathbf{m}^\alpha / m^\alpha$, three phase velocities must be found for one quasilongitudinal and two quasitransverse waves. We must distinguish between two cases: (1) all wave velocities (eigenvalues) are different, and (2) two velocities for the transverse waves coincide with each other, *i.e.*, $V_{ft} = V_{st}$. The second case can occur for wave propagation along the so-called acoustic axes of the anisotropic material.

We can determine the polarization vector following reference [6] using Fedorov's original method [3]. If all wave velocities are different, the following relationship exists for the polarization unit vector **p**

$$p_i p_\ell = W_{i\ell} / W_{jj}, \tag{4.33}$$

where $W_{i\ell} = |T_{i\ell}| T_{i\ell}^{-1}$ is the so-called "adjugate" matrix to the Christoffel matrix $T_{i\ell}$ that was previously defined by Eq. (4.25). The resulting equations for the polarization vectors take the same form as obtained in Chapter 2 by a less formal approach. The $W_{i\ell}$ matrix can be directly found from

$$W_{i\ell} = \frac{1}{2} \epsilon_{ijk} \epsilon_{\ell mn} T_{jm} T_{kn}. \tag{4.34}$$

Here, ϵ_{ijk} is the so-called permutation tensor, *i.e.*, $\epsilon_{123} = \epsilon_{231} = \epsilon_{312} = 1$, $\epsilon_{132} = \epsilon_{213} = \epsilon_{321} = -1$, and all other elements are zero. In our case, $n_2 = m_2 = 0$ and the matrix $T_{i\ell}$ is given by Eq. (4.26). For a generally anisotropic medium, from Eq. (4.34) we get

$$\begin{aligned} &W_{11} = T_{22}T_{33} - T_{32}T_{23}, \quad W_{12} = T_{23}T_{31} - T_{21}T_{33}, \\ &W_{22} = T_{33}T_{11} - T_{13}T_{31}, \quad W_{31} = T_{12}T_{23} - T_{13}T_{22}, \\ &W_{33} = T_{11}T_{22} - T_{21}T_{12}, \quad W_{23} = T_{31}T_{12} - T_{32}T_{11}. \end{aligned} \tag{4.35}$$

We can see from Eq. (4.33) that

$$p_i p_i = p_1^2 + p_2^2 + p_3^2 = W_{ii} / W_{jj} = 1, \tag{4.36}$$

i.e., the magnitude of the polarization vector **p** is unity as expected.

There are several other ways to find the polarization vector. For example, it can be found from any two equations of the three-equation system (Eq. 4.22), which is generally called the Christoffel equation, together with the normalization condition $p_i p_i = 1$ as we did in Chapter 2 or we can use Musgrave's [2] method. However, equations obtained in these ways will lead to division by zero for planes of symmetry. If we avoid the singularity by conversion of these equations as was done in Section 2.3, we obtain Eqs. (4.33), (4.35), and (4.36). The advantage of the above solution is the straightforward nature of the calculation that can be readily exploited for numerical analysis.

Propagation along acoustic axes and in planes of isotropy

If the velocities of the two transverse waves coincide, Eq. (4.33), which is based on the adjugate matrix, cannot be used for calculating the polarization vector [3]. Wave directions with this property are called acoustic axes. For transversely isotropic materials (hexagonal symmetry), the acoustic axes generate a continuous line and for isotropic materials any direction is an acoustic axis. As we have shown in Section 2.8, in a cubic medium, the acoustic axes coincide with the three principal directions and the four body-diagonals (seven axes in all). In general, the set of acoustic axes is included in the (much larger) set of directions for which there exist pure transverse modes (although not necessarily all the modes along an acoustic axis are pure modes). A tetragonal medium may have not more than nine acoustic axes, a trigonal not more than ten, a rhombic not more than 14, and a monoclinic not more than 16 [3]. Therefore, for most symmetry groups there are only a few directions for which we need to use equations different from Eq. (4.33) to determine the polarization directions.

For acoustic axes, the polarization vector of the quasitransverse wave is not uniquely determined by the wave direction itself; it can have any direction in the plane perpendicular to the polarization vector of the quasilongitudinal wave. However, it is always possible to find a polarization direction perpendicular to the wave normal and therefore this wave will be pure transverse.

If, for a given propagation direction n_i, the Christoffel matrix in Eq. (4.22) has two equal eigenvalues $V_{ft} = V_{st} = V_t$, then it is a so-called uniaxial tensor whose axis coincides with the direction of the polarization vector of the quasilongitudinal wave V_ℓ. The polarization vector $\mathbf{p}^\ell$ of the quasilongitudinal wave propagating along the acoustic axis can be found from the decomposition of the uniaxial tensor [3]

$$\frac{c_{ijk\ell}}{\rho} n_j n_k - V_t^2 = Q_{i\ell} = (V_\ell^2 - V_t^2) p_i^\ell p_\ell^\ell. \tag{4.37}$$

The polarization vector p^{ℓ}_{β} of the quasilongitudinal wave can be calculated from Eq. (4.37) as

$$p^{\ell}_{\beta} p^{\ell}_{\beta} = Q_{\beta\beta}/Q_{ii}, \tag{4.38}$$

where the summation convention does not apply for the Greek subscript β, but it does for i. However, if we extend the summation convention for β we find that the magnitude of the polarization vector is unity, as it should be.

If an acoustic axis coincides with the direction of the reflected or transmitted longitudinal wave, then Eq. (4.38) directly gives its polarization direction. If this occurs for a reflected or transmitted transverse wave, then we must first find the polarization vector $\mathbf{p}^{\ell}$ of the quasilongitudinal wave propagating in this direction from Eq. (4.38) although this wave is actually not excited at the interface because it does not satisfy Snell's law. The sought polarization direction of the transverse wave then must lie in the plane whose normal coincides with the polarization direction of the quasilongitudinal wave. To find the polarization vector $\mathbf{p}^t$ of the transverse wave that is actually excited at the interface, let us write it as a linear combination of two orthogonal unit vectors $\mathbf{p}'$ and $\mathbf{p}''$, which both lie in the plane that is normal to $\mathbf{p}^{\ell}$

$$\mathbf{p}^t = B'\mathbf{p}' + B''\mathbf{p}''. \tag{4.39}$$

B' and B'' are presently unknown constants that can be found later from the boundary conditions. First, let us assume that $\mathbf{p}^{\ell}$ is not parallel to the wave normal $\mathbf{n}$ (*i.e.*, it is not a pure longitudinal mode). Then, $\mathbf{p}'$ can be selected

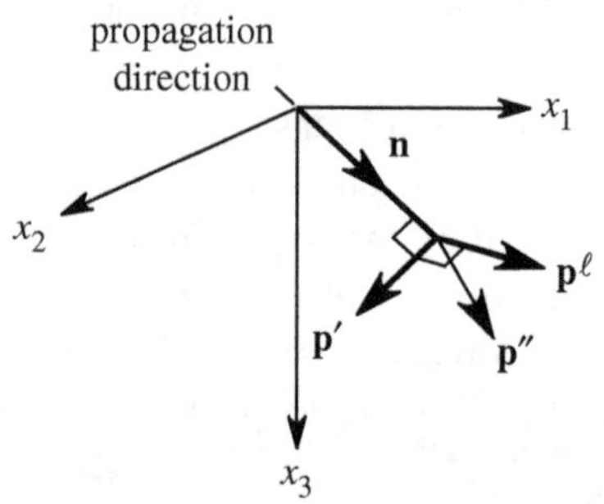

Figure 4.8. Polarization directions for propagation along an acoustic axis that does not support a pure longitudinal mode.
Note: If the reader prefers not to program acoustical axis propagation as a special case, the computations can be performed using the general algorithm described in previous sections. Instead of selecting $\phi = 0$ (propagation along the axis), compute the propagation in directions slightly deviated (say by 0.001°) from the acoustical axis to avoid on-axis velocity degeneracy. For the isotropic case, slight anisotropy may be introduced by small variations in the elastic stiffnesses (*e.g.*, typically less than 0.01%).

normal to the plane including $\mathbf{p}^\ell$ and $\mathbf{n}$ as is shown in Fig. 4.8 so that

$$\mathbf{p}' = \mathbf{n} \times \mathbf{p}^\ell / \mid \mathbf{n} \times \mathbf{p}^\ell \mid . \tag{4.40}$$

Finally, $\mathbf{p}''$ can be obtained from the condition of orthogonality to both $\mathbf{p}^\ell$ and $\mathbf{p}'$

$$\mathbf{p}'' = \mathbf{p}^\ell \times \mathbf{p}'. \tag{4.41}$$

When $\mathbf{p}^\ell$ is parallel to the wave normal $\mathbf{n}$ (*i.e.*, it is a pure longitudinal mode), we can choose

$$\mathbf{p}' = \mathbf{j} \quad \text{and} \quad \mathbf{p}'' = \mathbf{n} \times \mathbf{j}, \tag{4.42}$$

where $\mathbf{j}$ is the unit vector in the x_2 direction, *i.e.*, normal to the plane of incidence.

4.3.5 *Example: Refracted waves in isotropic solids with incident waves from a fluid*

In the beginning of Section 4.3.3, we showed that for the isotropic case the slowness vectors for the reflected and refracted waves can be found immediately from Snell's law. It is useful to see how the simple isotropic solution follows from the general theory. For isotropic materials, the elastic constants were previously given by Eq. (1.38)

$$C_{ij} = \begin{bmatrix} C_{11} & C_{12} & C_{12} & 0 & 0 & 0 \\ C_{12} & C_{11} & C_{12} & 0 & 0 & 0 \\ C_{12} & C_{12} & C_{11} & 0 & 0 & 0 \\ 0 & 0 & 0 & C_{66} & 0 & 0 \\ 0 & 0 & 0 & 0 & C_{66} & 0 \\ 0 & 0 & 0 & 0 & 0 & c_{66} \end{bmatrix}, \tag{4.43}$$

where $C_{66} = (C_{11} - C_{12})/2$. From Eq. (4.27), the elements of the T_{ij} matrix in the Christoffel equation are

$$T_{11} = C_{11}M^2 - \rho + C_{66}\beta^2, \tag{4.44}$$

$$T_{12} = T_{21} = 0, \tag{4.45}$$

$$T_{23} = T_{32} = 0, \tag{4.46}$$

$$T_{13} = T_{31} = (C_{12} + C_{66})M\beta, \tag{4.47}$$

$$T_{22} = C_{66}M^2 - \rho + C_{66}\beta^2, \tag{4.48}$$

$$T_{33} = C_{66}M^2 - \rho + C_{11}\beta^2. \tag{4.49}$$

For easier readability, in Eqs. (4.44–4.49) we substituted $M = m_1^0$ for the known tangential component of the slowness vectors and $\beta = m_3^\alpha$ for the unknown normal components. Thus, Eq. (4.31) becomes

$$\begin{vmatrix} T_{11} & 0 & T_{13} \\ 0 & T_{22} & 0 \\ T_{13} & 0 & T_{33} \end{vmatrix} = T_{22}(T_{11}T_{33} - T_{13}^2) = 0. \tag{4.50}$$

The first two roots can be easily obtained from $T_{22} = 0$ as

$$\beta^2 = \frac{1}{V_t^2} - M^2 \tag{4.51}$$

or

$$m_3^{1,2} = \pm\left(\frac{1}{V_t^2} - \frac{\sin^2\theta_i}{V_f^2}\right)^{1/2}, \tag{4.52}$$

where we expressed the common tangential component M of the slowness vectors with the sound velocity $V^0 = V_f$ in the fluid and the angle of incidence $\theta^0 = \theta_i$. The additional four roots of Eq. (4.50) can be found by solving $T_{11}T_{33} - T_{13}^2 = 0$. After some algebraic manipulation, we get

$$[C_{66}(M^2 + \beta^2) - \rho][C_{11}(M^2 + \beta^2) - \rho] = 0. \tag{4.53}$$

The first factor in Eq. (4.53) is identical to Eq. (4.51) and gives two additional roots of the same values, *i.e.*,

$$m_3^{3,4} = m_3^{1,2}, \tag{4.54}$$

as one would expect in the isotropic case. The remaining two roots come from equating the second factor in Eq. (4.53) to zero, which yields

$$m_3^{5,6} = \pm\left(\frac{1}{V_\ell^2} - \frac{\sin^2\theta_i}{V_f^2}\right)^{1/2}. \tag{4.55}$$

The solutions $m_3^{5,6} = 0$ and $m_3^{1,2,3,4} = 0$ give the first (longitudinal) and second (shear) critical angles, respectively, as follows

$$\theta_{cr1} = \arcsin\frac{V_f}{V_\ell} \quad \text{and} \quad \theta_{cr2} = \arcsin\frac{V_f}{V_t}. \tag{4.56}$$

Of course, the critical angles are real only when the corresponding sound velocities in the solid are higher than the sound velocity in the fluid, which is almost always true for the longitudinal mode and most of the time true for the shear mode. Above the first critical angle, the roots of Eq. (4.55) are imaginary, and above the second critical angle, the roots in Eq. (4.54) are also imaginary, forming evanescent waves. For use in the boundary conditions, the signs of the physically acceptable roots must be chosen according to the following:

i) *before* the critical angle: energy must flow away from the interface (in isotropic media the energy flow direction coincides with the wave direction),
ii) *after* the critical angle: the wave is evanescent and the amplitude of the wave must decay away from the interface.

4.4 Determination of Reflection and Transmission Coefficients

In order to successfully address the challenges encountered in ultrasonic nondestructive characterization of composite materials, the previously considered fluid/isotropic solid model must be extended to anisotropic solids. The basic structural unit of an advanced composite material is a unidirectional lamina. In such a lamina, fibers are often oriented in a single direction, parallel to the lamina surface. A multidirectional composite material consists of several unidirectional laminae of different fiber orientation. If the thickness of the unidirectional layers (consisting of one or more similarly oriented laminae) is much greater than the ultrasonic pulse length, the transmission from layer to layer can be considered as transmission through an interface between two anisotropic half-spaces with different fiber orientations. Therefore, a study of ultrasonic wave propagation through the interface between differently oriented laminae has quite general implications. Other important cases to consider are wave propagation through the interface between a coupling fluid and a unidirectional composite lamina (immersion inspection) and wave propagation through the interface between a coupling solid and a unidirectional composite lamina (*e.g.*, contact inspection using an angle-beam wedge or buffer).

Figure 4.9 illustrates the case under consideration, where the ultrasonic wave is incident from the upper half-space onto an anisotropic composite medium. The elastic properties are supposed to be already transformed from the material coordinate system (x_1', x_2', x_3') into the physical coordinate system (x_1, x_2, x_3) that is aligned with both the interface plane (x_1, x_2) and the plane of incidence (x_1, x_3). [**N.B.** This convention on the primed and unprimed coordinate systems is just the opposite of that from the earlier chapters.] The following calculations will be carried out in this physical coordinate system. The plane of incidence, in general, may be rotated from the fiber direction by an arbitrary angle of ϕ. The upper medium may be liquid, isotropic solid, or another anisotropic solid with different fiber orientation and material properties.

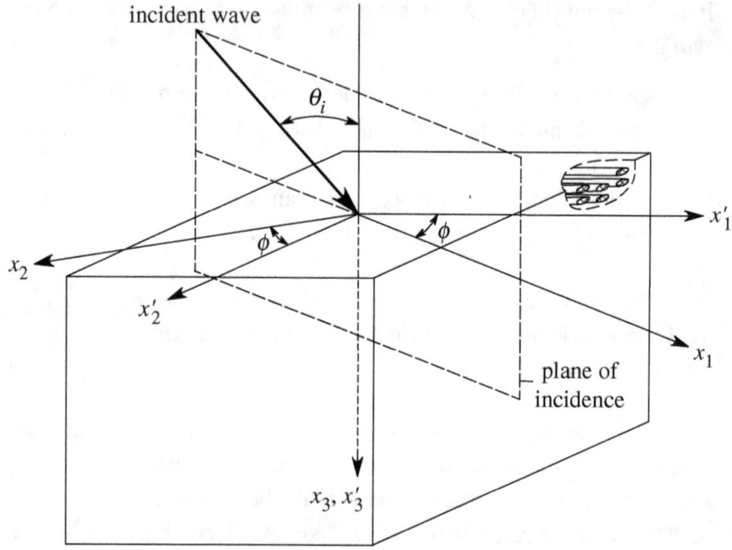

Figure 4.9. Schematic diagram of wave scattering at a composite interface (ϕ denotes the fiber orientation with respect to the plane of incidence, θ_i is the incident angle).

4.4.1 Boundary conditions, amplitude coefficients

Solid/solid interface with welded boundary conditions

The unknown displacement amplitudes A^α of the reflected and refracted waves can be found from the boundary conditions (Eqs. 4.12) and (4.13) that require that both displacement and interface tractions be continuous through the interface. Neglecting the common temporal $\exp(-i\omega t)$ and spatial $\exp(i\omega m_\ell^\alpha x_\ell)$ terms, the boundary conditions can be written in the form of six linear algebraic equations:

$$A^0 p_j^0 = -\sum_{\alpha=1}^{3} A^\alpha p_j^\alpha + \sum_{\alpha=4}^{6} A^\alpha p_j^\alpha \tag{4.57}$$

and

$$m_m^0 c_{i3jm}^I A^0 p_j^0 = -\sum_{\alpha=1}^{3} m_m^\alpha c_{i3jm}^I A^\alpha p_j^\alpha + \sum_{\alpha=4}^{6} m_m^\alpha c_{i3jm}^{II} A^\alpha p_j^\alpha. \tag{4.58}$$

The first three equations ($j = 1, 2, 3$) correspond to displacement continuity and the next three ($i = 1, 2, 3$) to stress continuity. For any given incident displacement amplitude A^0, there are six unknown displacement amplitudes A^α ($\alpha = 1, 2, \ldots, 6$) that can be found by solving Eqs. (4.57) and (4.58). It is convenient to normalize the unknown displacement amplitudes to the incident displacement amplitude and define the following scattering coefficients Γ^α

$$\Gamma^\alpha = \frac{A^\alpha}{A^0}, \tag{4.59}$$

that represent either reflection ($\alpha = 1, 2, 3$) or transmission ($\alpha = 4, 5, 6$) coefficients in terms of the displacement amplitudes.

Note that even in an isotropic upper solid half-space, owing to the anisotropy of the lower half-space, three elastic waves will be reflected in general. For example, an SH wave, which is polarized perpendicular to the plane of incidence, will be produced even when the incident wave is polarized in the plane of incidence, *i.e.*, it is either a longitudinal wave or an SV wave. Since SH and SV waves have the same velocity in an isotropic material, they will have the same reflection angle and will interfere, with each other. As a result of this interference, one can consider the reflected transverse field as a simple transverse wave with polarization vector not necessarily in the incident plane.

Solid/solid interface with slip boundary conditions

Another example is the slip boundary conditions which assume vanishing of shear tractions in the plane of the interface. For example, slip boundary conditions represent a good approximation for a thin layer of fluid couplant between a transducer wedge and the specimen to be inspected in angle-beam contact inspection. The boundary conditions require the continuity of the normal displacement

$$A^0 p_3^0 = -\sum_{\alpha=1}^{3} A^\alpha p_3^\alpha + \sum_{\alpha=4}^{6} A^\alpha p_3^\alpha \tag{4.60}$$

and normal stress

$$m_m^0 c_{33jm}^I A^0 p_j^0 = -\sum_{\alpha=1}^{3} m_m^\alpha c_{33jm}^I A^\alpha p_j^\alpha + \sum_{\alpha=4}^{6} m_m^\alpha c_{33jm}^{II} A^\alpha p_j^\alpha \tag{4.61}$$

through the interface, but place no restriction on the two tangential displacement components that lie in the plane of the interface. Furthermore, the tangential tractions in both the upper and lower media must vanish on the interface, *i.e.*,

$$m_m^0 c_{i3jm}^I A^0 p_j^0 = -\sum_{\alpha=1}^{3} m_m^\alpha c_{i3jm}^I A^\alpha p_j^\alpha \tag{4.62}$$

and

$$\sum_{\alpha=4}^{6} m_m^\alpha c_{i3jm}^{II} A^\alpha p_j^\alpha = 0 \tag{4.63}$$

for $i = 1, 2$.

We have mentioned above in connection with the case of welded boundary conditions that for an incident longitudinal wave in an isotropic wedge, due to the anisotropy of the lower half-space, three elastic waves will be reflected in general, including an SH wave which is polarized perpendicular to the plane of incidence. However, in the case of slip boundary conditions, the horizontally polarized shear wave reflection vanishes since it would be the only wave in the upper half-space capable of producing tangential stress (σ_{32}) at the interface normal to the plane of incidence. Therefore, the boundary condition of Eq. (4.62) is inherently satisfied for $i = 2$ and should be disregarded when solving for the rest of the reflection and transmission coefficients.

Fluid/solid interface

Numerically, the case of a fluid/solid interface can be treated using a general program written for the solid/solid case taking the real density and longitudinal velocity of the fluid and assuming a negligibly small hypothetical shear modulus. To consider the fluid case separately, the boundary conditions must be changed. In this case, only four unknown amplitudes are left: one for the reflected compressional wave and three for the transmitted waves. Note that due to anisotropy of the solid, all three transmitted waves will be excited for arbitrary orientations of the solid, *i.e.*, when the plane of incidence does not coincide with a plane of symmetry of the solid. Of course, when the plane of incidence coincides with a plane of symmetry, only two waves will be transmitted with polarizations in this plane.

The boundary conditions at a solid/fluid interface require that the normal displacement and normal stress be continuous through the interface and that the tangential shear stresses in the solid vanish on the interface. The tangential components of the displacement will be discontinuous through the interface. The continuity of the normal displacement and normal stress requires that

$$A^0 p_3^0 = -A^1 p_3^1 + \sum_{\alpha=4}^{6} A^\alpha p_3^\alpha \tag{4.64}$$

and

$$\frac{\lambda_f}{V_f} A^0 = -\frac{\lambda_f}{V_f} A^1 + \sum_{\alpha=4}^{6} m_m^\alpha c_{33jm}^{II} \, A^\alpha p_j^\alpha . \tag{4.65}$$

Here, we substituted the bulk modulus λ_f of the fluid for $c_{3311}^I = c_{3333}^I$ and $1/V_f$ for the magnitude of the slowness m^0 in the fluid into Eq. (4.61) to specialize it for the case of a fluid upper medium. Recognizing that

$V_f = \sqrt{\lambda_f/\rho_f}$, where ρ_f is the density of the fluid, we can also use the acoustic impedance $Z_f = V_f \rho_f = \sqrt{\lambda_f \rho_f}$ instead of λ_f/V_f in Eq. (4.65). The remaining two boundary conditions require that the tangential shear stresses in the solid vanish on the interface, while there are no restrictions on the two tangential displacement components that lie in the plane of the interface. The requirement of vanishing tangential stresses in the solid at the interface was previously given by Eq. (4.63) in connection with slip boundary conditions and is repeated here only for the sake of completeness

$$\sum_{\alpha=4}^{6} m_m^\alpha c_{i3jm}^{II} A^\alpha p_j^\alpha = 0 \tag{4.66}$$

for $i = 1, 2$.

4.4.2 Energy conversion coefficients

The previously derived reflection and transmission coefficients were obtained in terms of displacement amplitudes; therefore, they cannot be directly used to check their compliance with fundamental laws such as energy conservation and reciprocity. Although these laws are inherently satisfied by virtue of the the appropriate equilibrium equations used to derive the wave equation first and then the scattering coefficients at the interface, they still provide very useful means for checking the validity of the derivations and also for elucidating the results. In order to further characterize the phenomenon of acoustic wave scattering at a plane interface, let us consider the reflection and transmission coefficients in terms of energy flow. In order to distinguish them from the previously calculated displacement scattering coefficients Γ^α we are going to denote the corresponding energy scattering coefficients with γ^α.

In Section 2.5 we introduced the energy flux vector **P**, which is often called the acoustic Poynting vector. The energy scattering coefficients can be written as follows:

$$\gamma^\alpha = \overline{P}_3^\alpha / \overline{P}_3^0, \tag{4.67}$$

where P_3^α denotes the projection of the energy flux vector on the normal of the interface (x_3 in this case) and the bar indicates time averaging. The superscript α identifies the type of the reflected or transmitted mode ($\alpha = 1, 2, \ldots, 6$), and the superscript 0 indicates the incident wave. The γ^α coefficients characterize the redistribution of the incident energy among different reflected and refracted modes. These coefficients have clear physical meaning, and the requirement of energy balance provides an easy check on the calculations.

The energy flux vector **P** was given by Eq. (2.132) as

$$P_i = -\sigma_{ij}\dot{u}_j. \tag{4.68}$$

For harmonic vibrations, the normal component of the time-averaged energy flux vector $\overline{P}_3$ can be expressed from Eq. (2.145) as follows

$$\overline{P}_3 = \frac{1}{2}A^2\omega^2 c_{3j\ell k} m_\ell p_k p_j. \tag{4.69}$$

The direction of the time-averaged energy flux vector $\overline{\mathbf{P}}$ coincides with the direction of the group velocity $\mathbf{V}_g$ or ray direction and Eq. (2.146) can also be used to express $\overline{P}_i$ in terms of the corresponding group velocity vector component V_{gi}

$$\overline{P}_i = \frac{1}{2}A^2\omega^2 \rho V_{gi}. \tag{4.70}$$

The sought energy scattering coefficients γ^α can be expressed with the previously calculated displacement scattering coefficients Γ^α by substituting Eq. (4.70) into Eq. (4.67) as follows

$$\gamma^\alpha = (\Gamma^\alpha)^2 \frac{\rho^\alpha}{\rho^I}\frac{V_{g3}^\alpha}{V_{g3}^0} \tag{4.71}$$

where $\rho^\alpha = \rho^I$ ($\alpha = 0, 1, 2, 3$) denotes the density of the upper medium and $\rho^\alpha = \rho^{II}$ ($\alpha = 4, 5, 6$) denotes the density of the lower medium. Equation (4.71) leads immediately to the following important conclusion: when the energy flow (or group velocity) of a reflected or transmitted wave is oriented parallel to the interface $V_{g3}^\alpha = 0$, the corresponding energy scattering coefficient vanishes regardless of the direction of the wave vector. This is the case of a critical angle. A more complete discussion of the critical angle phenomenon can be found in Section 4.6.

The usefulness of the energy transmission coefficient is illustrated in Fig. 4.10 where A_i and A_t are the displacement amplitudes of the incident and transmitted waves. Let us assume that the transmitted wave is completely reflected from a large reference flaw in the lower half-space as shown in Fig. 4.10(a) and back propagates into the upper medium with a displacement amplitude A_b. In the figure, the displacement transmission coefficient T_{12} describes transmission from the upper medium (I) to the lower medium (II) and T_{21} describes transmission from the lower to the upper medium. The corresponding energy transmission coefficients are t_{12} and t_{21}, respectively. Owing to reciprocity, the one-way energy transmission coefficients are the same in both direction for any particular combination of incident and reflected or transmitted waves, *i.e.*, $t_{12} = t_{21}$. Furthermore, the one-way energy transmission coefficient is equal to the overall two-way amplitude transmission coefficient ($T_{12}T_{21}$) through the interface.

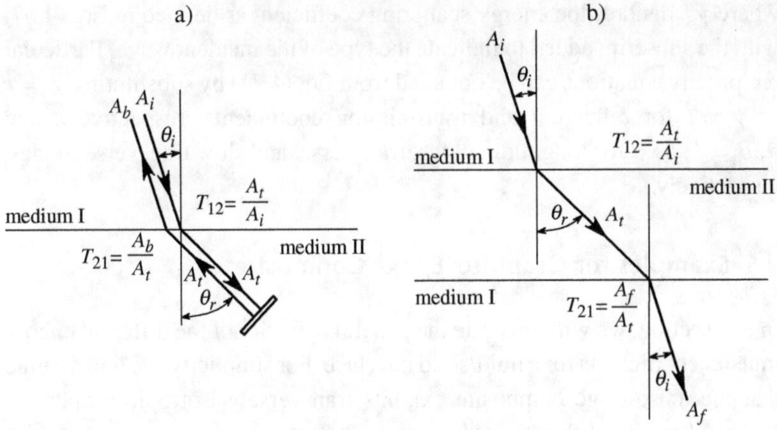

Figure 4.10. Schematic diagram of repeated transmission through an interface separating medium I (fluid or solid coupling medium) and medium II (composite specimen).

Therefore, the normalized flaw signal received by the transducer from a perfectly reflecting flaw behind the interface is

$$\frac{A_b}{A_i} = T_{12}T_{21} = t_{12} = t_{21}. \tag{4.72}$$

The amplitude transmission coefficients T_{12} and T_{21} are not equal. On the other hand, the one-way energy transmission coefficient is equal to the amplitude transmission coefficient through a plate between identical media

$$\frac{A_f}{A_i} = T_{12}T_{21} = t_{12} = t_{21}. \tag{4.73}$$

This case is illustrated in Fig. 4.10(b) through the example of a plate immersed in fluid.

Equations (4.72) and (4.73) imply that the energy transmission coefficient from medium I at incident angle θ_i^I to medium II at refraction angle θ_r^{II} is the same as the energy transmission coefficient from medium II at incident angle $\theta_i^{II} = \theta_r^{II}$ to medium I at refraction angle $\theta_r^I = \theta_i^I$. This is due to the law of Reciprocity, which is a result of material linearity, and energy conservation therefore holds for both isotropic and anisotropic media. Reciprocity relationships exist not only for transmitted modes of the same kind, but in a most general way for any combination of both reflected and transmitted waves regardless of mode conversion [6]. These relations can be formally stated in the following universal form

$$\gamma_\beta^\alpha(\theta_\beta \to \theta_\alpha) \equiv \gamma_\alpha^\beta(\theta_\alpha \to \theta_\beta), \tag{4.74}$$

where γ_β^α denotes the energy scattering coefficient as defined in Eq. (4.67) with the subscript added to indicate the type of the incident wave. Particular reciprocity equations can be obtained from Eq. (4.74) by substituting $\gamma = r$ or $\gamma = t$ for reflection and transmission coefficients, respectively, and $\alpha, \beta = \ell, ft, st$ for longitudinal, fast transverse, and slow transverse modes, respectively.

4.5 Examples for Graphite/Epoxy Composite

In this section, we will illustrate the general behavior of the different energy transfer coefficients for a fluid/solid interface. For simplicity, we will assume that the anisotropic composite exhibits transversely isotropic properties. It should be noted that the transversely isotropic symmetry represents no loss of generality and similar calculations can be done by using the same algorithm for orthotropic laminae as well. In general, three elastic waves of different velocities can propagate in an anisotropic medium along an arbitrary direction: a quasilongitudinal, a fast quasitransverse, and a slow quasitransverse wave and the polarization directions of these modes are mutually orthogonal. However, one interesting feature of wave propagation in transversely isotropic materials is that one of the transverse modes becomes a pure wave, *i.e.*, its polarization direction is orthogonal to the wave direction (see Fig. 2.23 and Eq. (2.208) of Chapter 2). Since the fiber reinforcement stiffens the material with respect to the matrix in essentially all structural composites, and the plane of isotropy is normal to the fiber direction in unidirectional composites, the true transverse mode is inevitably the slower one of the possible two transverse modes.

The schematic diagram of the geometrical arrangement was previously shown in Fig. 4.9. The surface plane of the composite includes the fiber direction and the plane of incidence is rotated by an azimuthal angle ϕ with respect to the fiber orientation. For $\phi = 0°$, the pure slow transverse wave is polarized perpendicular to the plane of incidence independent of the angle of propagation in this plane, *i.e.*, it is always polarized perpendicular to fibers and parallel to the surface, therefore it is called an SH wave. The second transverse wave is an SV-type quasitransverse wave and its velocity is generally higher than the velocity of the SH wave. Only in the direction of the fibers at refraction angle ($\theta_r = 90°$) do both speeds have the same value. This coincidence is caused by isotropy in the plane orthogonal to the fiber direction. In this plane of incidence ($\phi = 0°$), mode conversion from an incident wave in fluid occurs only to the two transmitted waves, namely the quasilongitudinal and the fast quasitransverse waves.

A similar situation occurs when the plane of incidence is normal to the fiber orientation $\phi = 90°$ (this is the plane of isotropy). The true transverse wave is polarized in this plane, therefore this time it exhibits SV behavior.

The SH wave is polarized in the fiber direction and it is faster than the SV wave. The velocity of this fast transverse wave polarized along the fibers is equal to the velocity of the slow transverse wave propagating along the fibers and polarized perpendicular to the fibers. If the rotation angle of the incidence plane is changed continuously from the fiber direction, the slow pure transverse wave which is SH in the fiber direction ($\phi = 0°$) becomes SV at $\phi = 90°$ while at the same time the fast quasitransverse wave changes from SV polarization to SH polarization.

In the following section, we present numerical results to illustrate the energy scattering coefficients at a fluid/composite interface. At an arbitrary orientation of the incident plane with respect to the fiber direction ($\phi \neq 0°$ or 90°), neither wave has either SV or SH character and both have displacement components in and out of the plane of incidence as well as the plane of the surface. However, we can still readily distinguish them as fast and slow transverse modes.

4.5.1 *Fluid/composite interface*

The anisotropic solid used in these calculations was the same transversely isotropic unidirectional graphite fiber-reinforced epoxy matrix composite we used in Chapters 2 and 3 as an example. The elastic constants of this material were listed in Table 2.1 (the fluid is water with $V_f = 1.5 \times 10^3$ m/s and $\rho_f = 10^3$ kg/m^3). Figure 4.11 shows the calculated energy reflection coefficient r_ℓ and transmission coefficients t_ℓ, t_{ft}, t_{st} for the transmitted quasilongitudinal, fast quasitransverse, and slow quasitransverse waves, respectively. Two angles of rotation of the incident plane relative to the fiber direction are presented in solid $\phi = 45°$ and dashed $\phi = 70°$ lines, while the angle of incidence θ_i was varied in the 0°–90° range.

Owing to the above-described deviation of the polarization directions of all three waves from the plane of incidence for an arbitrary rotation angle ϕ, the incident wave from the fluid is converted at the interface into three transmitted waves. It can be seen that for both incident planes, there exist three critical angles for the quasilongitudinal, the fast quasitransverse, and the slow true transverse waves. It is interesting to note that at $\phi = 70°$, the first and third critical angles are considerably higher than at $\phi = 45°$ since the velocities of the quasilongitudinal and true transverse modes monotonically decrease with increasing ϕ, while the second critical angle is higher for $\phi = 45°$ as the velocity of the fast quasitransverse mode increases in this angular range with increasing ϕ (see, *e.g.*, Figs. 3.23–3.25).

Generally, in anisotropic media, the wave and group velocities do not coincide, and the ultrasonic ray deviates from the wave normal not only in the plane of incidence, but also from this plane itself, as it was illustrated in Fig. 4.5. In composites, both in-plane (α) and out-of-plane

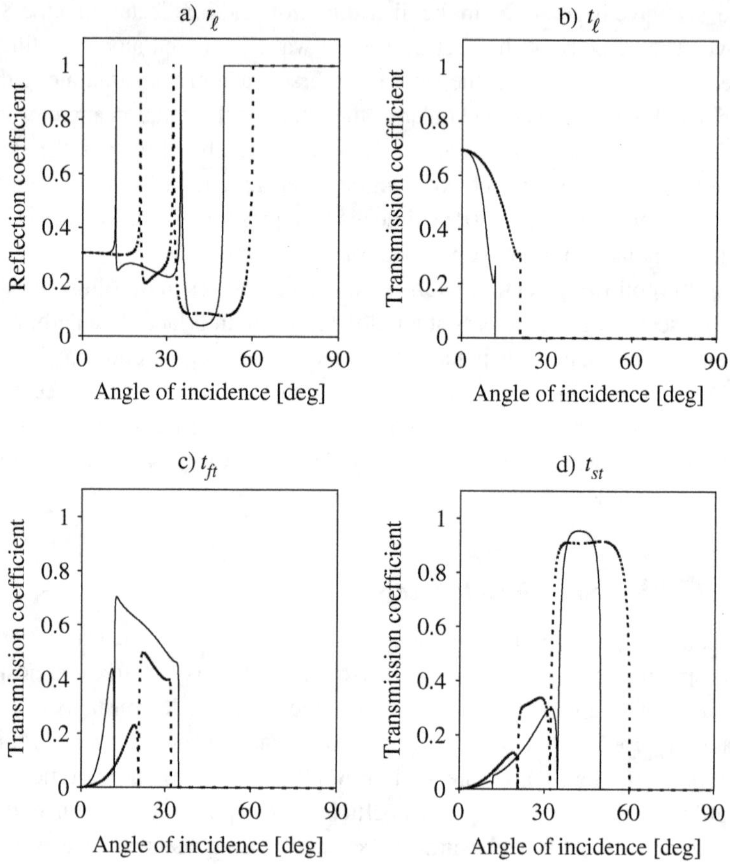

Figure 4.11. Calculated energy reflection coefficient r_ℓ and transmission coefficients t_ℓ, t_{ft}, t_{st} for the transmitted quasilongitudinal, fast quasitransverse, and slow quasitransverse waves, respectively, at a water/composite interface (solid lines $\phi = 45°$ and dashed lines $\phi = 70°$).

(β) deviation angles can be very large. This fact was illustrated for the same unidirectional graphite/epoxy composite material in Fig. 3.20 for rotation angle $\phi = 45°$. The out-of-plane deviation angle (dashed curve) is particularly large ($\beta > 40°$) whenever the refracted wave is significantly tilted with respect to the plane of isotropy, *i.e.*, $| \theta_r |> 30°$. It was also shown that the in-plane deviation angle (solid curve) equals zero both at a refraction angle of $\theta_r = 90°$, *i.e.*, in the plane of the surface, and at $\theta_r = 0°$, *i.e.*, in the plane of isotropy that lies normal to the surface. However, in between, the in-plane deviation angle α could also reach very high values in excess of 45°.

Figure 4.12 shows the in-plane angle of deviation of the group velocity from the wave normal for the quasilongitudinal and the fast quasitransverse

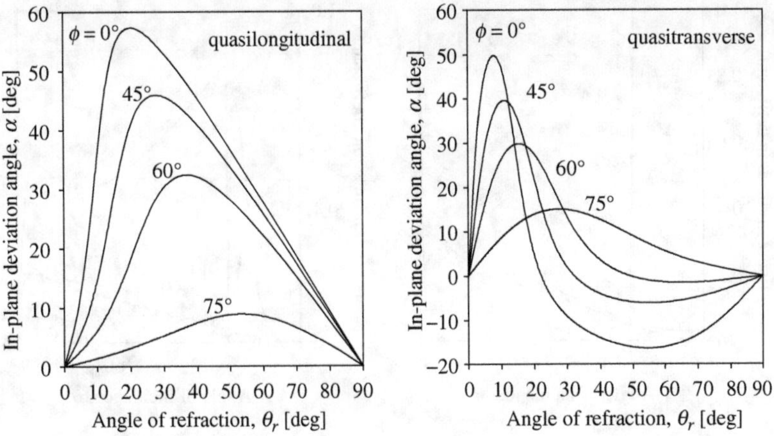

Figure 4.12. In-plane angle of deviation of the group velocity from the wave normal for the quasilongitudinal and the fast quasitransverse waves.

waves at four different values of ϕ. From this figure, one can see that the quasilongitudinal refracted ray always tends to deviate toward the material surface ($\alpha > 0°$), especially at small refraction angles ($\theta_r \approx 20°$). These very large deviations in composite materials were observed experimentally [13, 14]. While the quasilongitudinal ray tends not to penetrate the composite material by deviating toward the surface, the fast quasitransverse wave behaves differently. At incident plane orientation angles $\phi < 65°$, the in-plane deviation angle becomes negative for larger angles of refraction, *i.e.*, the ray direction is skewed toward the normal of the interface plane. Still, at small angles of refraction, the ray direction is very strongly skewed toward the surface, thereby limiting penetration through thick specimens. In this respect, the slower true transverse wave behaves similarly to the quasilongitudinal mode as its in-plane deviation angle is always positive, *i.e.*, it always skews toward the surface.

4.5.2 Isotropic wedge/composite interface

In order to realistically model angle-beam contact transducers, slip boundary conditions must be prescribed at the interface between the isotropic solid wedge and the anisotropic composite specimen to be inspected. These conditions, as discussed in Section 4.4.1, require that the normal displacement and stress be continuous through the interface and that the tangential components of the stress vanish on both sides. Physically, such boundary conditions can be produced by a thin, low-viscosity liquid layer at the interface [10, 15].

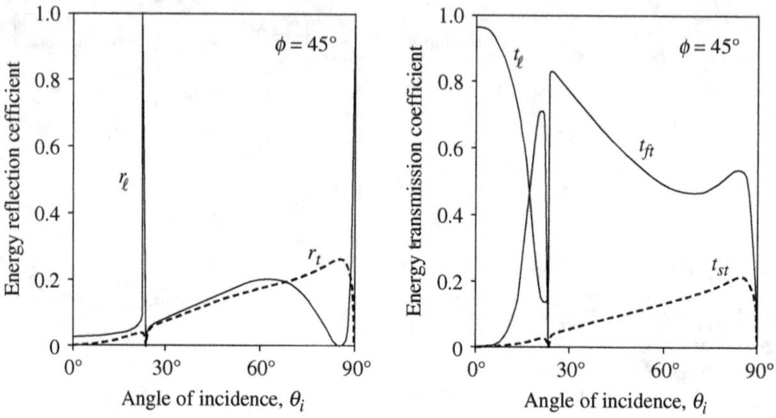

Figure 4.13. Energy reflection and transmission coefficients for a perspex/composite interface with slip boundary conditions.

Table 4.1 Material properties of the transversely isotropic composite and isotropic perspex used in the illustrations of Figs. 4.13–4.17.

Stiffness [10^9 N/m^2]	Composite	Wedge
C_{11}	162	8.79
C_{22}	17.0	8.79
C_{33}	17.0	8.79
C_{12}	5.47	3.97
C_{13}	5.47	3.97
C_{23}	7.0	3.97
C_{44}	3.3	2.41
C_{55}	6.01	2.41
C_{66}	6.01	2.41
ρ[10^3 kg/m^3]	1.61	1.2

The typical case of energy reflection and transmission coefficients for an angle-beam contact transducer with perspex wedge is illustrated in Fig. 4.13. The material properties of the transversely isotropic composite (taken from [16]) and of the perspex wedge are listed in Table 4.1. The plane of incidence is rotated by $\phi = 45°$ from the fiber orientation and a longitudinal incident wave is considered. First, note the very high transmission t_ℓ into the longitudinal mode at normal incidence as a result of the material parameters being very similar in this direction (the epoxy matrix is very similar to perspex and the fiber reinforcement causes little stiffening normal to the fibers). For the parameters used in this example, only one critical angle may be observed at $\theta_{cr} \approx 23°$ for the quasilongitudinal wave.

As in the case of a fluid/composite interface, at the critical angle, the corresponding reflection coefficient reaches a sharp maximum of unity, and the transmission coefficients are zero. This is a general rule that applies to both liquid/solid and solid/solid interfaces with slip boundary conditions as long as the surface (interface) of the anisotropic lower medium coincides with one of its planes of symmetry (and losses are negligible in both media). In these boundary conditions, the two tangential stresses (σ_{13} and σ_{23}) on the surface of the lower solid must be cancelled by the three transmitted waves themselves without any contribution from either the incident or the reflected waves. For slip boundary conditions, the requirement of vanishing of tangential stresses is separately applied to these waves for each semi-space). This action is possible only if the three transmitted waves maintain a given amplitude ratio among themselves, which is determined by the elastic properties of the second medium. Of course, this ratio is changing with the incident angle θ_i because the directions of the refracted waves change according to Snell's law, but it is not affected otherwise by the first medium.

Let us consider the case of a critically refracted longitudinal wave. Whenever the interface is a plane of symmetry for the lower medium, the critically refracted longitudinal wave produced along the interface cannot provide any tangential stress for this balance; therefore, the two transmitted transverse waves must identically vanish, *i.e.*, $t_{ft} = t_{st} = 0$. As the group velocity of the transmitted longitudinal wave is parallel to the surface, the energy transmission coefficient t_ℓ must also be zero (see Eq. (4.71)) though its displacement transmission coefficient is not zero and it still contributes to the continuity of the normal stress and displacement. Of course, the tangential stresses must also disappear on the other side of the interface. Assuming that the incident medium is isotropic, then by virtue of symmetry, this leads to $r_t = 0$, $r_\ell = 1$ so that the incident and reflected longitudinal waves cancel each other's tangential stress components but not the normal ones.

The results of the calculations for the same pair of materials but with welded boundary conditions are shown in Fig. 4.14. This situation can be approximated by applying a thin layer of special high-viscosity coupling fluid usually used for normal-incidence shear wave inspection only. In this case, the energy reflection coefficient $r_\ell(\theta_{cr}) \approx 0.1$ and does not approach unity at the critical angle. Another interesting difference from the case of slip boundary condition is that the reflected SH wave appears in the reflected field as we predicted earlier in our analytical discussion. It should also be noted that the slow transverse wave essentially disappears in the transmitted field and starts to play a significant role in the energy balance only at large angles of rotation ($\phi > 80°$) where the polarization of the fast transverse wave turns into a horizontal direction; therefore, it becomes uncoupled to the incident longitudinal wave. It is very interesting, however, that in the case of slip boundary conditions that exclude the generation of

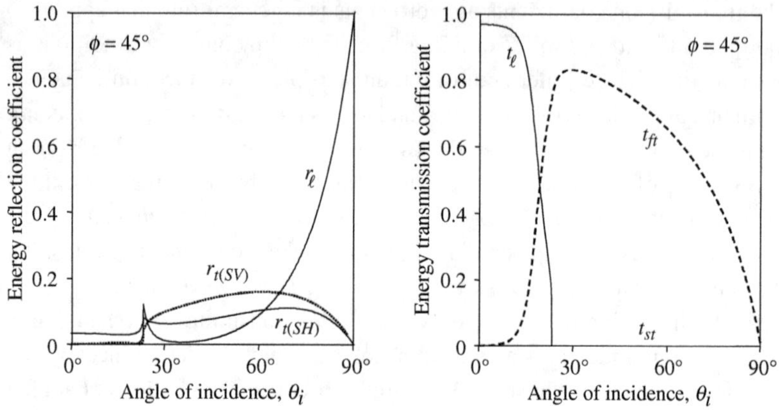

Figure 4.14. Energy reflection and transmission coefficients for a perspex/composite interface with welded boundary conditions.

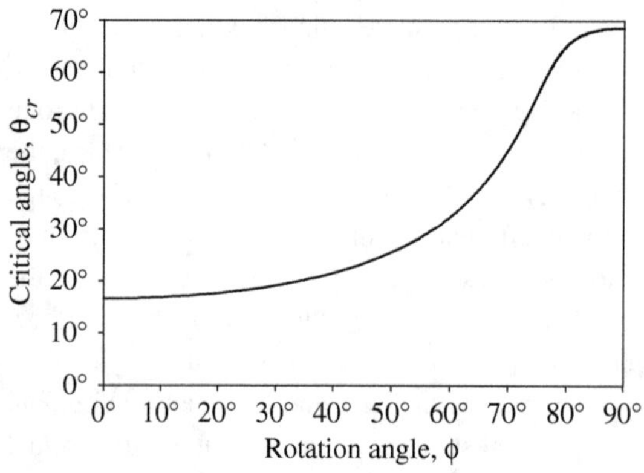

Figure 4.15. Dependence of the only (longitudinal) critical angle θ_{cr} on the rotation angle ϕ a perspex/composite interface.

horizontally polarized transverse reflection, the need to cancel the tangential stress on the surface of the composite did require a given ratio of all three transmitted modes and the slow transverse transmission was fairly substantial (see Fig. 4.13).

It is known that angle-beam wedge transducers are easiest to use above the longitudinal critical angle, where the transmitted shear wave is strong and less influenced by interference from other spurious beams. The dependence of the only (longitudinal) critical angle θ_{cr} on the rotation angle ϕ is shown in Fig. 4.15 for the same perspex/composite interface. As the longitudinal velocity in the plane of the surface decreases from a high value of $V_\ell \approx 9.5$ km/s along the fiber direction ($\phi = 0°$) to a low value of $V_\ell \approx 2.9$ km/s normal to it ($\phi = 90°$), the critical angle increases from about 17° to 70°.

4.5.3 Composite/composite interface

In this section we consider a few typical examples of composite/composite interfaces. Two examples for acoustic wave scattering at a composite/composite interface are shown in Figs. 4.16 and 4.17. Figure 4.16 shows the energy reflection and transmission coefficients for a composite/composite interface with welded boundary conditions. In this case, the incident plane is coincident with the fiber direction in the upper medium ($\phi_1 = 0°$) and rotated by $\phi_2 = 45°$ in the lower medium (longitudinal incident wave is assumed). Except for their orientation, the two composite half-spaces are identical (their material properties were previously given in Table 4.1). This case represents an example of an interface between two laminae of different orientations in a multiple-ply laminate. First, note that at normal incidence full transmission occurs without any mode conversion. There is, however, a relatively high and monotonically increasing reflection coefficient for the longitudinal wave at incident angles above 15° accompanied by a very strong skewing of the reflected ultrasonic beam toward the fiber direction, but there is no deviation of the reflected beam from the plane of incidence. Energy conversion into either of the reflected quasitransverse waves is very small. All three waves appear in the transmitted field with the quasilongitudinal wave being the strongest and the fast quasitransverse wave being the weakest.

The inverse situation is shown in Fig. 4.17. Here, the upper medium is rotated by $\phi_1 = 45°$ and in the lower medium, the fibers lie in the incident plane ($\phi_2 = 0°$). A critical angle phenomenon is observed in this case for

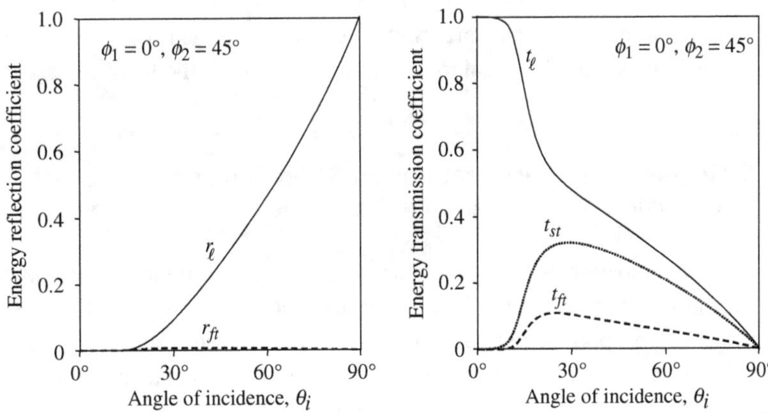

Figure 4.16. Energy reflection and transmission coefficients for a composite/composite interface with welded boundary conditions when the incident plane is coincident with the fiber direction in the upper medium and rotated by $\phi_2 = 45°$ in the lower medium (longitudinal incident wave).

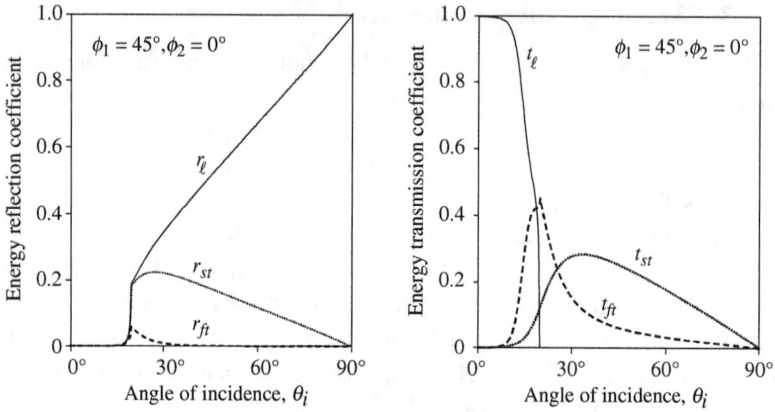

Figure 4.17. Energy reflection and transmission coefficients for a composite/composite interface with welded boundary conditions when the incident plane is coincident with the fiber direction in the lower medium and rotated by $\phi_1 = 45°$ in the upper medium (longitudinal incident wave).

the transmitted quasilongitudinal wave slightly below $\theta_i = 20°$. Just as in the previous case, very small reflection and mode conversion occur for the quasilongitudinal wave when it crosses the interface at an angle of incidence below $\theta_i = 15°$. On the other hand, beam deviation from the wave normal is very significant even at these small angles. At first sight, one might expect that the longitudinal transmission coefficients (t_ℓ) should be identical in the cases shown in Figs. 4.16 and 4.17 by virtue of reciprocity. Earlier, in Section 4.4.2, we showed that in a fundamental way they are identical, though in these figures they do not appear to be so because reciprocity works only if the propagation paths are truly inverted, *i.e.*, in the reverse case, the incident angle should be equal to the refraction angle in the forward case.

4.6 Geometrical Considerations on Reflection and Refraction, Grazing and Critical Angles

Before we start our discussion of the geometrical interpretation of wave scattering at an interface between generally anisotropic media, let us first prove a very important relationship between the slowness surface and the ray direction, namely that the ray direction is exactly normal to the slowness surface at every point. In Chapter 2 we found that the projection of the group velocity vector $\mathbf{V}_g$ on the wave direction $\mathbf{n}$ is equal to the phase velocity V (see Eq. 2.157)

$$V_{gi} n_i = V. \tag{4.75}$$

At the same time, the slowness vector **m** was defined as a vector parallel to the wave direction with a magnitude equal to the inverse of the phase velocity

$$m_i = \frac{1}{V} n_i, \tag{4.76}$$

which leads to

$$\mathbf{V}_g \cdot \mathbf{m} = 1. \tag{4.77}$$

Taking the exact differential of the left side of Eq. (4.77) yields

$$d\mathbf{V}_g \cdot \mathbf{m} + \mathbf{V}_g \cdot d\mathbf{m} = 0, \tag{4.78}$$

where

$$d\mathbf{m} = \frac{\partial \mathbf{m}}{\partial n_i} dn_i. \tag{4.79}$$

Here, the partial derivative of m_j with respect to n_i can be calculated as follows

$$\frac{\partial m_j}{\partial n_i} = \frac{\partial}{\partial n_i}\left(\frac{n_j}{V}\right) = \frac{\delta_{ij}}{V} - \left(\frac{n_j}{V^2}\right)\frac{\partial V}{\partial n_i}. \tag{4.80}$$

Substituting Eq. (4.80) into Eq. (4.79), multiplying by $\mathbf{V}_g$, and using Eq. (4.75) yields

$$\mathbf{V}_g \cdot d\mathbf{m} = -V_{gj}\frac{\partial m_j}{\partial n_i} dn_i = -V_{gj}\left(\frac{\delta_{ij}}{V} - \frac{n_j}{V^2}\frac{\partial V}{\partial n_i}\right) dn_i = -\frac{1}{V}\left(V_{gi} - \frac{\partial V}{\partial n_i}\right) dn_i. \tag{4.81}$$

The right side of Eq. (4.81) is obviously zero since from Eq. (4.75) $\partial V / \partial n_i = V_{gi}$, therefore our final result is that

$$\mathbf{V}_g \cdot d\mathbf{m} \equiv 0. \tag{4.82}$$

This means that $\mathbf{V}_g$ is always perpendicular to $d\mathbf{m}$, *i.e.*, to the slowness surface itself, which is the important geometrical relationship between the slowness surface and the ray direction we wanted to prove.

Now, let us consider the slowness surface of two anisotropic solids shown in Fig. 4.18. This type of polar diagram is very useful in the determination of the reflection and refraction slowness vectors by graphical means. Let us assume that a quasilongitudinal wave impinges at the interface between these two solids at an angle of incidence θ_i and draw a line in the incident wave direction through the origin of the slowness diagram of the first medium. The slowness vector $\mathbf{m}^0$ of the incident wave is aiming at the origin from the intersection point of this line with the slowness curve of the quasilongitudinal mode. According to Snell's law, the incident as well as all the reflected and transmitted waves will have the same tangential

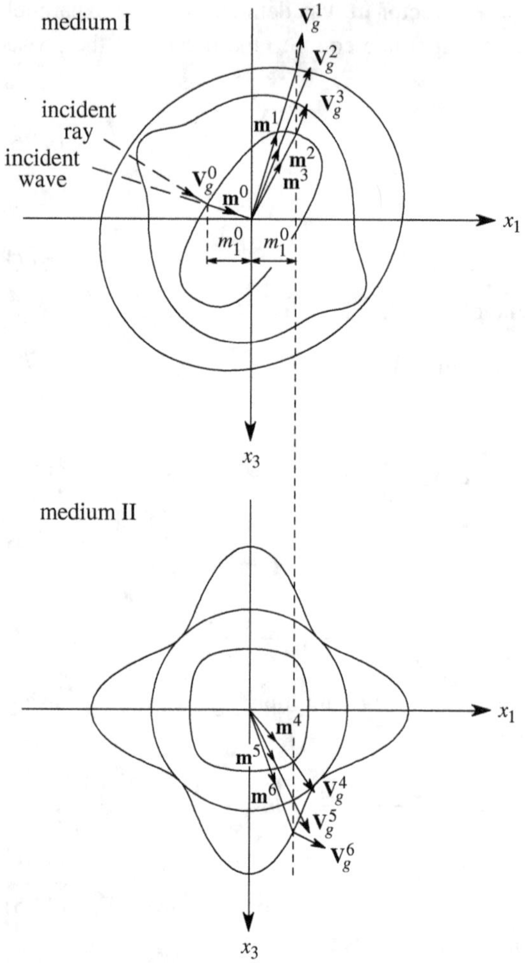

Figure 4.18. Example of the analysis of wave reflection and refraction at an interface between two anisotropic materials based on the slowness surfaces. $\mathbf{m}^0$ and $\mathbf{V}_g^0$ are the slowness and group velocity vectors for the incident quasilongitudinal wave, respectively, and $\mathbf{V}_g^\alpha$ ($\alpha = 1, 2, \ldots, 6$) are the group velocity vectors of the reflected and refracted waves. m_1^0 denotes the slowness vector component of the incident wave in the plane of the surface, which, according to Snell's law, is common for all reflected and refracted waves.

slowness vector component m_1^0, which is also indicated in Fig. 4.18. Now, let us draw a vertical line on the opposite side of the origin at a distance m_1^0 from it. The intersection of this line with the reflected and refracted slowness surfaces will determine the slowness vectors of the reflected and refracted waves as shown in Fig. 4.18. The group velocity vector (or ray direction) is normal to the slowness surface at these intersection points while the magnitude of the group velocity can be determined from the fact that the projection of the group velocity to the wave propagation direction is equal to the phase velocity, *i.e.*, to the inverse of the slowness (the length of the group velocity vectors in Fig. 4.18 are not drawn to exact scale to reflect their actual magnitude).

If no intersection is found with a certain slowness surface, it means that the incident angle is higher than the critical angle for the corresponding wave

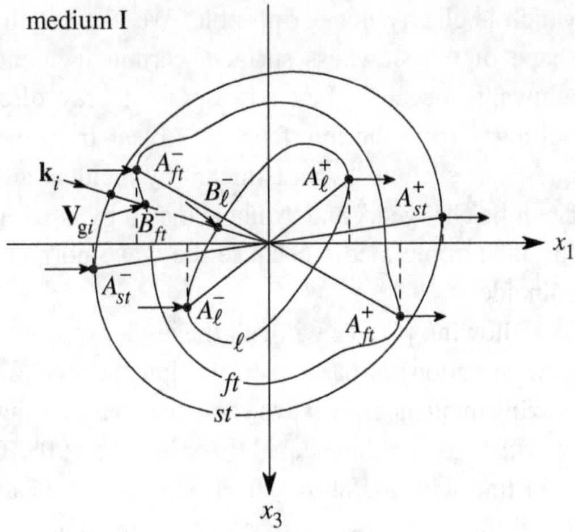

Figure 4.19. Two-dimensional graphical representation of grazing angle/critical angle phenomena. Points identified by A correspond to wave directions where the group velocity is parallel to the interface and the associated arrows indicate rays propagating in either the positive or negative direction along the interface.

and that particular mode will be evanescent. It can be seen in Fig. 4.19 that the critical angle for each wave is reached when the normal to the slowness surface becomes parallel to the interface (the projection of this point of the slowness surface on the abscissa axis is farthest from the origin). For simplicity, only the 2-D cross-section of the slowness surface of the upper medium is shown from Fig. 4.18. As we proved at the beginning of this section, the ray direction is exactly normal to the slowness surface, therefore at the critical angle the ray direction is parallel to the surface. It should be emphasized that the slowness surface is actually 3-D and the surface normal will generally also have an out-of-the plane (of incidence) component that is not perceivable in this 2-D representation.

The maximum ray angle of incidence Θ_i at which the incident acoustic beam can reach the interface is clearly 90° and this situation is called grazing incidence. However, the maximum wave angle of incidence θ_i could be either lower or higher than 90°. For example, in the particular case shown in Fig. 4.19, the grazing longitudinal incident wave has a wave angle of incidence as high as $\approx$ 130° as determined by points A_ℓ^- and A_ℓ^+ on its slowness curve. In comparison, the grazing fast and slow transverse incident waves have wave angles of incidence $\approx$ 70° (points A_{ft}^- and A_{ft}^+) and $\approx$ 100° (points A_{st}^- and A_{st}^+), respectively.

It is also very interesting that there are wave directions at which the phase velocity is directed toward the interface ($n_3 > 0$) while the group velocity propagates away from it ($V_{g3} < 0$). For example, if the incident wave previously illustrated in Fig. 4.18 were changed to a fast quasitransverse mode (this point is marked in Fig. 4.19 as point B_{ft}^-) from the quasilongitudinal mode (B_ℓ^-) without changing the (wave) angle of incidence θ_i, the ray angle of incidence Θ_i would have to be larger than 90°, which is clearly not permissible. We can conclude that depending on the shape of the slowness surface, certain incident wave directions are not allowable because they correspond to ray directions aimed not toward but away from the interface. It is clear from these arguments that simple questions such as "what is meant by incident angle," "through what range can it be changed," and "what is meant by grazing incidence" are not trivial for anisotropic media because the wave normal and ray directions do not coincide.

Following [6], we will call the angles at which the ray direction (energy flow direction) is parallel to the interface ($\Theta_i = 90°$) grazing angles. At grazing incidence, the wave vector can have an angle θ_i with the normal to the interface greater than, equal to, or less than 90°. This choice of terminology is in line with the above discussion of the meaning of the critical angle. The difference between ray and wave incident angles is further illustrated in Fig. 4.20. The incident acoustic beam ($\mathbf{V}_{gi}$) is directed toward the interface but its wave vector $\mathbf{k}_i$, which is normal to the wavefront, is directed away from the interface subtending an angle of incidence θ_i that is greater than 90°.

The permissible range of (wave) incident angles θ_i (where $\theta_i = \theta_1$) that corresponds to positive group velocity along the x_3 axis ($V_{g3} > 0$) is very sensitive to the orientation of the anisotropic material. As an example, Fig. 4.21 illustrates the permissible sectors of the slow quasitransverse incident wave for an anisotropic composite at three orientations as shaded areas. In the the first case on the left ($\theta_1 = 0°$), the physical coordinate system (x_1, x_3) is aligned with the material symmetry directions and the permissible range of incident angles comprises three discontinuous sectors. In the second case, in the middle ($\theta_1 = 12.5°$), there are still three separate permissible sectors, but the forbidden zone between them is substantially changed. Finally, in the third case on the right side ($\theta_1 = 27.5°$), the permissible

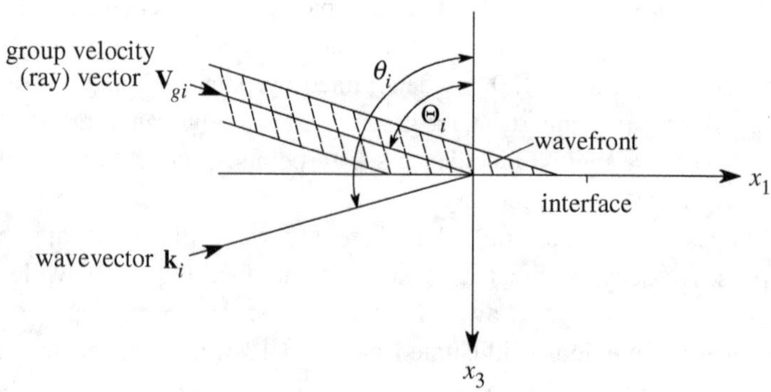

Figure 4.20. Schematic representation of the difference between ray and wave incident angles in an anisotropic medium.

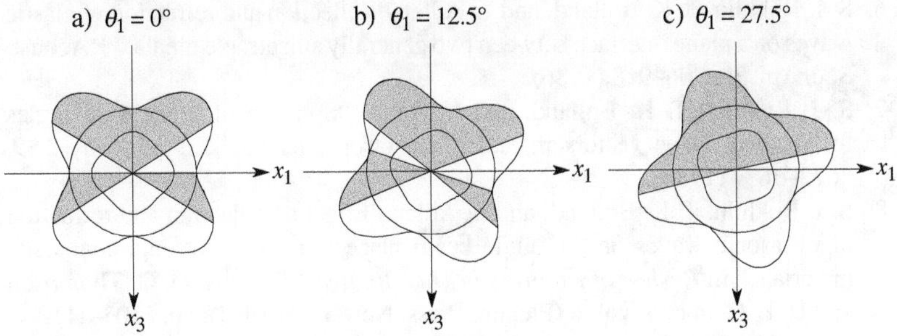

Figure 4.21. Schematic illustration of the permissible range (shaded areas) of the slowness diagram for the slow quasitransverse incident wave, which corresponds to positive group velocity along the x_3 axis ($V_{g3} > 0$) at three different orientations of an anisotropic material rotated by an angle θ about the x_2 axis (the slowness curves of the two other modes are shown only for comparison).

regime for θ_i is continuous over a 180°-range, though this range is not the same as the ±90°-range for Θ_i between grazing incidence from opposite directions.

The absolute values of the phase and group velocities do not change if the direction of wave propagation is reversed. Therefore, the slowness surface of a generally anisotropic material is symmetric relative to the coordinate origin and for each point on any slowness surface, a symmetrical point can be found. At these two points, the ray (group velocity) and wave vectors have the same magnitude but opposite directions. So, for each incident grazing direction θ_{cr}, the opposite direction $\theta_{cr} \pm 180°$ is also an incident grazing angle. The domain of incident angles for any anisotropy spans 180° regardless of whether the domain is split or not. For each impermissible (unshaded) sector, a corresponding equal permissible (shaded) sector may be found symmetrically positioned relative to the coordinate origin. Numerous examples of elastic wave scattering at interfaces separating anisotropic media with splitting of the permissible angular domains and analysis of critical angles are given in [7].

Bibliography

1. J. Krautkramer and H. Krautkramer, *Ultrasonic Testing of Materials*, 4th ed (Springer-Verlag, Berlin-New York, 1990).
2. M. J. P. Musgrave, *Crystal Acoustics* (Holden Day, San Francisco, 1970).
3. F. I. Fedorov, *Theory of Elastic Waves in Crystals* (Plenum, New York, 1968).
4. B. A. Auld *Acoustic Fields and Waves in Solids*, 2nd ed (Krieger Publishing, Malabar, 1990), Vols. I and II.
5. E. G. Henneke, II, "Reflection-refraction of a stress wave at a plane boundary between anisotropic media," J. Acoust. Soc. Am. **51**, 210–217 (1972).

6. S. I. Rokhlin, T. K. Bolland, and L. Adler, "Reflection and refraction of elastic waves on a plane interface between two generally anisotropic media," J. Acoust. Soc. Am. **79**, 906–918 (1986).
7. S. I. Rokhlin, T. K. Bolland, and L. Adler, "Splitting of domain of angles for incident wave vectors in elastic anisotropic media," J. Appl. Phys. **59**, 3672–3677 (1986).
8. S. I. Rokhlin, T. K. Bolland, and L. Adler, "Effect of reflection and refraction of ultrasonic waves on the angle beam inspection of anisotropic composite materials," in *Review of Progress in Quantitative NDE*, eds. D. O. Thompson and D. E. Chimenti, vol. **6** (Plenum Press, New York, 1987), pp. 1103–1110.
9. A. H. Nayfeh, *Wave Propagation in Layered Anisotropic Media* (Elsevier Science, Amsterdam, 1995).
10. S. I. Rokhlin and Y. J. Wang, "Analysis of boundary conditions for elastic wave interaction with an interface between two solids," J. Acoust. Soc. Am. **89**, 503–515 (1991).
11. S. I. Rokhlin and W. Huang, "Ultrasonic wave interaction with a thin anisotropic layer between two anisotropic solids. II: Second order asymptotic boundary conditions," J. Acoust. Soc. Am. **94**, 3405–3420 (1993).
12. P. Lanceleur, H. Ribeiro, and J.-F. De Belleval, "The use of inhomogeneous waves in the reflection–transmission problem at a plane interface between two anisotropic media," J. Acoust. Soc. Am. **93**, 1882–1892 (1993).
13. R. D. Kriz and W. W. Stinchcomb, "Elastic moduli of transversely isotropic fibers and their composites," Exper. Mech. **19**, 41–49 (1979).
14. W. R. Rose, S. I. Rokhlin, and L. Adler, "Evaluation of anisotropic properties of graphite-epoxy composite plates using Lamb waves," in *Review of Progress in Quantitative NDE*, eds. D. O. Thompson and D. E. Chimenti, vol. **6** (Plenum Press, New York, 1987), pp. 1111–1118.
15. S. I. Rokhlin and D. Maron, "Study of adhesive bonds using low-frequency obliquely incident ultrasonic waves," J. Acoust. Soc. Am. **80**, 585–590 (1986).
16. R. D. Kriz and H. M. Ledbetter, "Elastic representation surfaces of unidirectional graphite/epoxy composites," in *Proceed. of 2nd US Japan Conf. on Composite Materials* (ASTM, Philadelphia, 1985), pp. 661–675.
17. L. Wang, A. I. Lavrentyev, and S. I. Rokhlin, "Beam and phase effects in angle-beam-through-transmission method of ultrasonic velocity measurement," J. Acoust. Soc. Am. **113**, 1551–1559 (2003).

5

Guided Waves in Plates and Rods

5.1 Introduction

In this chapter we consider elastic wave modes which propagate in composites with finite boundaries. There are those waves that exist between the two plane parallel boundaries of a homogeneous anisotropic solid. We consider that well-known problem, as well as waves in an elastic anisotropic rod, specifically an individual graphite fiber. Composite laminates seen in applications are essentially all multilayered structures, and in many cases can be considered periodically layered. So, we also take up the subject of guided waves in layered plates in later chapters.

In a plate geometry, as illustrated in Fig. 5.1, we choose the propagation direction to be parallel to the x_1 axis and the x_3 axis to be normal to the plate surfaces. This geometry is particularly significant for composite materials since, by design, laminates are often locally planar in nature. While the solutions we find are appropriate for flat plates, with some modifications they describe wave motion in gently curved structures as well. Clear and mathematically straightforward descriptions of the characteristics of plate waves exist for isotropic media [1, 2]. The results obtained for isotropic media are not, however, directly applicable to most composites. We begin by considering the behavior of waves in a uniaxial composite laminate. In later chapters we generalize the calculation to layered orthotropic media, concentrating on the results and physical interpretation rather than the algebraic details.

To begin a description of waves in plates, let us consider the possible polarizations of particle motion. Let the plate surfaces lie in the (x_1, x_2) plane of mirror symmetry with the origin dividing the plate thickness in half, as shown in Fig. 5.1. Then, we will at first assume the wave to be uniform in

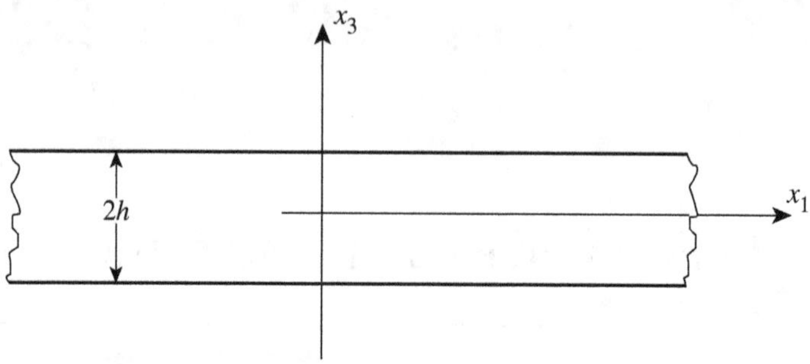

Figure 5.1. Geometry of plate structure showing coordinate axes; x_2 axis points inward from origin.

the x_2 direction and propagating in the x_1 direction, and (x_1, x_3) is the plane of symmetry. Particle motion can occur along any axis. Note that in this restricted symmetry, shear partial waves polarized along the x_2 axis will have no component of particle motion normal to the plate surfaces. Partial waves are a concept introduced by Rayleigh [3] to acknowledge that a superposition of both shear and longitudinal particle motion is generally needed to produce plate waves polarized in the vertical plane. From our earlier discussion of scattering at planar interfaces, it is clear that in isotropic media, SH waves (shear waves, polarized horizontally, *i. e.*, parallel to the plate surfaces) could not convert to other modes in reflection from the plate surfaces. Therefore, we anticipate that this polarization, even in the case of a uniaxial fiber layup, should lead to waves which are distinct from other normal modes of the plate, in the case of straight-crested waves propagating in the x_1 direction and for an excitation that is also invariant in the x_2 direction. For generally anisotropic media, or when the propagation direction is not along any symmetry axis in the composite, satisfying the plate boundary conditions leads to a more complicated circumstance. In this case, plate modes will be composed of quasilongitudinal and quasishear partial waves, polarized in both the vertical and horizontal planes. The implication is that all three partial waves will be coupled for such plate modes. With propagation along a symmetry axis in orthotropic media, the vertically and horizontally polarized wave motion will decouple for straight-crested waves. The former type is similar to Lamb waves observed in isotropic media, while the latter one yields SH plate waves.

5.2 Guided Waves in a Uniaxial Laminate

5.2.1 Preliminaries

We begin by noting that the presence of anisotropy means that resolution of the wave motion into longitudinal and shear pure wave components occurs only under special circumstances. For this reason, the discussion of guided waves in a uniaxial laminate is couched in terms of displacements, rather than wave potentials, as is typically done in the case of isotropic materials [1,2].

This situation parallels the generalizations treated in Chapters 3 and 4, and demonstrates again the complications of dealing with anisotropic materials. The three possible wave modes will, in general, be coupled when the wave strikes a boundary, and they uncouple (into SH waves and waves polarized in the vertical plane) only for wave propagation along material symmetry axes, as we saw earlier. We have two choices in deciding how to calculate dispersion. On the one hand, we could leave the laminate material properties in transversely isotropic (TI) form and deal with the wave coupling through the additional dimensionality of the equations of motion [4]. Alternatively, we could transform the elastic stiffness matrix to the form suitable for an orthotropic material rotated through some arbitrary angle, and then analyze the problem in the plane of the propagation direction and the plate normal. We assume that the fibers always lie in the plane of the plate surface (composite laminates are seldom made any other way), but we allow that the fiber direction can have any orientation with respect to the plate wave propagation direction. We choose the latter option, analysis of propagation with a rotated stiffness matrix in the invariant (x_1, x_3) plane, because of its convenience when we later analyze the problem of fluid-coupled plate waves and because of its generality and our experience with this analysis from Chapter 4.

We will transform the elastic properties of the material to yield an effective elastic stiffness matrix appropriate to the chosen direction of propagation. Restating the equation of motion of the anisotropic linear elastic solid,

$$\rho_s \frac{\partial^2 u_i}{\partial t^2} = c_{ijk\ell} \frac{\partial^2 u_k}{\partial x_j \partial x_\ell}, \qquad i = 1, 2, 3, \tag{5.1}$$

where the $c_{ijk\ell}$ is the rotated elastic stiffness tensor for the uniaxial plate material (discussed in Chapter 3), and ρ_s is the mass density of the solid plate. For propagation in the (x_1, x_3) plane of symmetry, the tensor of elastic stiffnesses $c_{ijk\ell}$ can also be expressed from $\bar{c}_{ijk\ell}$. Using the familiar subscript contractions $ii \to i$ and $ij \to 9-(i+j)$, $i \neq j$, the fourth rank elastic stiffness tensor can be conveniently expressed as a 6×6 stiffness matrix, $\overline{C}_{IJ}$,

$$\overline{C}_{IJ} = \begin{pmatrix} \overline{C}_{11} & \overline{C}_{12} & \overline{C}_{13} & 0 & 0 & 0 \\ \overline{C}_{12} & \overline{C}_{22} & \overline{C}_{23} & 0 & 0 & 0 \\ \overline{C}_{13} & \overline{C}_{23} & \overline{C}_{33} & 0 & 0 & 0 \\ 0 & 0 & 0 & \overline{C}_{44} & 0 & 0 \\ 0 & 0 & 0 & 0 & \overline{C}_{55} & 0 \\ 0 & 0 & 0 & 0 & 0 & \overline{C}_{66} \end{pmatrix}, \tag{5.2}$$

for an orthotropic material (nine independent elastic stiffnesses). If we are considering a single ply or a uniaxial laminate, then transverse isotropy is appropriate, and the following simplifications result. The number of independent constants is five, and if the fiber direction is coincident with x_1, $\overline{C}_{22} = \overline{C}_{33} = \overline{C}_{23} + 2\overline{C}_{44}$, $\overline{C}_{12} = \overline{C}_{13}$, and $\overline{C}_{55} = \overline{C}_{66}$, which implies that the x_2 and x_3 directions are equivalent and interchangeable. If the plate wave vector is along x_1 and we assume a uniform wave in the x_2 direction, only $u_1(x_1, x_3)$ and $u_3(x_1, x_3)$ will be nonzero, as in the isotropic case. The resulting wave motion is very much like Lamb waves in isotropic media, except that additional stiffness terms enter the characteristic equations and the plane of incidence is not isotropic. In addition, there will be an uncoupled SH wave having only one nonzero displacement component $u_2(x_1, x_3)$.

In general, however, this relatively simple situation does not persist if the plate wave vector ξ_p is neither along the fiber direction nor perpendicular to it. Instead, for an arbitrary in-plane angle, all possible particle motion in the plate (both contained in the incident plane and normal to it) will be coupled upon reflection from the plate boundaries, and the corresponding equations of motion assume a more complicated form. We have chosen to accommodate this additional coupling in the case of arbitrary propagation direction by transforming the material properties to their effective values referred to the plane of propagation, as mentioned above. This operation is equivalent to rotating the $\bar{c}_{ijk\ell}$ stiffness tensor by applying well-known orthogonal rotation operators which transform the tensor to $c_{ijk\ell}$. The rotation operator about the x_3 axis β_{im} is just the transformation equations of the axes expressed as a second rank tensor that rotate the coordinate system through an angle of positive ϕ,

$$\beta_{im} = \begin{pmatrix} \cos\phi & \sin\phi & 0 \\ -\sin\phi & \cos\phi & 0 \\ 0 & 0 & 1 \end{pmatrix}. \tag{5.3}$$

We need one rotation tensor to transform each rank of the elastic tensor. To transform from $\bar{c}_{mnpq}$ to $c_{ijk\ell}$, we just apply

$$c_{ijk\ell} = \beta_{im}\beta_{jn}\beta_{pk}\beta_{q\ell}\bar{c}_{mnpq}, \tag{5.4}$$

where the choice of subscripts is arbitrary. For orthotropic media, there are also the well-known Bond transformation matrices for operation on the stiffness matrix itself, as discussed in detail at the end of Chapter 1, or also shown in [5]. The result of applying these rotations about the x_3 axis to $\overline{C}_{IJ}$ is

$$C_{IJ} = \begin{pmatrix} C_{11} & C_{12} & C_{13} & 0 & 0 & C_{16} \\ C_{12} & C_{22} & C_{23} & 0 & 0 & C_{26} \\ C_{13} & C_{23} & C_{33} & 0 & 0 & C_{36} \\ 0 & 0 & 0 & C_{44} & C_{45} & 0 \\ 0 & 0 & 0 & C_{45} & C_{55} & 0 \\ C_{16} & C_{26} & C_{36} & 0 & 0 & C_{66} \end{pmatrix}, \tag{5.5}$$

where the new matrix C_{IJ}, including the new nonzero entries, is derived from the original by a transformation of the elements of $\overline{C}_{IJ}$. The circumstance is illustrated schematically in Fig. 5.2.

For the sake of completeness, we treat the more general problem, and then show how to recover the restricted results as special cases. This approach parallels our exposition in the preceding chapter, as well. Due to the high interest in guided waves in composite plates over the past 20 years, this problem has received attention from many workers. Li and Thompson [4] have solved for the modes of an anisotropic plate and calculated a grid of bounds. Mal [6] has addressed this problem from a formal standpoint. Datta *et al.* [7] and Hosten [8] have also analyzed this problem. Nayfeh and Chimenti [9, 10] have developed solutions for plates of monoclinic symmetry in the context of calculations on ultrasonic wave reflection and transmission in composites. Earlier, Jones [12], Kosevich and Syrkin [13], Kulkarni and Pagano [14], Solie and Auld [15], and others [16–25] contributed to aspects of this problem. In our treatment we will follow the general problem statement outlined by Kulkarni and Pagano [14] and later extended by Nayfeh and Chimenti [9] for treatment of wave reflection from a fluid-coupled orthotropic plate.

We proceed from a consideration of the displacements and stresses and their boundary conditions on the plate surfaces. This analysis serves as a precursor to the derivation of the characteristic equations for guided waves in the plate. We begin with a plate in vacuum and later extend the analysis to a plate immersed in liquid. For anisotropic media, we should speak more

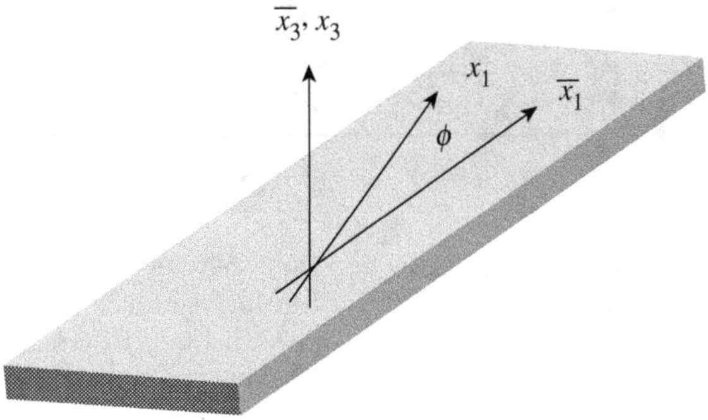

Figure 5.2. Schematic drawing of a uniaxial composite laminate plate showing direction of original $\bar{x}_1$ and transformed x_1 axes, rotated through an angle ϕ; the x_3 and $\bar{x}_3$ axes are unchanged. The fiber direction is $\bar{x}_1$.

properly of guided *plate waves*. Let us reserve Lamb waves to denote the excitations, polarized in the vertical plane, of a traction-free isotropic plate. This convention makes sense, too, since this is exactly the problem Rayleigh [3] and Lamb [26] dealt with. For guided waves in immersed anisotropic plates, we will refer to the radiation-damped versions as *leaky plate waves*. Later, we will use these results to solve for liquid-coupled guided wave modes by analyzing the problem of plane wave reflection from an orthotropic plate immersed in an inviscid liquid.

5.2.2 Plate Wave Solutions

To cast the problem conveniently, first we write the displacements in a form suggestive of the plate wave problem we want to analyze. Referring to Fig. 5.1, our plate lies with its surfaces parallel to the (x_1, x_2) plane, the origin of coordinates at the center, and the x_3 axis is along the plate normal. Now, consider two groups of three partial waves, one quasilongitudinal (QL) and two quasishear (QS) waves ($\overline{x}_1$ is the fiber direction) incident onto both the upper and lower plate surfaces with the wave vectors all lying in the (x_1, x_3) plane, as illustrated in Fig. 5.3. From Snell's law generalized to anisotropic media Eq. (4.14), we expect for lossless solids that the x_1 projection of the wave vector will be equal for all six modes. Following a familiar program, we identify the x_1 partial wave vector component in the plate as ξ_p, the plate wave vector, when ξ_p satisfies the appropriate characteristic equations derived from the boundary conditions.

Analogous to the solution of Christoffel's equation for slowness vector component m_3^α, as we did in Chapter 4 for a single surface, we are looking for the x_3 component of the partial wave vectors whose x_1 component $\xi_p$we

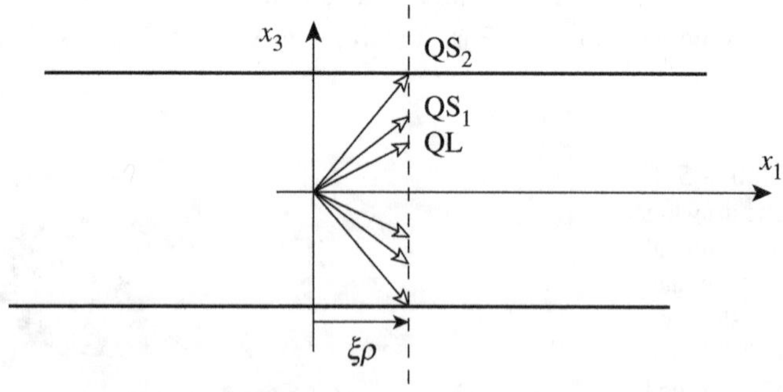

Figure 5.3. Plate geometry showing the quasilongitudinal and quasishear partial waves in the laminate; not all partial waves propagate under all conditions, but they all generally exist, even if only as evanescent waves. The common plate wave vector ξ_p is shown at the bottom.

assume is already determined (and is common to all partial waves). We have seen in Chapter 4 how the partial differential equation of motion Eq. (5.1) can be reduced to an algebraic equation by the assumption of a harmonic plane wave solution. Let us express the postulated solution for the particle displacement vector as

$$\mathbf{u}^k = \mathbf{U}^k \exp[i(\xi_p x_1 + \alpha^k x_3 - \xi_p V_p t)], \tag{5.6}$$

where V_p is the phase velocity of the guided plate wave. The U_i^k are components of the displacement amplitude vectors $\mathbf{U}^k$ of the partial waves; these amplitudes are related to the notation for the polarization vector $\mathbf{p}^k$ used in previous chapters through $\mathbf{U}^k = \mathrm{U}^k\, \mathbf{p}^k$. In Eq. (5.6), the subscript k refers to one of the six partial wave modes, one QL and two QS waves for each of two x_3 wave vector components. The guided wave phase velocity V_p equals the trace velocity V_{tr} in the x_1 direction and can be related to the slowness $\mathbf{m}^k$ of Chapter 4 through $m^k \sin\theta^k = 1/V_p$, where θ^k is the refraction angle on the plate surface of the kth partial wave. Then, $\xi_p = \omega/V_p$ is the wave vector projection onto the x_1 axis, α^k is the x_3 component of the partial wave vector, and the exponential term $\exp(-i\xi_p V_p t)$ is equivalent to the familiar exponential factor $\exp(-i\omega t)$. Substitution of Eq. (5.6) into Eq. (5.1) and using the C_{IJ} stiffness matrix in Eq. (5.5) yields the matrix equation

$$\begin{pmatrix} A + C_{55}\bar{\alpha}_k{}^2 & C_{16} + C_{45}\bar{\alpha}_k{}^2 & (C_{13} + C_{55})\bar{\alpha}_k \\ C_{16} + C_{45}\bar{\alpha}_k{}^2 & B + C_{44}\bar{\alpha}_k{}^2 & (C_{36} + C_{45})\bar{\alpha}_k \\ (C_{13} + C_{55})\bar{\alpha}_k & (C_{36} + C_{45})\bar{\alpha}_k & D + C_{33}\bar{\alpha}_k{}^2 \end{pmatrix} \begin{pmatrix} U_1^k \\ U_2^k \\ U_3^k \end{pmatrix} = \begin{pmatrix} 0 \\ 0 \\ 0 \end{pmatrix}, \tag{5.7}$$

where we expect six solutions for $\bar{\alpha}_k (= \alpha^k/\xi_p)$ from the homogeneous form of the above equation, as we saw in Chapters 1 and 2. The variables A, B, and D are given by

$$\begin{aligned} A &= C_{11} - \rho_s V_p^2 \\ B &= C_{66} - \rho_s V_p^2 \\ D &= C_{55} - \rho_s V_p^2, \end{aligned} \tag{5.8}$$

corresponding to a pure longitudinal and two pure shear modes, where ρ_s is the mass density of the solid plate. The determinant of the coefficient matrix in Eq. (5.7) must vanish, leading to a sixth-degree polynomial equation for $\bar{\alpha}_k$. The determinant of Eq. (5.7) is equivalent to Eq. (4.24); it can be obtained from Eq. (4.24) with proper normalization and simplifications owing to the monoclinic symmetry expressed in Eq. (5.5).

By our choice of material property matrix Eq. (5.2) and the rotation operations on it, we have implied the existence of a mirror symmetry plane coplanar with the plate surfaces (*i. e.*, the fibers are always parallel to the plate surfaces). Therefore, the above relation leads to a sixth-order equation for α^k in which the odd-power terms are absent. There will be, as before, three solutions and their negative values for α^k, each pair corresponding to one of the three possible modes of vibration in the solid anisotropic medium. That is, $\alpha^{(2m)} = -\alpha^{(2m-1)}$, $m = 1, 2, 3$. These solutions relate to three identical pairs of partial waves, half propagating with components oriented along the $+x_3$ direction and half along the $-x_3$ direction. Each pair has the same plate wave vector value because the x_1 component ξ_p for all wave vectors is invariant. Were we to define a material coordinate system for even a uniaxial composite laminate, represented as a homogeneous transversely isotropic solid, in which the $+x_3$ and $-x_3$ directions were not equivalent, then the entire stiffness matrix of Eq. (5.5) would be completely populated with nonzero elements [10], and the resulting characteristic equation for α would yield six distinct solutions. At that point, an analytical solution for wave propagation in the plate would be, practically speaking, out of reach [10].

Solving the resulting system Eq. (5.7) of linear algebraic equations for the displacement amplitudes of each of the six partial waves, and their corresponding α_k, yields the following solutions for the normalized displacement amplitudes

$$\begin{aligned} \overline{U}_1^k &= 1, \qquad k = 1, 2, \ldots 6, \\ \overline{U}_2^k &= \frac{K_{11}K_{23} - K_{12}K_{13}}{K_{13}K_{22} - K_{12}K_{23}} \\ \overline{U}_3^k &= \frac{K_{11}K_{22} - K_{12}^2}{K_{12}K_{23} - K_{13}K_{22}}, \end{aligned} \tag{5.9}$$

where we have normalized each displacement by the value of the x_1 displacement amplitude U_1^k for each of the six modes, $k = 1, 2, \ldots 6$. (The normalized displacement vectors $\bar{\mathbf{U}}^k$ are analogous to the unit polarization vectors $\mathbf{p}^k$.) The factors K_{ij} in Eq. (5.9) are given by

$$\begin{aligned} K_{11} &= A + C_{55}\bar{\alpha}_k{}^2 \\ K_{12} &= C_{16} + C_{45}\bar{\alpha}_k{}^2 \\ K_{13} &= (C_{13} + C_{55})\bar{\alpha}_k \\ K_{22} &= B + C_{44}\bar{\alpha}_k{}^2 \\ K_{23} &= (C_{36} + C_{45})\bar{\alpha}_k \\ K_{33} &= D + C_{33}\bar{\alpha}_k{}^2. \end{aligned} \tag{5.10}$$

Introducing simplifications, $\bar{\alpha}_k$ occurs in Eq. (5.9) for U_2^k only to the second order; therefore, this displacement amplitude has the same sign for propagation in the $+x_3$ or $-x_3$ directions, whereas U_3^k changes sign.

The displacements in Eq. (5.9) are valid for arbitrary in-plane angle ϕ. In the special case that the plate wave vector ξ_p lies along the fiber direction $\bar{x}_1$, the expressions for the displacements simplify considerably. The appropriate stiffness matrix is like the one in Eq. (5.2), and because of the coincidence of ξ_p and a principal material axis, the only constants that enter the calculation are C_{11}, C_{13}, C_{33}, and C_{55}, assuming plane wave excitation. In fact, this assertion is also nearly true for acoustic beams sampled in the incident plane, as we shall see later in the discussion of reflection coefficients. For propagation in the plane containing the fibers, the displacements are

$$
\begin{aligned}
\overline{U}_1^k &= 1, \qquad k = 1, 2, \ldots 4, \\
\overline{U}_2^k &= 0 \\
\overline{U}_3^k &= -\frac{K_{11}}{K_{13}} = -\frac{A + C_{55}\bar{\alpha}_k{}^2}{(C_{13} + C_{55})\bar{\alpha}_k}.
\end{aligned}
\tag{5.11}
$$

where the particle motion normal to the (x_1, x_3) plane vanishes, as expected, since this wave can no longer couple to the SH-like displacements. Physically, the two halves of the x_2 axis are now elastically equivalent, and particle motion in the (x_1, x_3) plane cannot deviate from this plane since the direction of such a deviation would be indeterminate. In this case, the SH wave is an independent excitation of the plate and decouples from the vertically polarized plate waves. For the plate waves polarized in the vertical plane, the $\bar{\alpha}_k$ are given by [11]

$$
\bar{\alpha}_{1,3}^2 = \frac{-S \pm \sqrt{S^2 - 4RT}}{2R} \tag{5.12}
$$

$$
R = C_{33}C_{55}, \qquad S = AC_{33} + DC_{55} - (C_{13} + C_{55})^2, \qquad T = AD,
$$

where the terms A, B, and D are defined in Eq. (5.8). Another implication of symmetry axis propagation is the reduction in number of partial waves. As straight-crested waves imply the absence of particle motion out of the incident plane, the slow pure transverse wave—propagation or particle motion in the x_2 or x_3 directions—cannot be excited for plate waves propagating in either the x_1 or the x_2 directions. In addition to the waves polarized in vertical plane, shear (pure transverse) guided waves can propagate in the x_2 direction with x_1 polarization and in the x_1 direction with x_2 polarization. Finally, if ξ_p is normal to the fibers, the x_1 and x_3 directions are equivalent, and the material is effectively isotropic in this plane.

Then, the displacements reduce to values equivalent to those for isotropic media,

$$\begin{aligned} \overline{U}_1^k &= 1, \quad k = 1, 2, \ldots 4, \\ \overline{U}_3^1 &= \bar{\alpha}_1 \\ \overline{U}_3^3 &= -1/\bar{\alpha}_3, \end{aligned} \tag{5.13}$$

and the $\alpha^{1,3}$ are given by $\alpha^1 = \sqrt{k_\ell^2 - \xi_p^2}$ and $\alpha^3 = \sqrt{k_s^2 - \xi_p^2}$.

Returning to the general case, the program now is to apply these intermediate results to the problem of waves in a plate; in particular, our objective is to find ξ_p the yet unknown propagation wave vector, common to all partial waves. The procedure is to write the boundary conditions for the traction-free anisotropic plate surfaces,

$$\begin{aligned} T_i = \sigma_{ij} n_j = 0, \quad \text{or} \\ \sigma_{33} = \sigma_{13} = \sigma_{23} = 0 \quad (x_3 = \pm h), \end{aligned} \tag{5.14}$$

because the unit vector n_j points along the x_3 axis. The traction vector can be expressed (for each of the k partial waves), without the common time and x_1-dependent harmonic terms and after normalization of the displacement amplitude,

$$\mathbf{T}^k = \begin{pmatrix} T_1^k \\ T_2^k \\ T_3^k \end{pmatrix} = i\xi_p U_1^k e^{i\xi_p \bar{\alpha}_k x_3} \begin{pmatrix} C_{55}\bar{\alpha}_k & C_{45}\bar{\alpha}_k & C_{55} \\ C_{45}\bar{\alpha}_k & C_{44}\bar{\alpha}_k & C_{45} \\ C_{13} & C_{36} & C_{33}\bar{\alpha}_k \end{pmatrix} \begin{pmatrix} 1 \\ \overline{U}_2^k \\ \overline{U}_3^k \end{pmatrix}, \tag{5.15}$$

where summation over the six solutions ($k = 1 \ldots 6$) is implied in each of the three terms, and U_1^k is the amplitude of the x_1 displacement component for each of the k partial waves. Forcing the tractions to zero at the plate surfaces leads to very complicated intermediate algebraic expressions, largely owing to the presence of anisotropy. The substantial additional complexity is also seen in the plate characteristic functions, where three terms appear because all three partial waves—one QL and two QS—are needed to describe the wave motion. As a result, a system of algebraic equations is obtained from the boundary conditions. To have nontrivial solutions for displacement amplitudes $\mathbf{U}^k$, the determinant of the system must be zero, leading to the equations for the wave vector ξ_p, the so-called *dispersion relations*. Owing to symmetry, the determinant can be factored into symmetric and antisymmetric 3×3 sub-determinants. As a result, these sub-determinants lead, with simplifications, to the symmetric and antisymmetric plate-wave dispersion functions for the orthotropic plate in the geometry of Fig. 5.2.

These functions are given by

$$S\xi_p = \mathcal{D}_1 \cot(\alpha^1 h) + \mathcal{D}_3 \cot(\alpha^3 h) + \mathcal{D}_5 \cot(\alpha^5 h) = 0$$

$$A\xi_p = \mathcal{D}_1 \tan(\alpha^1 h) + \mathcal{D}_3 \tan(\alpha^3 h) + \mathcal{D}_5 \tan(\alpha^5 h) = 0, \qquad (5.16)$$

where h is the plate half-thickness.

The $\mathcal{D}_i$ $(i = 1, 3, 5)$ stand for complicated expressions, each one in general dependent on the elastic properties C_{IJ}, the normalized displacements $\overline{U}_j^k$, the partial wave vector components α^k, and the propagation constant ξ_p. The $\mathcal{D}_i(C_{IJ}, \overline{U}_j^k, \alpha^k, \xi_p)$ are calculated from terms in the Laplace expansion on the top row of the following determinant,

$$\begin{vmatrix} D_{11} & D_{13} & D_{15} \\ D_{21} & D_{23} & D_{25} \\ D_{31} & D_{33} & D_{35} \end{vmatrix}. \qquad (5.17)$$

For example, $\mathcal{D}_3 = D_{13}(D_{31}D_{25} - D_{21}D_{35})$. The matrix elements D_{jk} are

$$\begin{pmatrix} D_{1k} \\ D_{2k} \\ D_{3k} \end{pmatrix} = \begin{pmatrix} C_{13} & C_{36} & C_{33}\bar{\alpha}_k \\ C_{55}\bar{\alpha}_k & C_{45}\bar{\alpha}_k & C_{55} \\ C_{45}\bar{\alpha}_k & C_{44}\bar{\alpha}_k & C_{45} \end{pmatrix} \begin{pmatrix} 1 \\ \overline{U}_2^k \\ \overline{U}_3^k \end{pmatrix}, \qquad (5.18)$$

where $\bar{\alpha}_k (= \alpha^k/\xi_p)$ is the normalized x_3 wave vector, and the subscript $k = 1, \ldots, 6$ runs over six partial waves, with the stipulation that because of mirror symmetry in the (x_1, x_2) plane, $\alpha_{2m} = -\alpha_{2m-1}$, $m = 1, 2, 3$. The normalized displacements can be solved in terms of elastic properties and wave vector components using Eqs. (5.9) and (5.10),

$$\overline{U}_2^k = \frac{\left(C_{11} - \rho_s(\omega/\xi_p)^2 + C_{55}\bar{\alpha}_k^2\right)(C_{36} + C_{45})\bar{\alpha}_k - (C_{13} + C_{55})\bar{\alpha}_k\left(C_{16} + C_{45}\bar{\alpha}_k^2\right)}{(C_{13} + C_{55})\bar{\alpha}_k\left(C_{66} - \rho_s(\omega/\xi_p)^2 + C_{44}\bar{\alpha}_k^2\right) - (C_{16} + C_{45}\bar{\alpha}_k^2)(C_{36} + C_{45})\bar{\alpha}_k}$$

$$\overline{U}_3^k = \frac{\left(C_{11} - \rho_s(\omega/\xi_p)^2 + C_{55}\bar{\alpha}_k^2\right)\left(C_{66} - \rho_s(\omega/\xi_p)^2 + C_{44}\bar{\alpha}_k^2\right) - (C_{16} + C_{45}\bar{\alpha}_k^2)^2}{(C_{16} + C_{45}\bar{\alpha}_k^2)(C_{36} + C_{45})\bar{\alpha}_k - (C_{13} + C_{55})\bar{\alpha}_k\left(C_{66} - \rho_s(\omega/\xi_p)^2 + C_{44}\bar{\alpha}_k^2\right)}. \qquad (5.19)$$

Had we not restricted the composite geometry to fibers parallel to the plate surfaces only, with the attendant mirror symmetry, then the concept of distinct and independent symmetric and antisymmetric plate-wave functions would no longer exist. The equations of motion would then be prohibitively complicated [27]. An analytical formulation along the lines of the one developed above can also be undertaken for guided waves in more complicated anisotropic media, namely piezoelectric materials.

In that case, coupling of mechanical and electrical waves will occur, leading to additional degrees of freedom represented by the electrical potentials. This coupling is inherent in the piezoelectric constitutive relations. If in addition, the piezoelectric plate is immersed in a fluid such as water (with a static dielectric constant of 79), then both the mechanical and electrical wave fields will be altered by the presence of the fluid. The complete formulation for a single-layer monoclinic piezoelectric plate immersed in a dielectric fluid has a total of 12 degrees of freedom, accompanied by 12 elastic or electrical boundary conditions [28–30]. The problem of guided waves in fluid-immersed composite plates is taken up in the following section.

The transcendental equations in Eq. (5.16) have as their solutions a discreet set of wave vectors ξ_p for a given frequency and plate thickness with only a finite number of real roots for propagating modes. These equations form a complete description of uniform, straight-crested wave motion in the traction-free anisotropic plate. Their solution set includes both the propagating and nonpropagating modes. For a typical polymer-matrix composite, the propagating plate modes in the fiber direction of

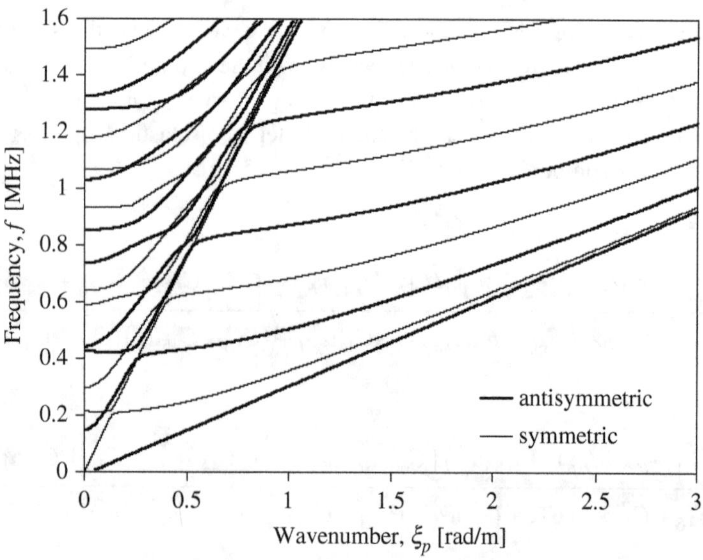

Figure 5.4. Plate-wave dispersion curves for graphite-epoxy composite, plotted as wavevector versus frequency. The thick curves are antisymmetric wave modes, and the thin are symmetric. The strong shift in the curves close to the frequency axis corresponds to the longitudinal critical angle, below which the behavior is simpler because the propagation is dominated by shear waves only. No modes appear at high frequency below the line corresponding to the shear wavespeed.

a uniaxial laminate exhibit dispersion behavior as shown in Fig. 5.4. The calculation assumes that the material is similar to AS4 fiber in 3501 resin at a fiber volume fraction of 60%. A major difference between the behavior of composites and isotropic materials lies in the form of the curves in Fig. 5.4. The rather steep transition to shear partial wave motion in the higher order, or overtone, modes at the longitudinal critical angle is characteristic of materials having a high degree of elastic anisotropy. In Fig. 5.4, this is the concentration of curves along a line from the origin to a frequency of 1.6 MHz at a wavenumber of 1 rad/m. This behavior, which is a consequence of the large anisotropy in the effective longitudinal wavespeeds, $V_{11} = \sqrt{C_{11}/\rho_s}$ and $V_{33} = \sqrt{C_{33}/\rho_s}$, is generally not observed for Lamb waves in isotropic media, where these two speeds are equal. For comparison, Fig. 5.5 shows the same plate-wave modes for a plate of isotropic polycrystalline aluminum. The transition from combined longitudinal and shear partial wave motion to shear partial waves beyond the longitudinal critical angle (below the longitudinal wavespeed) is much more gradual than in the case of the composite. For plate-wave propagation in arbitrary directions in a composite laminate, the mode cutoffs become much more complicated functions of the elastic properties, given by the anisotropic wave velocity surface in the appropriate plane.

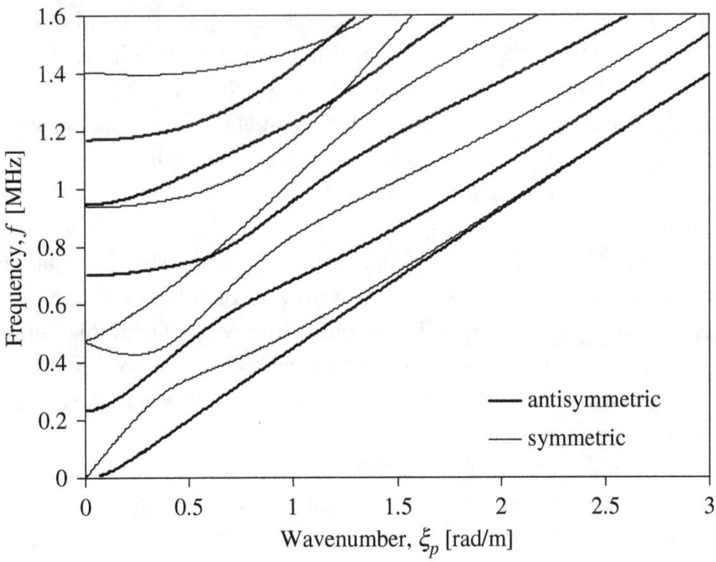

Figure 5.5. Plate-wave dispersion curves for aluminum, plotted as wavevector versus frequency. The thick curves are antisymmetric, and the thin are symmetric wave modes. The appearance of the longitudinal critical angle is much gentler here than for the composite. Below this line, the wave motion is dominated by shear partial waves.

5.3 Leaky Guided Waves in a Fluid-Loaded Plate

This problem of guided wave propagation, when the plate is immersed in a fluid, has a long and distinguished history, at least in part because of its importance to the radiation of sound from the hulls of ships. Cremer [31] suggested the existence of a "coincidence effect," whereby the mechanical impedance of the plate vanishes if the trace velocity of the incident acoustic wave equals the velocity of "free waves" in the plate. Cremer made no distinction, however, between free waves in the presence or absence of the fluid because the fluid he considered was air. When the loading effect of the fluid is negligible, the trace velocity for coincidence is real $V_{\text{tr}} = V_f / \sin\theta$, and the reflection coefficient reaches a minimum. Schoch [32] put the problem on a firmer theoretical footing with his analysis of reflection and transmission in the fluid-coupled isotropic plate. He continued, however, to associate total transmission with the equality of the real acoustic trace velocity and the modes of the fluid-coupled plate, not recognizing that even for small values of the density ratio $\rho = \rho_f / \rho_s$, these two ways of viewing coincidence can be quite different in certain parts of the guided wave spectrum. Schoch did note that for typical values of ρ, the Lamb modes (in vacuum) and the fluid-coupled plate modes would be in close proximity, the latter differing primarily by the existence of a small imaginary part. Schoch's more rigorous viewpoint leads to the idea of a *complex trace velocity* given by $V_{\text{tr}} = \omega / \xi_{p0}$, where the angular frequency ω is real and the plate wave vector ξ_{p0} is complex. When this condition is satisfied, the reflection coefficient vanishes identically, as we shall see next.

Later, Plona *et al.* [33] observed that calculated plate-mode dispersion for the case of ρ near unity leads to behavior rather different from that predicted for Lamb modes in vacuum. In a discussion of the relationship between poles and zeroes of the reflection coefficient for a fluid-coupled plate, Pitts *et al.* [34] showed that the real parts of the two fundamental plate poles (S_0 and A_0 modes) departed, in a narrow range of frequency–thickness product, from their corresponding zeroes. Moreover, beyond this range of fd (where $d = 2h$) the zeroes leave the real axis. The work of Rokhlin *et al.* [35] expands on this point to provide a full exposition of the character of the plate-wave poles under fluid loading.

When a plate immersed in a fluid is subjected to an incident plane wave of ultrasound, the plate responds with both transmitted and reflected wave amplitudes characteristic of the plate and fluid material properties, the angle of incidence of the plane wave, and its wavelength compared to the plate thickness. In an ideal fluid, the sound wave is purely longitudinal in nature, whereas in the plate quasilongitudinal and quasitransverse modes will generally be present. Partial waves will be reflected internally from both plate surfaces and will interfere at the distant plate surface. When that interference yields out-of-plane particle motion identical in amplitude to that at the insonified surface, the plate's influence on the transmitted sound wave intensity is negligible and total transmission results. This circumstance corresponds to the upper and lower plate surfaces moving simultaneously toward and/or away from the plate midline. But these are just the conditions of plate-wave propagation.

This fortuitous circumstance has been exploited many times to deduce the velocity dispersion of Lamb (or more correctly plate) waves by observing the ultrasonic reflected field, or simply the minimum reflected amplitude (corresponding to real trace velocity coincidence) versus angle, from fluid-loaded plates. With the velocity dispersion in hand, it is a task of nonlinear optimization to deduce elastic stiffnesses. For most materials of interest, the occurrence of total transmission and guided wave propagation in the plate are very nearly coincident. If however, the loading of the fluid becomes, in some sense, heavy, then this approximation is no longer useful and substantial deviations can result. The effect of the fluid can be seen in curves of the reflection coefficient minima as functions of incident angle and frequency. Figure 5.6 from a 1986 paper by Chimenti and Nayfeh [37] shows the trace velocity dispersion of leaky plate waves in a uniaxial graphite-epoxy plate

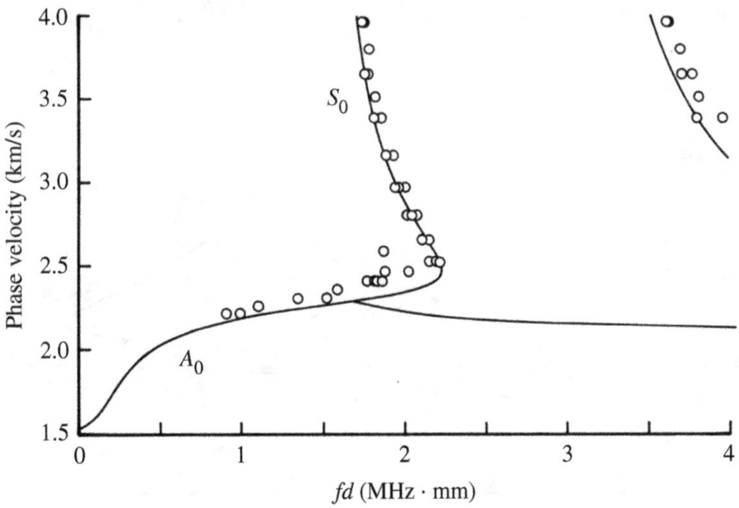

Figure 5.6. Leaky plate-wave trace velocity dispersion curves for uniaxial graphite-epoxy composite, plotted as phase velocity versus frequency–thickness product. The open circles are experimental measurements, and the solid curves are predictions of the reflection coefficient zeroes from an analytical model. The branches labeled A_0 and S_0 have particle displacements that are appropriate for each of these modes. The two branches meet as a single curve near phase velocity $V_p = 2.5$ km/s. From these distorted curves, a Rayleigh wave branch extends to the right (after Chimenti and Nayfeh [37]).

immersed in water. The behavior shown in the A_0 and S_0 modes here is quite different from that expected for negligible fluid loading, such as one would observe for a steel plate in water. Instead, the mode curves have been distorted and joined, yet the model calculation accurately predicts this behavior. What happens when the plate mass density is not much larger than that of the fluid is that the fluid begins to modify the boundary conditions at the plate–fluid interface. Evidence of this effect is seen in Fig. 5.7, where the same situation as in Fig. 5.6 is modeled with varying fluid density.

The suggestion that the fluid-to-plate mass density ratio holds the key to understanding this anomalous finding led Rokhlin *et al.* [35] to re-analyze the conditions that might give rise to this behavior. Their conclusion is that the entire complex wave mode spectrum is distorted as the fluid–solid density ratio grows. Eventually, when the fluid–solid density ratio is quite high, the interfacial boundary conditions effectively change from continuity in tractions and normal displacements to the so-called mixed boundary conditions, where normal interfacial displacements vanish because the fluid has become so "heavy." This situation is illustrated in Fig. 5.8 where we see both the real and imaginary parts of the wave vector plotted as a function of frequency for varying fluid–solid density ratios. From frame (a) to (b) of this figure, there is little change in the S_0 or S_1 modes in the

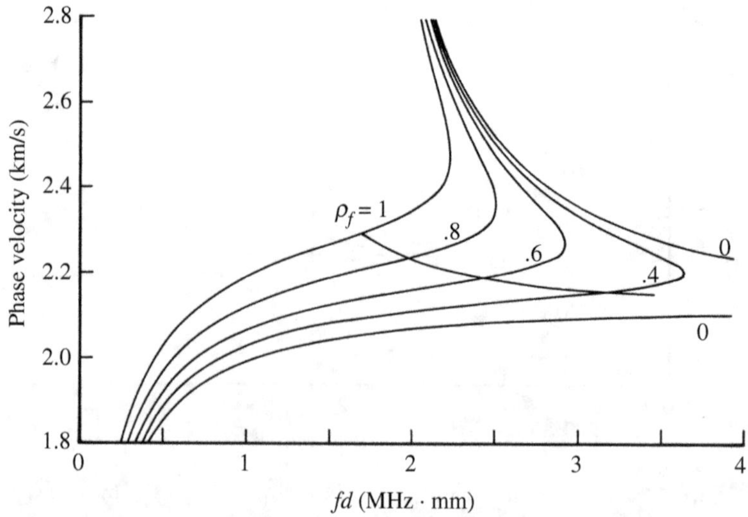

Figure 5.7. Leaky plate-wave trace velocity dispersion curves for a fluid-immersed uniaxial graphite-epoxy composite, plotted as phase velocity versus frequency–thickness product. The fluid density ρ_f (here in units of 10^3 kg/m^3) is allowed to vary from 1.0 to 0.0. The branches of the two fundamental plate wave modes are seen to revert to the behavior expected for negligible fluid loading as the fluid density decreases (after Chimenti and Nayfeh [37]).

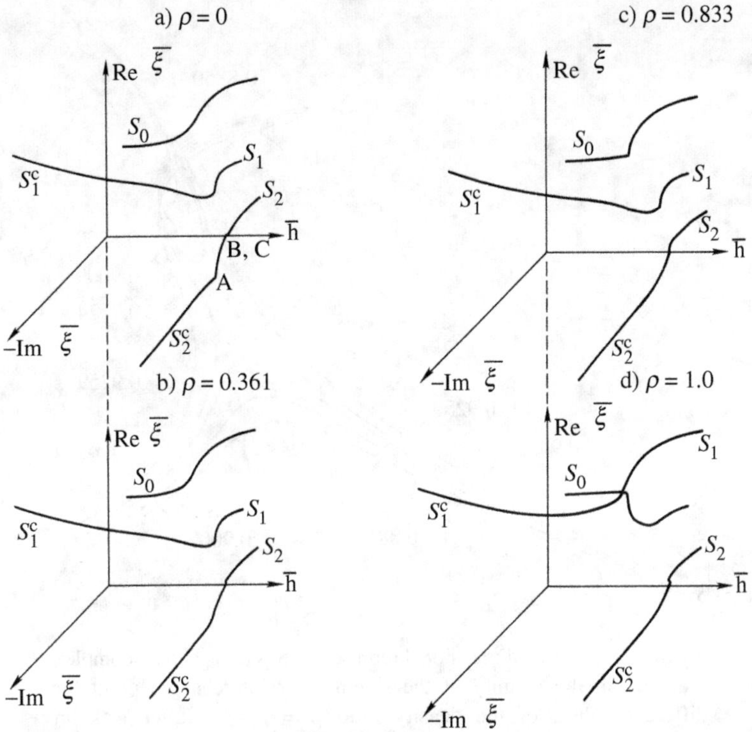

Figure 5.8. Three-dimensional representations of the first three symmetric modes of an isotropic plate, where $\bar{\xi} = \xi_p/\xi_s$ is the plate wavenumber normalized by the shear wavenumber, and $\bar{h} = \omega h/V_s$ is the normalized frequency or normalized thickness. Both the real and imaginary values of $\bar{\xi}$ are plotted in pseudo-3D. The terminology S_1^c denotes a complex mode branch. In frame (a) the fluid–solid density ratio of 0, equivalent to a plate in vacuum. In frame (b) the density ratio ρ is 0.361, or that for aluminum in water. In frames (c) and (d), the fluid density grows compared to that for the solid, and the guided wave spectrum topology is strongly modified (after Rokhlin *et al.* [35]).

two plots. The S_2 mode, is modified near its cut-off frequency from the upper to the lower plot. Frames (c) and (d) of Fig. 5.8, however, shows much stronger modification of the mode structure. In (d) of this series, the liquid is as dense as the solid. Between frames (c) and (d) there is an exchange between the S_0 and S_1 modes. The S_0 mode connects to the high-frequency branch of the S_1 mode, and the complex S_1^c branch is connected to the S_0 mode.

In Fig. 5.9, the critical mode behavior for the S_0 and S_1 modes is calculated just at the transition point. As the density ratio changes from $\rho = 0.9259$ to 0.9174, the mode exchange occurs. This behavior is typical of other mode branch exchanges as the fluid–solid density ratio increases. A more

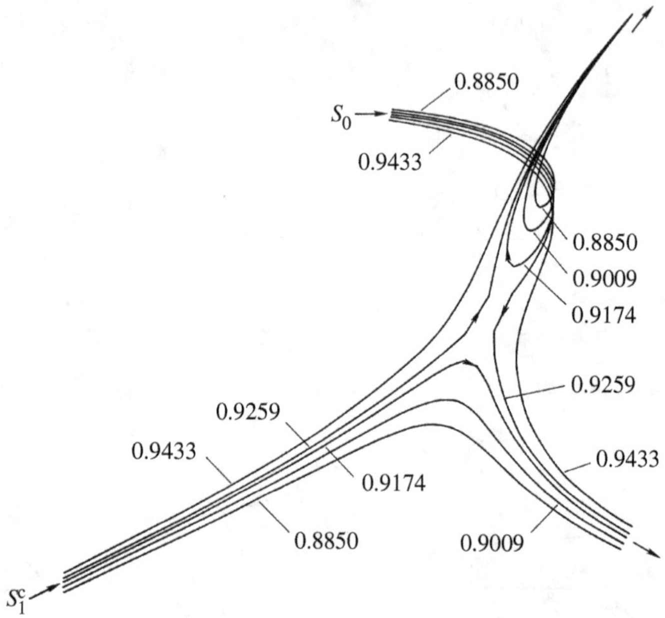

Figure 5.9. Three-dimensional representations of S_0 and S_1 complex branches in the vicinity of the singular (or interchange) point for different values of the density ratio $\rho(=\rho_f/\rho_s)$ (after Rokhlin *et al.* [35]).

detailed version of this exchange behavior between the S_0 and S_1 modes is seen in Fig. 5.10, where the S_0 mode and the complex portion of the S_1 mode, denoted S_1^c, are represented. The S_1^c branch is a portion of the mode spectrum where the imaginary part is not principally connected with wave energy leakage into the fluid. In the absence of a fluid, this branch would still be complex. In this figure, the case of aluminum in water is shown in the dashed curves for a density ratio of $\rho = 0.361$, whereas the heavier fluid loading cases are shown in dotted ($\rho = 0.833$) and solid ($\rho = 1.0$) curves. The vertical axis records the real and imaginary parts of the plate wavevector ratio, normalized by the shear wavevector. The horizontal axis is shear wavevector times plate thickness. Normal behavior of the two modes is seen for a density ratio of $\rho = 0.361$. The dotted curves, for a density ratio of 1.0, still follow generally the paths of the untransformed mode trajectories, but the solid curves for a density ratio of $\rho = 0.833$ demonstrate the complete mode exchange between the S_0 and the complex branch of the S_1^c mode. As the fluid density increases, this mode transformation phenomenon spreads to other higher-order modes as well.

These effects for an isotropic plate at a more advanced stage of transformation can be seen in Fig. 5.11, where the density ratio is $\rho = 2.165$.

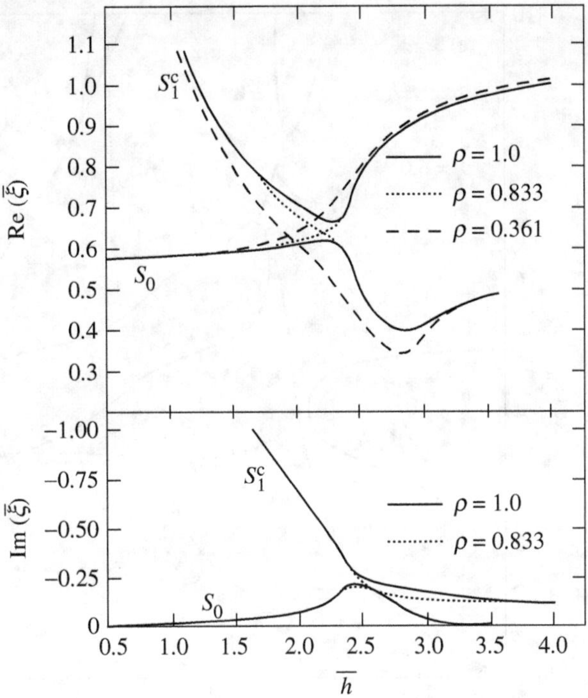

Figure 5.10. The projection of the S_0 and S_1^c branches on the real and imaginary normalized wavenumber planes for different density ratios. Projections onto the real plane clearly demonstrate anomalous behavior of S_0 and S_1 modes for equal densities of fluid and plate (solid line) (after Rokhlin *et al.* [35]).

In this calculation, the real and imaginary parts of the trace velocities V_{tr} for the reflection coefficient zeroes are plotted versus the frequency–thickness product *fd*. Near $V_{tr} = 8$ km/s and $fd = 2.5$ MHz·mm, the upper portion of the originally separate A_1 and S_1-like branches now join to form an open loop *ac* extending to higher trace velocity, while at the same time the lower portions of these two branches also form an open loop *bd* to the right, accompanied by a double-valued complex branch *e* whose real part curves sharply toward zero. The (negative) imaginary part of this region of the spectrum is shown in Fig. 5.11(b). For most branches computed, the imaginary part is zero. The exceptions are the dotted curves segments in Fig. 5.11(a) and their correspondingly labeled branches in the lower frame of the figure. For clarity, the imaginary A_1^c branch, labeled *f*, is omitted in Fig. 5.11(b). From the convergence of the two fundamental branches near $\Re(V_{tr}) = 4$ km/s and $fd = 2.0$ MHz·mm is initiated a complex conjugate pair, denoted 0. The segment denoted *M* in Fig. 5.11(a), while not part of the reflection coefficient zero spectrum, is still a loci of minima on the real *fd* axis. A similar transformation occurs at larger fluid densities in the second-order branches in Fig. 5.11 at the point where the complex branch *f* begins.

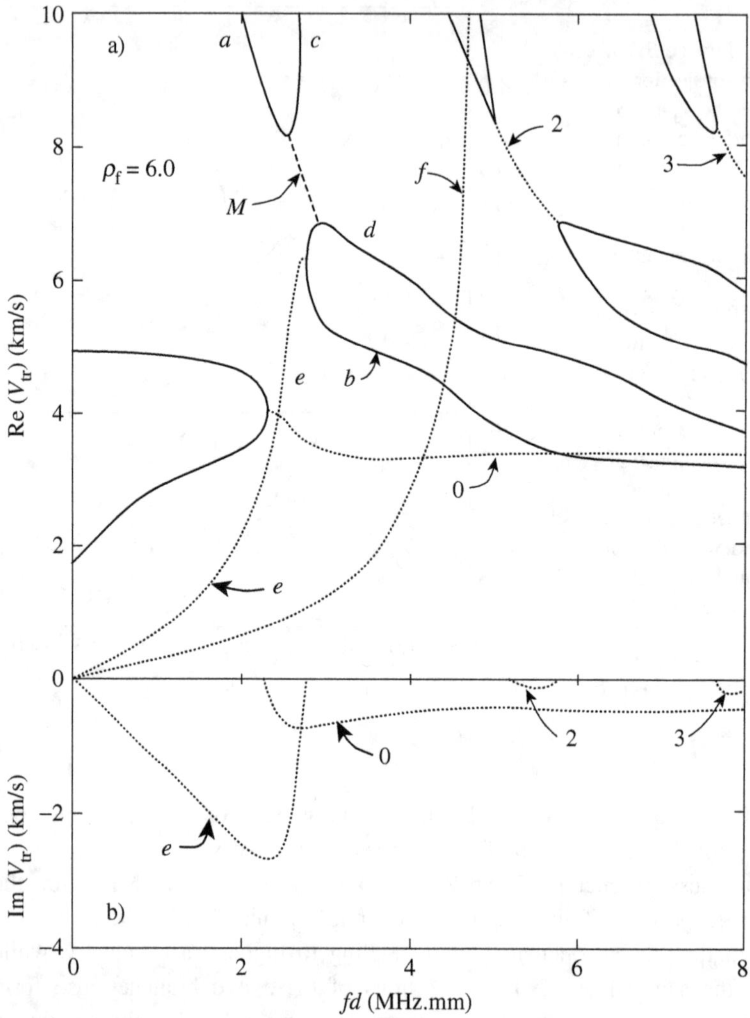

Figure 5.11. The projection of the real and imaginary parts of the trace velocities, corresponding to reflection coefficient zeroes, of guided wave mode branches for a density ratio $\rho(=\rho_f/\rho_s) = 2.165$. The real reflection coefficient zero spectrum is shown in the top portion (a), and the imaginary spectrum in the lower portion (b). The real and imaginary branch labels are explained in the text (after Chimenti and Rokhlin [38]).

As mentioned earlier, all these modifications in the reflection coefficient spectrum occur at much lower fluid density when the solid is a graphite-epoxy composite, where this behavior was first observed [37]. For there to be little or no mode transformation of the zero spectrum of a composite plate in a fluid, the fluid must be very tenuous, as shown in Fig. 5.12. Most of the mode transformation seen in this figure for a composite in water mirrors the

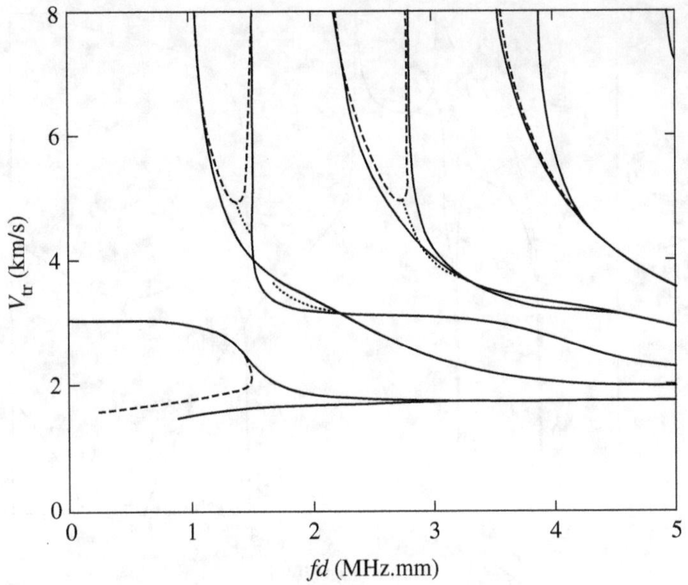

Figure 5.12. Reflection coefficient zero spectrum of guided waves in uniaxial graphite-epoxy for a density ratio $\rho = 0.004$ (solid curves) and in water $\rho = 0.361$ (dashed curves). The plane of the incident wave vector is normal to the fiber axis. The stage of mode transformation for a composite in water is similar to that shown in Fig. 5.11 for aluminum immersed in a fluid of much higher density (after Chimenti and Rokhlin [38]).

calculation of aluminum in a fictitious fluid of density equal to 6000 kg/m^3. The conventional approximation of water as a tenuous fluid appears to be quite valid for a steel plate in water or even an aluminum plate in water, but not at all for a water-immersed composite plate. A graphic illustration of the mode transformation of a composite plate in a fluid of varying density at a single value of frequency times thickness $fd = 1.5$ MHz·mm is shown in Fig. 5.13. Here, the trace velocity V_{tr} is plotted versus the log of the fluid density. Even considerably below the density of water, substantial changes occur in the reflection spectrum, indicating the sensitive nature of this fluid–solid combination. Just below water density, the two fundamental zero branches, A_0 and S_0, coalesce into a complex zero connecting to inverted curves at somewhat higher fluid density. Each of the mode curves is labeled with its symmetry and order at a point where the fluid plays no role in spectral modification. Unusual behavior is seen in the S_1 and S_2 modes. At the value of *fd* chosen for the calculations in this plot, these curves have very high slope (Fig. 5.12) and are therefore extremely sensitive to small variations in fluid density. Their corresponding zeroes do not appear in Fig. 5.13 at all.

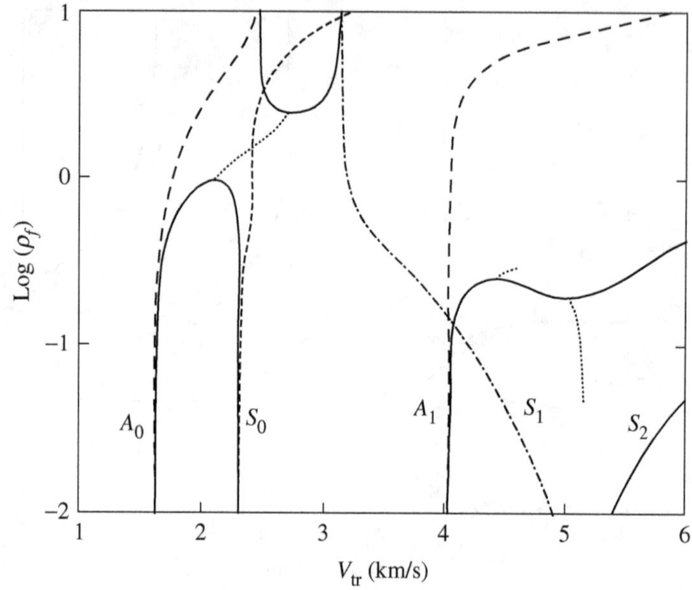

Figure 5.13. Fluid density (in units of 10^3 kg/m^3) expressed as $\log(\rho_f)$ vs V_{tr} for a graphite-epoxy plate with $fd = 1.5$ MHz·mm, showing pole and zero modifications in the first few spectral branches. Considerable deformation in the zero spectrum occurs below the value of fluid density for water, $\log(\rho_f) = 0$. Dotted curves are complex zero branches; real zero branches are solid curves (after Chimenti and Rokhlin [38]).

In the limit of this phenomenon, the density of the fluid becomes high enough to inhibit normal displacements at the plate surface. In that case, the traditional traction-free boundary conditions (or when immersed in a fluid, the stress and displacement continuity conditions) are replaced by a different set of conditions, the so-called *mixed boundary conditions* [1]. When mixed boundary conditions apply, the shear traction, σ_{13} in Fig. 5.1 still vanishes, but not σ_{33}. Instead, the normal displacement u_3 vanishes at the surface. The unusual consequence of this new set of conditions is that the shear and longitudinal partial waves in the (isotropic) plate are no longer coupled at the plate surfaces [1], implying a much simpler guided wave spectrum. In the absence of partial wave coupling at the interfaces, the guided waves (for isotropic media, or orthotropic media along a symmetry axis) decompose into a set of guided shear wave modes and an independent set of guided longitudinal wave modes. These two independent spectra are the limit of guided plate waves for increasing fluid density. It is an accident of nature that much of the transformation of the guided wave modes and corresponding reflection coefficient zeroes is already well underway for the technologically interesting case of a fiber-reinforced plastic plate in water.

5.4 Fluid–Solid Plate Reflection Coefficient

It is instructive to consider the difference between Rayleigh surface waves and Lamb plate waves when seeking to appreciate the relationship between reflection coefficient poles and zeroes. For a surface wave on an ideal solid half-space loaded by an ideal fluid half-space, the magnitude of the plane-wave reflection coefficient beyond the shear critical angle must be unity. The reason for this is that all energy from the incident wave must reappear in the fluid, because no energy is dissipated in the solid, nor can it be carried infinitely far from the interface, according to the Sommerfeld radiation condition, when the partial waves are both evanescent. (If we include material attenuation in the solid, then typically a reduction in amplitude at the Rayleigh wave location is observed in the reflection coefficient; this phenomenon is not significant for a graphite-epoxy composite.) In that case, the Rayleigh wave pole, or zero of the reflection coefficient denominator, will lie above the real axis in the complex wave vector plane, as illustrated in Fig. 5.14(a) and (b) showing the poles and zeros in the complex plane for a half-space and a plate, respectively. It is useful to note that this complex plane representation does not show the branch points; a full complex plane picture for the reflection coefficient would show these added details and can be used for the evaluation of the integral transform solution in ultrasonic beam reflection, but its consideration is beyond our current focus. The positive imaginary part of the reflection coefficient pole insures that the wave energy coupled from the fluid to the surface wave will reappear in the fluid. The surface wave zero, on the other hand, cannot occur at an experimentally accessible location on the real wave vector axis, because of the requirement that the reflection coefficient magnitude be unity beyond the shear critical angle. Instead, the zero lies as far below the real wave vector axis as the

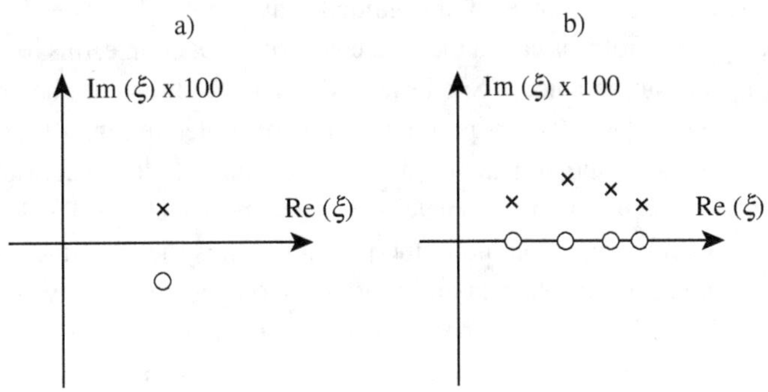

Figure 5.14. Representations of poles (×) and zeroes (o) of the fluid–solid reflection coefficient for a half-space (a) and a plate (b) immersed in an (semi-)infinite, ideal fluid. The vertical scale is expanded for clarity.

pole lies above it. As the incident angle is varied so that the incident wave vector on the real axis passes between the pole and zero of the surface wave reflection coefficient, only the phase of the reflected wave is affected, not its magnitude. By contrast, in Fig. 5.14(b) the situation of a plate of finite thickness immersed in an ideal fluid is illustrated. In this case, the reflection coefficient behaves quite differently. Instead of the unity magnitude required by the half-space, there will be an entire spectrum of zeroes located mostly on the real axis and therefore accessible in an experiment. As we have seen, there can also be complex plate wave zeroes, but this is not the rule. The poles are located above the real wave vector axis, and energy converted to a guided wave will reappear in the fluid, apparently leaking from the guided wave as it propagates. As the coupling to the fluid depends on the nature of the surface displacements and because these will vary with mode and frequency, the poles are illustrated in Fig. 5.14 having different imaginary values. The only other effect that can lift the reflection coefficient zeroes from the real axis are material losses in the plate itself. In most technologically important materials, these material damping effects are either small or negligible at frequencies of interest in material characterization. When this is not the case, a more rigorous analysis must be employed [40]. The coupling of the guided wave to a wave field in the fluid suggests that energy leaks from the guided wave into the fluid. A graphic representation of a leaky guided plate wave is shown in Fig. 5.15. In this photograph, made stroboscopically with a time delay of 20 μs from the excitation toneburst, we see the five or six cycles of the reflected wave and a trailing leaky wave field above the plate. There is also a leaky wave field below the plate, and both leaky wave fields are attached to the A_1 guided wave propagating in the 7-mm thick glass plate. The wavefronts of the guided plate wave are exactly commensurate with the wavefronts of the leaky waves in the water. That is, the advance of the leaky wavefront projected along the glass–water interface is identical to the advance of the guided wavefronts. The speed of the advancing wavefronts is called the "trace velocity." Of course, this behavior is all simply a consequence of Snell's law. Much of the ultrasonic phenomena related to leaky plate waves is graphically demonstrated in Fig. 5.15. Leaky Lamb waves are inhomogeneous waves; their amplitudes decrease with propagation distance along the plate owing to energy leakage. The leakage decreases with propagation along the plate, and thus the amplitude of the wave radiated into the fluid is an inhomogeneous plane wave, where the wave amplitude in a direction perpendicular to the plate surface is actually *increasing*. The reason for this phenomenon is that upper rays (farther from the plate) were radiated at an earlier time, when the guided plate-wave amplitude was higher.

To analyze the plate in a fluid, we repeat the calculation of Section 2 of this chapter, but extend the calculation to a fluid medium located both

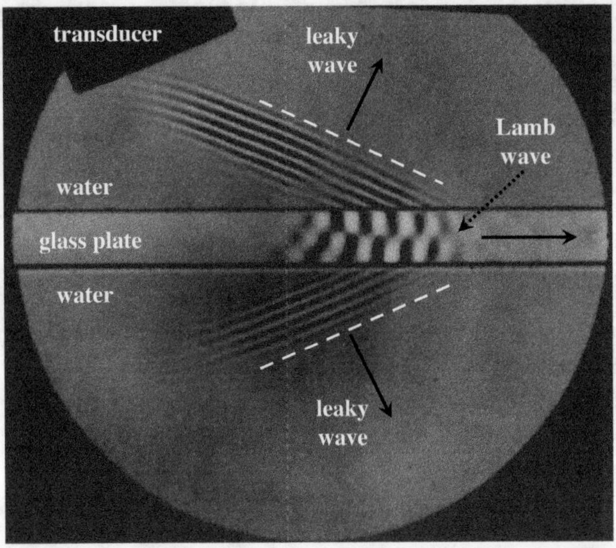

Figure 5.15. Photo-elastic image of an ultrasonic guided leaky wave in a glass plate. The guided wave, as well as the leaky wave field above and below the plate are clearly visible. Note the wavefront correspondence between the guided plate wave and the leaky waves in the water. The frequency is 0.80 MHz and the transducer incident angle is 23.24° (after Courouble and Moufle [39]).

above and below the plate. In the fluid, the corresponding α wave vector component will be given very simply by a geometrical expression similar to the one for isotropic media,

$$\alpha_f = (\omega^2/V_f^2 - \xi_p^2)^{\frac{1}{2}}. \tag{5.20}$$

The quantity V_f is the compressional wavespeed in an ideal, lossless fluid, $V_f = \sqrt{\lambda_f/\rho_f}$. For the geometry of Fig. 5.1, but with a fluid above and below the plate, the boundary conditions on the plate change from the traction-free conditions of Eq. (5.14) to include the continuity of normal tractions and displacements just inside ($^<$) the upper and lower plate surfaces ($x_3 = \pm h$) and just outside ($^>$) these surfaces, but inside the fluid ($x_3 = \pm h$) and the vanishing of shear tractions at the upper and lower plate surfaces,

$$\begin{gathered} \sigma_{33}^<(x_3 = \pm h) = \sigma_{33}^>(x_3 = \pm h) \\ u_3^<(x_3 = \pm h) = u_3^>(x_3 = \pm h), \quad \sigma_{k3}^<(x_3 = \pm h) = 0, \quad k = 1,\ 2, \end{gathered} \tag{5.21}$$

where these conditions apply at $x_3 = \pm h$, the half-thickness.

Using the solutions of Eq. (5.16) and wave solutions in the fluid media, and applying the boundary conditions of Eq. (5.21), leads to a linear system of equations for the particle displacements, in this case of dimension 8 × 8. Solving for the reflected and transmitted complex wave amplitudes in the fluid medium gives, after much algebra, the reflection and transmission coefficients,

$$R = \frac{AS - Y^2}{(S - iY)(A + iY)}$$
$$T = \frac{iY(S + A)}{(S - iY)(A + iY)}, \tag{5.22}$$

where the terms S and A denote, as before, the symmetric and antisymmetric characteristic functions for the traction-free anisotropic plate.

The new term here, denoted by Y, is essentially real and contains the influence of the fluid,

$$Y = \left(\frac{\mathcal{D}_1}{D_{11}} U_{31} + \frac{\mathcal{D}_3}{D_{13}} U_{33} + \frac{\mathcal{D}_5}{D_{15}} U_{35} \right) \frac{\rho_f V_p^2}{\bar{\alpha}_f}, \tag{5.23}$$

where ρ_f is the fluid density, $\bar{\alpha}_f (= \alpha_f / \xi_p)$ is the normalized fluid wave vector component in the x_3 direction, and V_p is the phase velocity of the guided wave in the plate, which corresponds to the trace velocity V_{tr} discussed earlier in this chapter.

The reflection and transmission coefficients in Eq. (5.22) reveal some interesting aspects of waves in anisotropic and fluid-loaded plates. In the previous section, we considered the plate wave behavior *without* the fluid. The normal modes of the traction-free plate are found, as we saw earlier, by setting the symmetric and antisymmetric characteristic functions to zero, $S = 0, A = 0$.

The normal modes of the fluid-loaded plate are found by postulating the existence of a wave in the absence of an excitation. The scattering coefficients in Eq. (5.22) are just the ratio of the respective reflected or transmitted components to the incident wave. If we assume an exciting wave amplitude of zero, then, by implication, the scattering coefficients will be unbounded. For that to happen, the denominators of Eq. (5.22) must vanish. Since there are two multiplicative factors to consider, each may vanish separately, giving the characteristic equations for symmetric and antisymmetric waves in the anisotropic fluid-loaded plate,

$$\mathcal{D}_1 \cot(\alpha^1 h) + \mathcal{D}_3 \cot(\alpha^3 h) + \mathcal{D}_5 \cot(\alpha^5 h) = iY$$
$$\mathcal{D}_1 \tan(\alpha^1 h) + \mathcal{D}_3 \tan(\alpha^3 h) + \mathcal{D}_5 \tan(\alpha^5 h) = -iY. \tag{5.24}$$

Together, solutions of these equations describe the complete normal mode spectrum of waves in the fluid-loaded anisotropic plate. The fluid term iY perturbs the traction-free plate wave spectrum, lifting nearly all roots of the above equations off the real wave vector axis, as illustrated in Fig. 5.14.

Only on those branches for which the phase velocity lies below the fluid wavespeed, will the roots be entirely real. Complex roots imply that the particle displacements in the plate can be expressed as

$$\mathbf{u}^k = \mathbf{U}^k \exp[i(\{\xi_p' + i\xi_p''\}x_1 + \alpha^k x_3)], \tag{5.25}$$

where the prime and double prime refer to the real and imaginary parts of ξ_p, and the time dependence has been suppressed. This wave will be attenuated as it propagates. The fluid loads the plate inertially for plate wavespeeds lower than V_f and resistively for $V > V_f$. In the latter case, the plate wave experiences radiation damping by the fluid, and the plate wave energy is mode converted to an acoustic wave. The plate wave essentially "leaks" its energy to the fluid, giving rise to the colorful terminology for this excitation—*leaky waves*. The term ξ_p will be complex for nearly all plate modes of most composites, except possibly for portions of the lowest-order antisymmetric branch.

Beyond the vanishing of the denominator, the occurrence of a zero in the numerator of the scattering coefficient, in particular $R = 0$ in Eq. (5.22),

$$AS - Y^2 = 0, \tag{5.26}$$

also has interesting consequences. The solution set of this equation is the spectrum of reflection coefficient zeroes. If the fluid-to-solid density ratio ρ in Eq. (5.23) is small compared to unity, then following Schoch [32], the reflection coefficient zeroes will lie close to the real parts of the plate-wave modes ($AS \approx 0$), since Y (the fluid term) appears squared in Eq. (5.26). Schoch ignored the small differences introduced by the fluid term and instead took the real parts of the roots of Eq. (5.26) to be coincident with the guided wave modes. This approximation is generally acceptable for most materials of higher density, such as metals in water, and has been widely used to characterize plate waves in metallic plates [36]. In low-density materials, such as polymers [33] or polymer composites [37], however, the approximation can be substantially inaccurate, as we have seen earlier in Figs. 5.6 and 5.7.

5.5 Waves in Composite Rods

The problem of guided waves in rods has been amply addressed in the literature on elastic waves [41–43] for both isotropic and transversely isotropic media, with the symmetry axis along the fiber. We follow the treatment of Nagy [44] and Nagy and Kent [45] who have extended the analysis to account for immersion in an ideal fluid. Of course, there are few composite structures in the shape of rods, and those that exist may have a material symmetry more complicated than transverse isotropy.

The importance of this analysis lies in its application to the elastic characterization of individual fibers. Measuring the axial Young's modulus for almost any fiber is a straightforward process. Measuring Poisson's ratio, on the other hand, is increasingly difficult and fraught with uncertainty as the fiber diameter decreases. In the case of boron fibers or silicon carbide fibers, measuring transverse strain as the fiber is loaded axially may yield marginally useable results. For fibers as small as S-glass or graphite, however, it is essentially impossible to achieve reliable measurements through a direct observation of the change in fiber diameter with axial loading.

Acoustic waves in thin rods could possibly assist in this task. For thin rods, where the wavelength is much greater than the fiber diameter, only the lowest-order modes will propagate. Of these, the axisymmetric longitudinal mode has the approximate phase velocity

$$V_p \approx V_0[1 - \frac{1}{4}\nu^2(\zeta a)^2 + O(\zeta a)^4], \tag{5.27}$$

where ζ is the guided wavenumber in the z direction, a is the rod radius, ν is the Poisson ratio, and V_0 is the low-frequency asymptotic value of the phase velocity ($V_0 = \sqrt{E_{\text{ax}}/\rho}$). (Here, E_{ax} is the axial Young's modulus.) Of the other two possible modes, one is independent of Poisson's ratio, and the dependence of the other is ambiguous. Practically speaking, it is advantageous to measure the group velocity of the axisymmetric guided rod wave, instead of the phase velocity. Even then, the variation of the group velocity with Poisson's ratio over a range of frequency can be quite small. So, although there is a functional dependence, it can be quite challenging to use this variation to obtain an accurate estimate of Poisson's ratio. The only practical tool remaining in the assessment of Poisson's ratio for thin rods is the leaky wave attenuation.

Before we proceed to a discussion of that question, let us review the mechanics of guided waves in transversely isotropic free rods with the symmetry axis oriented along the rod. The rod is positioned so that its axis is in the z direction in the cylindrical coordinate system (r, θ, z). The stress–displacement relations for this geometry are more complicated than for the Cartesian case,

$$\begin{aligned}
\sigma_{rr} &= C_{11}\frac{\partial u_r}{\partial r} + C_{12}\frac{u_r}{r} + C_{13}\frac{\partial u_z}{\partial z} \\
\sigma_{\theta\theta} &= C_{12}\frac{\partial u_r}{\partial r} + C_{11}\frac{u_r}{r} + C_{13}\frac{\partial u_z}{\partial z} \\
\sigma_{zz} &= C_{13}\frac{1}{r}\frac{\partial (ru_r)}{\partial r} + C_{33}\frac{\partial u_z}{\partial z}
\end{aligned} \tag{5.28}$$

$$\sigma_{rz} = C_{44}\left(\frac{\partial u_r}{\partial z} + \frac{\partial u_z}{\partial r}\right),$$

where u_r and u_z are the radial and axial displacements, and C_{IJ} are the elastic stiffness matrix elements. Putting these values into the equations of motion for radial and axial wave motion yields wave equations in terms of displacements,

$$C_{11}\left[\frac{\partial^2 u_r}{\partial r^2} + \frac{1}{r}\frac{\partial u_r}{\partial r} - \frac{u_r}{r^2}\right] + (C_{13}+C_{44})\frac{\partial^2 u_z}{\partial r \partial z} + C_{44}\frac{\partial^2 u_r}{\partial z^2} = \rho\frac{\partial^2 u_r}{\partial t^2}$$

$$C_{13}\frac{\partial}{\partial z}\left(\frac{1}{r}\frac{\partial(r u_r)}{\partial r}\right) + C_{33}\frac{\partial^2 u_z}{\partial z^2} + C_{44}\frac{1}{r}\frac{\partial}{\partial r}\left[r\left(\frac{\partial u_r}{\partial z} + \frac{\partial u_z}{\partial r}\right)\right] = \rho\frac{\partial^2 u_z}{\partial t^2}. \tag{5.29}$$

The cylindrical symmetry of these equations suggests solutions in the form of Bessel functions,

$$\begin{aligned} u_r &= AJ_1(\gamma r)e^{i[\zeta z - \omega t]} \\ u_z &= BJ_0(\gamma r)e^{i[\zeta z - \omega t]}, \end{aligned} \tag{5.30}$$

where the J_0 and J_1 are zeroth- and first-order Bessel functions, ζ and γ are the axial and radial wavenumbers, ω is the angular frequency, and the coefficients A and B are unknown complex amplitudes. Inserting these solutions into the wave equations leads to Christoffel's equation for the coefficients A and B. As the system is homogeneous, the vanishing of the determinant of the Christoffel matrix yields a secular equation for the radial wavenumbers. From these, follow the ratios of the unknown coefficients A and B. These ratios can be expressed as

$$(B/A)_{1,2} = -\frac{\gamma_{1,2}^2 C_{11} + C_{44}\zeta^2 - \rho\omega^2}{i\zeta\gamma_{1,2}(C_{13}+C_{44})}, \tag{5.31}$$

where the two radial solutions from Christoffel's equation are $\gamma_{1,2}$. These two solutions can be employed to find radial and axial displacements

$$\begin{aligned} u_r &= [A_1 J_1(\gamma_1 r) + A_2 J_1(\gamma_2 r)]e^{i(\zeta z - \omega t)} \\ u_z &= [B_1 J_0(\gamma_1 r) + B_2 J_0(\gamma_2 r)]e^{i(\zeta z - \omega t)}. \end{aligned} \tag{5.32}$$

The traction-free boundary conditions require that both normal and transverse stresses vanish at the rod circumference $r = a$,

$$\begin{bmatrix} \sigma_{rr}(a) \\ \sigma_{rz}(a) \end{bmatrix} = \begin{bmatrix} d_1 & d_2 \\ e_1 & e_2 \end{bmatrix}\begin{bmatrix} A_1 \\ A_2 \end{bmatrix} = \begin{bmatrix} 0 \\ 0 \end{bmatrix}, \tag{5.33}$$

where

$$d_{1,2} = (C_{11}\gamma_{1,2} + C_{13}i\zeta(B/A)_{1,2}J_0(\gamma_{1,2}a) - (C_{11} - C_{12})\frac{J_1(\gamma_{1,2}a)}{a} \tag{5.34}$$

and

$$e_{1,2} = C_{44}(i\zeta - (B/A)_{1,2}\gamma_{1,2})J_1(\gamma_{1,2}a). \tag{5.35}$$

The vanishing of the determinant of Eq. (5.33)

$$d_1e_2 - e_1d_2 = 0 \tag{5.36}$$

forms the characteristic equation of the system, and its roots in ζ are the guided wave solutions of the transversely isotropic rod in vacuum.

Adding the influence of the fluid requires changing the boundary condition on normal stress to $\sigma_{rr}(a) = -p(a)$, so that it equals the pressure in the fluid at the rod circumference. With some additional algebra, the new boundary conditions become

$$\begin{bmatrix} \sigma_{rr}(a) + p(a) \\ \sigma_{rz}(a) \end{bmatrix} = \begin{bmatrix} d_1 + f_1 & d_2 + f_2 \\ e_1 & e_2 \end{bmatrix} \begin{bmatrix} A_1 \\ A_2 \end{bmatrix} = \begin{bmatrix} 0 \\ 0 \end{bmatrix}. \tag{5.37}$$

Here,

$$f_{1,2} = -\frac{\rho_f\omega^2}{\gamma_f}\frac{H_0(\gamma_f a)}{H_1(\gamma_f a)}J_1(\gamma_{1,2}a), \tag{5.38}$$

where ρ_f is the fluid density, $\gamma_f^2 = (\omega/V_f)^2 - \zeta^2$ is the wavenumber in the fluid, and H_0 and H_1 are zeroth- and first-order Hankel functions. The roots of

$$(d_1 + f_1)e_2 - e_1(d_2 + f_2) = 0 \tag{5.39}$$

are the solutions in ζ for the guided wave modes of the fluid-loaded rod. The loss of wave energy in the guided wave mode caused by radiation into the fluid is critical to the solution of the Poisson ratio determination problem in thin rods. As the axial wave motion at the rod–fluid interface does not couple energy into an ideal fluid, the entire effect is dependent on the radial motion, exactly the component related to the Poisson ratio.

For the fundamental axisymmetric longitudinal mode in the thin rod, the ratio of the maximum radial to axial displacements has been shown [45] to be proportional to $|u_r|/|u_z| \propto \nu a\omega$, where ν is the Poisson ratio, a is the rod radius, and ω is the frequency. The attenuation per wavelength caused by the leaky wave is proportional to the ratio between the energy lost per wavelength P_ℓ and the energy transmitted down the guide in one period P_{trans}. As $P_\ell \propto \rho_f |u_r|^2$ and $P_{\text{trans}} \propto \rho_s |u_z|^2$, the attenuation α can be expressed as

$$\alpha \propto \frac{(P_\ell / P_{\text{trans}})}{\lambda} \propto \rho a^2 \omega^3 \nu^2. \tag{5.40}$$

A more rigorous asymptotic analysis leads to a leaky attenuation of

$$\alpha \approx \frac{\pi \rho a^2 \omega^3 \nu^2}{2V_0^3}, \tag{5.41}$$

which has the same functional dependence as Eq. (5.40). The crucial dependence here is on the square of the Poisson ratio. By measuring the leaky wave attenuation along the rod (or fiber), it is possible to estimate the Poisson ratio much more accurately than a direct measurement of the transverse strain could ever hope to achieve. The approximation in Eq. (5.41) is valid only at very low frequency because of the small argument asymptotic behavior of Bessel–Hankel functions. To overcome this limitation, and extend the range of applicability of the formula to frequencies of experimental interest, requires a more careful numerical analysis, including first-order expansion terms for the Bessel functions. The result [45] yields

$$\alpha \approx \frac{\rho a^2 \omega^3 \nu^2}{V_0^3} \left(\frac{\pi}{2} - \frac{0.82 a \omega \sqrt{V_0^2 - V_f^2}}{V_0 V_f} \right), \tag{5.42}$$

where V_f is the wavespeed in the fluid. A measurement of leaky wave attenuation in a 28-cm long copper wire 0.5 mm in diameter is shown in Fig. 5.16. The attenuation has been measured over nearly 1.5 MHz, where its value changes from nearly zero to over 120 dB/m. This large range permits an excellent best fit of the data to a corresponding theory curve. On the figure are plotted three predictions, one for a Poisson ratio of $\nu = 0.4$ and two bracketing values of $\nu = (0.38, 0.42)$. The experimental data are the solid curve. The best fit to the data indicates a value for the axial Poisson ration of about 0.40, which is quite a bit higher than the bulk value for pure copper ($\nu_{\text{bulk}} = 0.34$). The extrusion process by which the wire is produced, however, has been shown to lead to significant anisotropy induced by crystalline texture. Larger diameter copper wires exhibit axial Poisson ratio measurements much closer to the bulk value. One additional possible concern is the contribution to the total wave attenuation from the viscous nature of the fluid, in this case water. The kinematic viscosity of water is relatively low, so in the frequency range represented by Fig. 5.16, the additional damping owing to the viscous liquid amounts to no more than about 2 dB/m, a small fraction of the leaky wave loss. As the rod or fiber becomes smaller, however, the viscous contribution grows in importance. Extracting the rod, or fiber, from the water bath just prior to the measurement

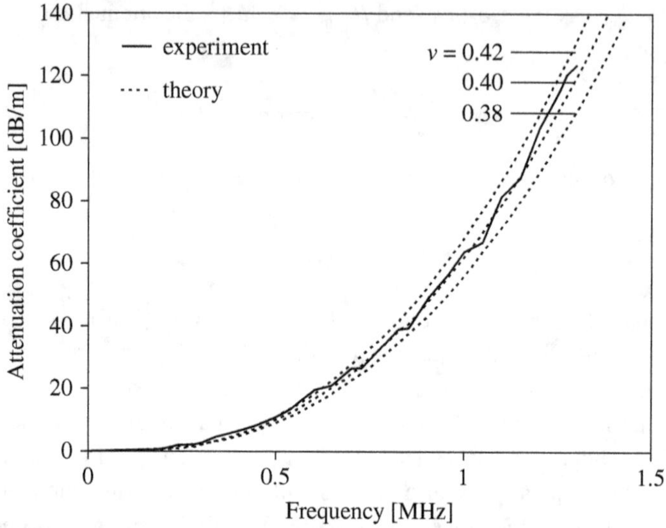

Figure 5.16. Measured and predicted leaky wave attenuation in a long copper wire 0.5 mm in diameter, plotted as a function of frequency. The three dashed predictions are for three different values of Poisson ratio ν, as recorded in the legend (after Nagy and Kent [45]).

leaves a thin liquid layer clinging to the fiber. Measurement on the wet but otherwise free fiber allows the measurement of the viscous loss combined with the intrinsic loss of the fiber itself. Then comparing the loss in the immersed fiber to this case allows the separate measurement of the leaky wave loss effect only.

The lowest-order dilatational modes in thin plates and rods are physically very much alike. In spite of their similar vibration patterns, the effect of fluid-loading on the phase velocity of the dilatational modes is quite different in thin plates and rods. At low frequencies, fluid-loading slightly increases the phase velocity of the fundamental symmetric Lamb mode propagating in a thin plate, while it decreases that of the fundamental axisymmetric mode of a thin rod. Although the magnitude of the relative change in the phase velocity due to fluid-loading is fairly small compared to the more significant attenuation effect in both cases, the very fact that the sign of the velocity change is opposite in plates and rods deserves some attention. The opposite velocity effect of fluid-loading on immersed rods and plates is caused by the different nature of radiation loading [46]. At low frequencies, the vibrating rod possesses a complex radiation impedance with a dominating reactive part in contrast to the essentially resistive radiation impedance of a plate. These conclusions were also confirmed experimentally by high-precision velocity measurements [47].

For a thin plate, the fluid-loading induced velocity change of the fundamental dilatational mode is

$$\frac{V_p - V_{p0}}{V_{p0}} \approx \frac{1}{32}\left[\frac{\rho d \nu \omega V_f}{V_s^2}\right]^2, \tag{5.43}$$

where V_p and $V_{p0}(= V_s\sqrt{2/(1-\nu)})$ are the phase velocities of the fundamental dilatational guided modes in the fluid-loaded and free plates, respectively, ρ is the density ratio between the fluid and the solid, d is the thickness of the plate, ν is Poisson's ratio, ω is the angular frequency, V_f is the sound velocity in the fluid, and V_s is the shear velocity in the solid. Similarly, for a thin rod, the fluid-loading induced velocity change of the fundamental dilatational mode is

$$\frac{V_r - V_{r0}}{V_{r0}} \approx \frac{\rho a^2 \omega^2 \nu^2}{2V_s^2(1+\nu)} \ln(a\alpha_f), \tag{5.44}$$

where V_r and $V_{r0}(= V_s\sqrt{2(1+\nu)})$ are the phase velocities of the fundamental dilatational guided modes in the fluid-loaded and free rods, respectively, a is the rod radius, and α_f is the radial wave vector component previously given in Eq. (5.20). At low frequencies, $a\alpha_f$ is less than unity, therefore the logarithmic function in the correction term for the rod makes it negative while the correction term for the plate is always positive.

It is relatively straightforward to extend the analytical treatment of guided wave propagation in rods and multilayered coaxial fibers immersed in a viscous fluid [48] or embedded in a solid matrix [49,50]. Viscous fluids can be modeled as hypothetical isotropic solids having rigidity $C_{44} = -i\omega\eta$, where η denotes the viscosity of the fluid, *i.e.*, the vorticity mode associated with the viscosity of the fluid is formally described as the shear mode in a hypothetical solid. Viscosity-free fluid-loading exerts its influence on guided wave propagation solely through the normal component of the surface vibration. In contrast, viscous fluid-loading couples to the solid waveguide mainly through tangential surface vibrations. Therefore, there exist sharp minima in the viscosity-induced attenuation of guided modes at particular frequencies when the generally elliptical polarization of the surface vibration becomes linearly polarized in the radial direction. The relative role of viscous drag on the fundamental dilatation mode increases over leaky effects at low frequencies where the surface motion becomes dominantly tangential. In a thin immersed rod, the viscosity-induced attenuation can be calculated as follows

$$\alpha_{vr} = \frac{1}{a}\sqrt{\frac{\omega\eta\rho}{2E}}, \tag{5.45}$$

where E denotes the axial Young's modulus [48]. In contrast to the leaky attenuation given in Eq. (5.42), the viscous attenuation increases rather than

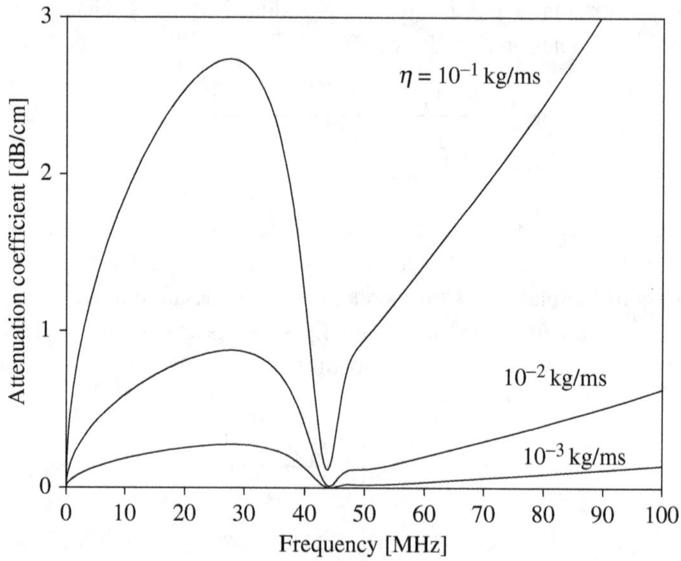

Figure 5.17. Viscosity induced attenuation of a 150-μm-diameter SiC fiber immersed in water-like fluids of different viscosities as a function of frequency (after Nagy and Nayfeh [48]).

decreases with decreasing fiber radius, therefore it becomes more important in thinner fibers.

As an example, Fig. 5.17 shows the viscosity-induced attenuation of the fundamental dilatational wave mode in a 150-μm-diameter SiC fiber immersed in water-like fluids of different hypothetical viscosities as a function of frequency. The lowest curve corresponds to ordinary water of $\eta = 10^{-3}$ kg/ms, while the other two fluids have water-like density and compressibility, but increased viscosity as indicated. These results illustrate that the location of the attenuation minimum is not affected by the viscosity of the fluid as it is associated with the changing vibration pattern of the lowest-order dilatational mode in that particular frequency region. At very low frequencies, the surface vibration of the fundamental dilatational mode is essentially linearly polarized in the axial direction, *i.e.*, there is but a negligible normal surface vibration component with respect to the tangential one. As the frequency increases, the Poisson effect produces more and more radial motion and the surface vibration becomes elliptically polarized with clockwise rotation. At very high frequencies, the same mode asymptotically approaches a Rayleigh-type surface mode that also exhibits elliptical surface vibration trajectory but with counterclockwise rotation. Obviously, there must be a point in the transition region where the surface vibration polarization becomes purely linearly polarized but this time in the

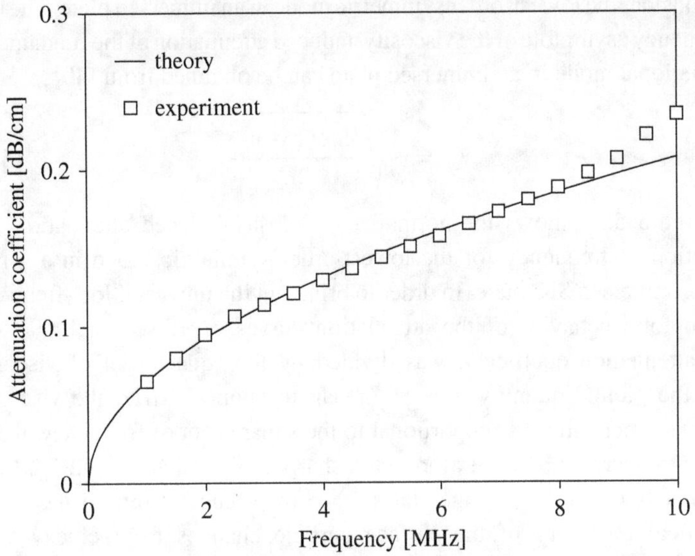

Figure 5.18. Measured and calculated viscosity-induced attenuations of the lowest-order dilatational mode in an SCS-6 fiber (after Nagy and Nayfeh [48]).

radial direction. At that particular frequency, there is only normal surface vibration as the tangential component completely vanishes.

Figure 5.18 shows the measured and calculated viscosity-induced attenuations of the lowest-order dilatational mode in an 80-cm-long SCS-6 fiber up to 10 MHz. This fiber is often used as a reinforcement in metal matrix composite materials. It is composed of a carbon core, a relatively thick SiC cladding, and a very thin carbon-rich coating layer, and its outside diameter is approximately 150 μm. The theoretical curve was obtained by matching the experimental data via varying the viscosity of the fluid. The best-fitting parameter was found to be $\eta = 1.23 \times 10^{-3}$ kg/ms, slightly higher than the tabulated viscosity ($\eta = 10^{-3}$ kg/ms) of water at room temperature. This discrepancy could be caused by the Photo-Flo 200 solution (made by the Eastman Kodak Company) which had to be added to the water to assure that the fluid wets the surface of the slightly hydrophobic fiber. As expected, the measured attenuation is proportional to the square root of frequency over most of the frequency range.

The same analytical approach of modeling viscous fluids as hypothetical solids having a purely imaginary rigidity $C_{44} = -i\omega\eta$ allows us to determine numerically the frequency-dependent velocity and attenuation of Lamb waves propagating in plates immersed in a viscous fluid. As an example, let us consider the lowest-order symmetric mode in an immersed plate. The low-frequency asymptote of the viscosity-induced attenuation of the fundamental dilatational mode in an immersed plate can be obtained from [40]

$$\alpha_{vp} = \frac{1}{d}\sqrt{\frac{\omega\eta\rho(1-\nu^2)}{2E}}. \tag{5.46}$$

Figure 5.19 shows the normalized viscosity-induced attenuation as a function of frequency for the lowest-order symmetric mode in a 1-mm-thick immersed SiC plate. In order to bring out the universal low-frequency asymptotic behavior of the attenuation curves expressed in Eq. (5.46), the attenuation coefficient was divided by the square root of viscosity, *i.e.*, the plotted quantity is $\alpha_{vp}/\sqrt{\eta}$. Up to about 2 MHz, the viscosity-induced attenuation is proportional to the square root of frequency, then it starts to decrease and goes through a sharp minimum at around 5.2 MHz. At that frequency the tangential surface displacement diminishes as the elliptical trajectory of the surface particle changes from clockwise to counterclockwise. Like in the previously considered case of a rod, the position of this minimum is essentially independent of viscosity and depends only on the properties of the solid plate.

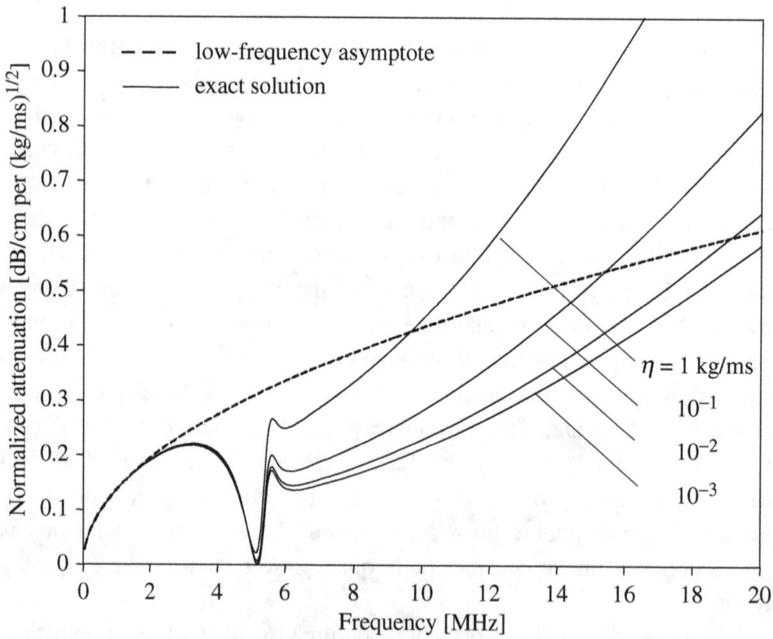

Figure 5.19. Normalized viscosity-induced attenuation ($\alpha_{vp}/\sqrt{\eta}$) versus frequency for the lowest-order symmetric mode in a 1-mm-thick immersed SiC plate (after Nagy and Nayfeh [40]).

Leaky guided modes propagating along embedded and clad fibers in a composite material can be used for characterizing the fiber–matrix interface [49–52]. For example, Drescher-Krasicka *et al.* applied this concept to study mechanical interface properties in an aluminum matrix model composite containing a single large-diameter silicon carbide rod [50]. This principle can be applied to real composites containing small-diameter fibers by using laser interferometric detection which offers diffraction-limited receiver apertures around a few microns [53]. Since the fiber diameter is usually of the order of 10–100 μm, a sharply focused optical beam is needed to obtain acoustic micrographs from epoxy, metal, and ceramic matrix composites with a few micron lateral resolution. These measurements are usually done at relatively low frequencies between 1 and 10 MHz so that a sufficient penetration depth can be maintained. Although the sensitivity of such inspection is still limited by the ultrasonic frequency, its lateral resolution is determined by the much smaller diffraction-limited spot size of the laser beam and individual fibers can be inspected.

Typically, the sound velocity is much lower in the matrix than in the fiber; therefore, the guided modes of well-bonded fibers are strongly attenuated by leaking their energy into the matrix as they propagate. Figure 5.20 shows surface vibration at the tip of a 500-μm-diameter, 12.5-mm-long glass fiber embedded in an epoxy matrix measured at 2 MHz for a well-bonded, a loose,

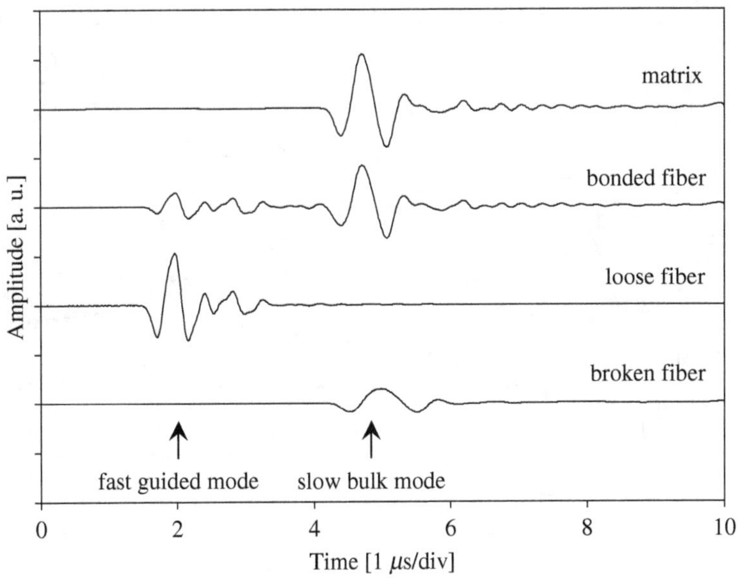

Figure 5.20. Measured surface vibration at the tip of a 500-μm-diameter, 12.5-mm-long glass fiber embedded in an epoxy matrix at 2 MHz for a well-bonded, a loose, and a broken fiber as well as on the matrix further away from the fiber (after Nagy [54]).

and a broken fiber as well as on the matrix further away from the fiber. Loose fibers can be readily identified from the strong early signal produced by the fast guided mode. In the case of a well-bonded fiber–matrix interface, the guided mode is strongly attenuated by the loading of the matrix depending on the fiber diameter and the interfacial stiffness of the interface. When an otherwise well-bonded fiber is broken not too deep below the surface, the fast guided mode is completely eliminated and even the slower bulk mode becomes much weaker.

Figure 5.21 shows the frequency-dependent phase velocity of this mode for a silicon carbide fiber embedded in an aluminum matrix. The dashed line represents the well-known dispersion curve of a completely free fiber. At low frequencies, the mode asymptotically approaches the so-called rod velocity. At higher frequencies, due to the Poisson effect, the phase velocity starts to decrease and, at very high frequencies, asymptotically approaches the Rayleigh velocity. The solid line represents the dispersion of the rigidly held fiber. The most interesting feature of this mode is that, at low frequencies, the phase velocity drops below the shear velocity of the matrix and approaches zero. This unusual behavior is caused by the increasing loading of the matrix via the transverse component of the surface vibration. When the frequency decreases, the fiber diameter becomes negligible with respect to the acoustic

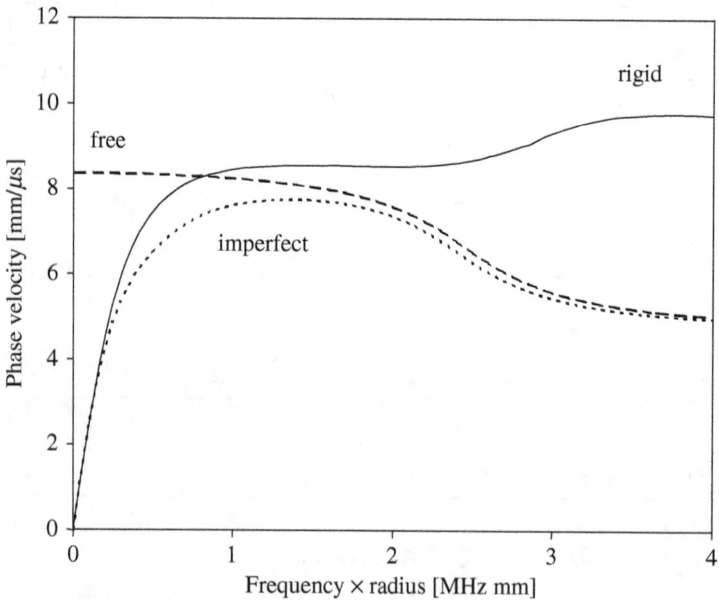

Figure 5.21. Transition of the lowest-order axisymmetric guided mode from "rigidly bonded" behavior at low frequencies to "free" behavior at high frequencies (silicon carbide fiber in aluminum matrix) (after Nagy [54]).

wavelength in the fiber and the surrounding matrix. As a result, the elastic singularity presented by the finite-diameter fiber embedded in the infinite matrix vanishes and the guided mode disappears with it. In comparison, the previously discussed loading effect on a thin fiber immersed in both viscosity-free and viscous fluids diminishes at low frequencies since the normal surface vibration (radial motion) and the viscous drag itself vanish, respectively. In a thin fiber embedded in an elastic solid, the remaining transverse surface vibration (axial motion) is also coupled to the surrounding medium; therefore, the loading effect of the matrix does not disappear but rather increases at low frequencies. In this region, the attenuation increases without limit and the guided mode cannot propagate any more.

Besides the free and rigidly bonded fibers, Fig. 5.21 also shows the lowest-order axisymmetric guided mode along an imperfectly bonded fiber (both the tangential and normal interfacial stiffnesses were assumed to be 10^{13} N/m^3). At very low frequencies, an imperfectly bonded interface of finite interfacial stiffness behaves as a perfect bond and the guided mode asymptotically approaches the previously discussed case of a rigidly bonded fiber. At very high frequencies, the same imperfect interface acts as a total disbond and the guided mode asymptotically approaches the case of a free fiber. In between, there is a smooth transition from the low-frequency "rigidly-bonded" behavior to the high-frequency "free" behavior. Naturally, the frequency at which this transition occurs is determined by the value of the interfacial stiffness. At a given inspection frequency, above a maximum interfacial stiffness, all boundaries appear to be perfectly bonded. Similarly, below a minimum interfacial stiffness, the fibers appear to be perfectly free. The interfacial stiffness range between these limiting values is the actual measuring range at that particular frequency.

Fiber–matrix interface characterization is made possible by the fact that in the case of a well-bonded fiber–matrix interface, the axial guided modes are slowed down and strongly attenuated by the loading of the matrix depending on the fiber diameter and the interfacial stiffness of the interface. Figure 5.22 shows the frequency-dependent propagation parameters of the fundamental dilatational leaky guided mode along a silicon carbide fiber in an aluminum matrix for different values of interfacial stiffness (the normalized attenuation is the ratio of the attenuation coefficient and frequency). At any particular frequency, the phase velocity increases while the attenuation decreases with decreasing interfacial stiffness. The frequency dependence of the attenuation coefficient is markedly different from the behavior of the leaky loss in an immersed rod where the attenuation increases with frequency. From a practical point of view, guided wave propagation in an embedded fiber is feasible only above a minimum frequency where the finite interfacial stiffness produces sufficiently weak coupling to the matrix so that the leaky loss is sufficiently low.

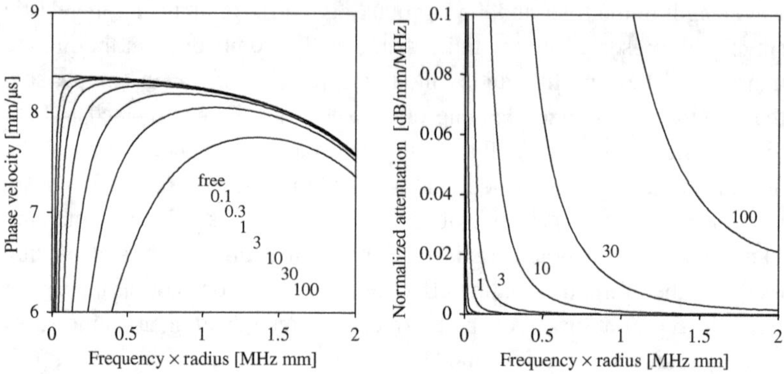

Figure 5.22. Frequency-dependent propagation parameters of the lowest-order axisymmetric guided mode along a silicon carbide fiber in aluminum matrix for different values of interfacial stiffness in units of 10^{11} N/m^3 (after Nagy [54]).

Interestingly, the guided wave approach is also suitable for a unified analytical treatment of wave propagation along the fiber direction in multilayered coaxial fibrous systems embedded in a host material [55]. In general, the host medium can be either fluid, solid, or vacuum representing a free rod. Alternatively, the host can be populated by a uniform distribution of the fibrous systems thus simulating actual unidirectional fibrous materials. Furthermore, each of the involved material components can possess transverse or full isotropy that insures axial symmetry along the fibers and the unified model also allows possible imperfections at material layer interfaces. In order to illustrate how this generalized guided wave approach can be exploited to model wave propagation in an infinite composite medium with uniform fiber reinforcement, let us consider a periodic distribution of fibers embedded in a matrix according to the hexagonal arrangement shown in Fig. 5.23. Due to symmetry, the composite can be approximated by concentric cylindrical regions consisting of the fiber assembly as the core that is surrounded by a coaxial matrix layer. The symmetry implies that the radial displacement and shear stress at the outer radius a of the concentric fiber-matrix system vanish. By requiring that the radial displacement u_r and shear stress σ_{rz} vanish at $r = a$, we can simulate the case of a periodic medium consisting of uniformly distributed fiber assemblies in the host material. With this choice of boundary conditions guided waves along the clad fiber well approximate bulk wave propagation along the fiber direction in an infinite periodically reinforced composite medium.

In order to illustrate the potential of bulk wave simulation via guided wave propagation under appropriate boundary conditions, let us consider an infinite composite medium consisting of a periodic distribution of SiC fibers embedded in a titanium matrix according to the hexagonal arrangement

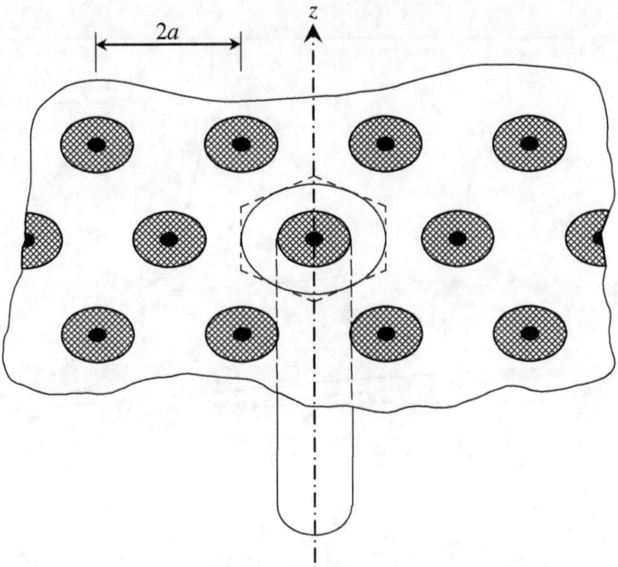

Figure 5.23. A schematic diagram of an infinite composite material with periodically distributed multilayered fibers.

shown in Fig. 5.23, where we assume that the separation between neighboring embedded fibers is twice the diameter of the individual clad fibers. Based on the simplified cylindrical unit cell, this ratio corresponds to a fiber volume fraction of 25%. Based on the actual hexagonal unit cell, however, the real volume fraction is $\approx 22.7\%$. As a result of the periodic distribution of the fibers, guided waves are attenuation-free bulk modes propagating along the fibers but not confined to them.

Figure 5.24 shows the dispersion curves of the two lowest-order modes in imperfectly bonded fibers for different values of interfacial stiffness. Two main effects are apparent from these results. First, the velocity of the lowest-order or "slow" mode, which is the only mode that can propagate at arbitrarily low frequencies, substantially decreases with decreasing interfacial stiffness. At very low frequencies, this mode propagates at a velocity determined by the law of mixture. In the case of loosely bonded fibers, this mode quickly drops to and then stays at the slightly lower longitudinal velocity of the matrix over a wide frequency range. Ultimately, the velocity of this mode further decreases as the frequency increases and asymptotically approaches the velocity of a Stoneley type interface wave. The higher-order or "fast" mode asymptotically approaches the axial vibration of the totally misbonded, or free, fiber as the interfacial stiffness decreases. This mode starts to propagate above a certain cutoff frequency which is mainly determined by the interfacial stiffness. Figure 5.25 displays

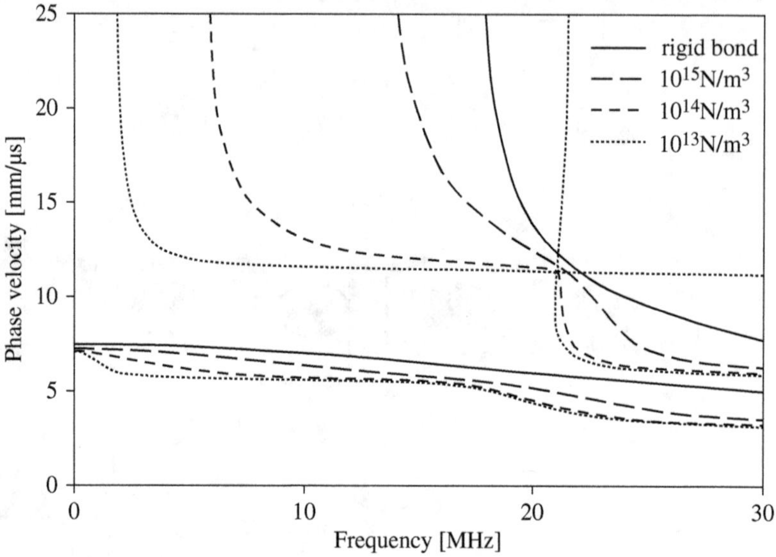

Figure 5.24. The two lowest-order modes in a titanium composite with textured SiC fibers of $\approx 25\%$ volume fraction for different interfacial stiffnesses in N/m^3 (after Nayfeh and Nagy [55]).

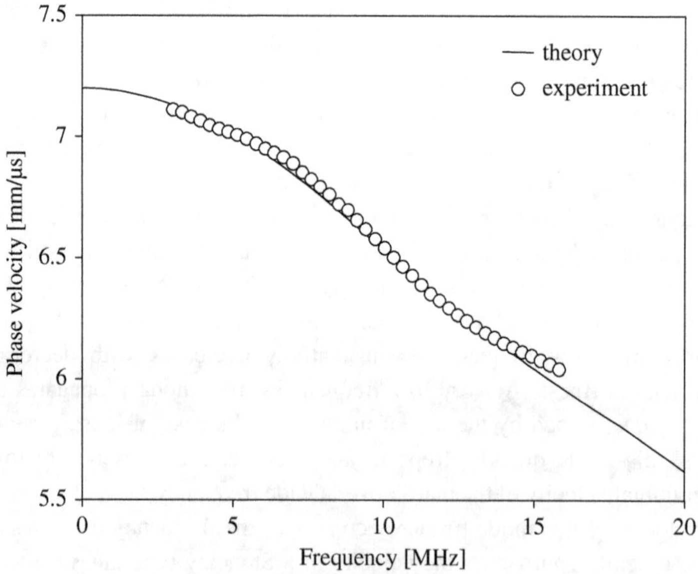

Figure 5.25. Comparison between theoretical prediction and the experimental data of Huang and Rokhlin [56] (after Nayfeh and Nagy [55]).

the good comparison between experimental data of Huang and Rokhlin and calculated results [55] for the fundamental mode based on the elastic properties and fiber dimensions given in [56, 57]. The fiber volume fraction of 25.4% was chosen as a best-fitting parameter, which is entirely consistent with the geometrical considerations mentioned above.

Finally, we should mention that there exists a closely related area of great interest, namely guided wave inspection of large-scale reinforced materials and structures, such as steel-reinforced concrete [58]. In this case, because of the significantly larger dimensions, higher-order guided modes can be used to minimize leaky losses. Pavlakovic *et al.* showed that higher-order axial guided modes propagating in a circular bar imbedded in a medium of lower impedance exhibit a series of more-or-less periodic attenuation minima [59, 60]. At these minima, the particle displacements and energy density of the particular mode are concentrated around the center of the rod and thereby leakage of energy into the surrounding medium is minimal. At the same frequencies, the group or energy velocity reaches local maxima, which further increases the detectability of these modes in practice. Unfortunately, the lower attenuation due to leakage also implies reduced sensitivity to interfacial properties; therefore, the primary application of this method is corrosion detection in rebars [61–63]. In comparison, low-frequency inspection based on the fundamental dilatational mode can be used to characterize interface conditions between the rebar and the surrounding concrete [64, 65].

Bibliography

1. J. D. Achenbach, *Wave Propagation in Elastic Solids* (Elsevier, Amsterdam, 1973).
2. K. F. Graff, *Wave Motion in Elastic Solids* (Dover Publications, New York, 1991).
3. J. W. S. Rayleigh, *The Theory of Sound* (Dover Press, New York, 1898).
4. Y. Li and R. B. Thompson, "Influence of anisotropy on the dispersion characteristics of guided ultrasonic plate modes," J. Acoust. Soc. Am. **87**, 1911–1931 (1990).
5. A. E. H. Love, *A Treatise on the Mathematical Theory of Elasticity*, 4th ed (Dover Press, New York, 1927).
6. A. K. Mal, "Wave propagation in layered composite laminates under periodic surface loads," Wave Motion **10**, 257–266 (1988).
7. S. K. Datta, R. L. Bratton, T. Chakraborthy, and A. H. Shah, "Wave propagation in laminated composite plates," J. Acoust. Soc. Am. **83**, 2020–2026 (1988).
8. B. Hosten, "Reflection and transmission of acoustic plane waves on an immersed orthotropic and viscoelastic solid layer," J. Acoust. Soc. Am. **89**, 2745–2752 (1991).
9. A. H. Nayfeh and D. E. Chimenti, "Ultrasonic wave reflection from liquid-coupled orthotropic plates with application to fibrous composites," J. Appl. Mech. **55**, 863–870 (1988).

10. A. H. Nayfeh and D. E. Chimenti, "Free wave propagation in plates of general anisotropic media," J. Appl. Mech. **56**, 881–886 (1989).
11. D. E. Chimenti and A. H. Nayfeh, "Ultrasonic reflection and guided wave propagation in biaxially laminated composite plates," J. Acoust. Sec. Am. **87**, 1409–1415 (1990).
12. A. T. Jones, "Exact natural frequencies and modal functions for a thick off-axis lamina," J. Compos. Mater. **5**, 504–520 (1971).
13. Yu. A. Kosevich and E. S. Syrkin, "Elastic waves in plates of highly anisotropic crystals," Sov. Phys.–Acoustics **31**, 365–367 (1985).
14. S. V. Kulkarni and N. J. Pagano, "Dynamic characteristics of composite laminates," J. Sound Vib. **123**, 127–143 (1972).
15. L. P. Solie and B. A. Auld, "Elastic waves in free anisotropic plates," J. Acoust. Soc. Am. **54**, 50–65 (1973).
16. I. Abubakar, "Free vibrations of a transversely isotropic plate," Quart. J. Mech. Appl. Math. **15**, 129–136 (1962).
17. E. R. Baylis and W. A. Green, "Flexural waves in fiber reinforced laminated plates," J. Sound Vib. **110**, 1–26 (1986).
18. F. C. Moon, "Wave surfaces due to impact on anisotropic plates," J. Compos. Mater. **6**, 62–79 (1972).
19. Zh. G. Nikiforenko, V. T. Bobrov, and I. I. Averbukh, "Propagation of Lamb waves in anisotropic sheets," Sov. J. Nondestr. Test. **8**, 543 (1972).
20. R. S. Wagers, "Plate modes in surface acoustic wave devices," in *Physical Acoustics* **14**, eds. W. P. Mason and R. N. Thurston (Academic Press, New York, 1977), pp. 49–78.
21. S. A. Markus, M. D. Kaplan, and S. V. Vermeenko, "Propagation of natural waves in orthotropic plates," Sov. J. Nondestr. Test. **21**, 739 (1985).
22. R. D. Mindlin, "Waves and vibrations in isotropic elastic plates," in *Structural Mechanics*, eds. J. N. Goodier and N. J. Hoff (Pergamon Press, New York, 1960).
23. R. D. Mindlin, "Influence of rotary inertia and shear in flexural motion of isotropic, elastic plates," J. Appl. Mech. **18**, 31–38 (1951).
24. R. D. Mindlin and M. A. Medick, "Extensional vibrations of elastic plates," J. Appl. Mech. **26**, 561–569 (1959).
25. B. Tang and E. G. Henneke, "Long wavelength approximation for Lamb wave characterization of composite laminates," Res. Nondestr. Eval. **1**, 51–64 (1989).
26. H. Lamb, "Waves in an elastic plate," Proc. Roy. Soc. (London) **A93**, 114–128 (1917).
27. A. H. Nayfeh, *Wave Propagation in Layered Anisotropic Media* (Elsevier Science, Amsterdam, 1995).
28. C-H. Wang and D. E. Chimenti, "Acoustic waves in a piezoelectric plate loaded by a dielectric fluid," Appl. Phys. Lett. **63**, 1328–1330 (1993).
29. C-H. Wang and D. E. Chimenti, "Guided plate waves in piezoelectrics immersed in a dielectric fluid: I. Analysis," J. Acoust. Soc. Am. **97**, 2103–2109 (1995).
30. C-H. Wang and D. E. Chimenti, "Guided plate waves in piezoelectrics immersed in a dielectric fluid: II. Experiments," J. Acoust. Soc. Am. **97**, 2110–2115 (1995).
31. L. Cremer, "Theory of sound isolation by thin walls at oblique incidence," (in German), Akust. Z. **7**, 81–104 (1942).
32. A. Schoch, "Sound transmission through plates," (in German), Acoustica **2**, 1–18 (1952).

33. T. J. Plona, M. Behravesh, and W. G. Mayer, "Rayleigh and Lamb waves at liquid–solid boundaries," Ultrasonics **13**, 171–177 (1975).
34. L. E. Pitts, T. J. Plona, and W. G. Mayer, "Theory of nonspecular reflection effects for an ultrasonic beam incident on a solid plate in a liquid," Trans. IEEE **SU-24**, 101–109 (1977).
35. S. I. Rokhlin, D. E. Chimenti, and A. H. Nayfeh, "On the topology of the complex wave spectrum in a fluid-coupled elastic layer," J. Acoust. Soc. Am. **85**, 1074–1080 (1989).
36. D. C. Worlton, "Experimental confirmation of Lamb waves at megacycle frequencies," J. Appl. Phys. **32**, 967–971 (1961).
37. D. E. Chimenti and A. H. Nayfeh, "Anomalous ultrasonic dispersion in fluid-coupled, fibrous composite plates," Appl. Phys. Lett. **49**, 492–493 (1986).
38. D. E. Chimenti and S. I. Rokhlin, "Relationship between leaky Lamb modes and reflection coefficient zeroes for a fluid-coupled elastic layer," J. Acoust. Soc. Am. **88**, 1603–1611 (1990).
39. B. Courouble and J. P. Moufle, "Evaluation of ultrasonic techniques by visualization of ultrasound in water and in transparent solids," in *Ultrasonics International 91* (Butterwork-Heinemann, Oxford, 1991), pp. 135–138.
40. A. H. Nayfeh and P. B. Nagy, "Excess attenuation of leaky Lamb waves due to viscous fluid loading," J. Acoust. Soc. Am. **101**, 2649–2658 (1997).
41. C. Chree, "On the longitudinal vibrations of aleotropic bars with one axis of material symmetry," Q. J. Math. **24**, 340–358 (1890).
42. R. W. Morse, "Compressional waves along an anisotropic circular cylinder having hexagonal symmetry," J. Acoust. Soc. Am. **26**, 1018–1021 (1954).
43. I. Mirsky, "Wave propagation in transversely isotropic circular cylinders, Part I: Theory," J. Acoust. Soc. Am. **37**, 1018–1026 (1965).
44. P. B. Nagy, "Longitudinal guided wave propagation in a transversely isotropic rod immersed in fluid," J. Acoust. Soc. Am. **98**, 454–457 (1995).
45. P. B. Nagy and R. M. Kent, "Ultrasonic assessment of Poisson's ratio in thin rods," J. Acoust. Soc. Am. **98**, 2694–2701 (1995).
46. W. Hassan and P. B. Nagy, "Why fluid-loading has an opposite effect on the velocity of dilatational waves in thin plates and rods?" J. Acoust. Soc. Am. **102**, 3478–3483 (1997).
47. W. Hassan and P. B. Nagy, "Experimental verification of the opposite effect of fluid loading on the velocity of dilatational waves in thin plates and rods," J. Acoust. Soc. Am. **105**, 3026–3034 (1999).
48. P. B. Nagy and A. H. Nayfeh, "Viscosity induced attenuation of longitudinal guided waves in fluid-loaded rods," J. Acoust. Soc. Am. **100**, 1501–1508 (1996).
49. J. A. Simmons, E. Drescher-Krasicka, and H. N. G. Wadley, "Leaky axisymmetric modes in infinite clad rods. I," J. Acoust. Soc. Am. **92**, 1061–1090 (1992).
50. E. Drescher-Krasicka and J. A. Simmons, "Leaky axisymmetric modes in infinite clad rods. II," J. Acoust. Soc. Am. **92**, 1091–1105 (1992).
51. J. A. Simmons, E. Drescher-Krasicka, H. N. G. Wadley, M. Rosen, and T. M. Hsieh, "Ultrasonic methods for characterizing the interface in composites," in *Review of Progress in Quantitative Nondestructive Evaluation*, eds. D. O. Thompson and D. E. Chimenti, Vol. **7** (Plenum, New York, 1988), pp. 893–901.
52. P. C. Xu and S. K. Datta, "Characterization of fiber–matrix interface by guided waves: Axisymmetric case," J. Acoust. Soc. Am. **89**, 2573–2583 (1991).

53. C. H. Yew and P. N. Jogi, "A study of wave motion in fiber-reinforced medium," Int. J. Solids Struct. **12**, 693–703 (1976).
54. P. B. Nagy, "Leaky guided wave propagation along imperfectly bonded fibers in composite materials," J. Nondestr. Eval. **13**, 137–145 (1994).
55. A. H. Nayfeh and P. B. Nagy, "General study of axisymmetric waves in layered anisotropic fibers and their composites," J. Acoust. Soc. Am. **99**, 931–941 (1996).
56. W. Huang and S. I. Rokhlin, "Frequency dependencies of ultrasonic wave velocity and attenuation in fiber composites. Theory and experiment," in *Review of Progress in Quantitative Nondestructive Evaluation*, eds. D. O. Thompson and D. E. Chimenti, vol. **14** (Plenum, New York, 1995), pp. 1233–1240.
57. W. Huang, S. I. Rokhlin, and Y. J. Wang, "Effect of fiber-matrix interphase on wave propagation along, and scattering from, multilayered fibers in composites – Transfer matrix approach," Ultrasonics **33**, 365–375 (1995).
58. W. B. Na, T. Kundu, and M. R. Ehsani, "A comparison of steel/concrete and glass fiber reinforced polymers/concrete interface testing by guided waves," Mater. Eval. **61**, 155–161 (2003).
59. B. N. Pavlakovic, M. J. S. Lowe, and P. Cawley, "The inspection of tendons in post-tensioned concrete using guided ultrasonic waves," Insight **41**, 446–452 (1999).
60. B. N. Pavlakovic, M. J. S. Lowe, and P. Cawley, "High-frequency low-loss ultrasonic modes in imbedded bars," J. Appl. Mech. **68**, 67–75 (2001).
61. H. Reis, B. L. Ervin, D. A. Kuchma, and J. T. Bernhar, "Estimation of corrosion damage in steel reinforced mortar using guided waves," J. Press. Vessel Techn. **127**, 255–261 (2005).
62. H. Reis, and B. L. Ervin, "Longitudinal guided waves for monitoring corrosion in reinforced mortar," Meas. Sci. Techn. **19**, 1–19 (2008).
63. H. Reis, B. L. Ervin, D. A. Kuchma, and J. T. Bernhar, "Monitoring corrosion of rebar embedded in mortar using high-frequency guided ultrasonic waves," J. Eng. Mech. **127**, 255–261 (2009).
64. W. B. Na, T. Kundu, and M. R. Ehsani, "Ultrasonic guided waves for steel bar concrete interface testing," Mater. Eval. **60**, 437–444 (2002).
65. W. B. Na and T. Kundu, "Inspection of interfaces between corroded steel bars and concrete using the combination of a piezoelectric zirconate–titanate transducer and an electromagnetic acoustic transducer," Exp. Mech. **43**, 24–31 (2003).

6

Elastic Waves in Multilayer Composites

6.1 Introduction

Expanding on the theme of bulk waves from the previous chapters, we will examine the problem of plane wave sound propagation in layered media. We assume we have an finite stack of planar layers with perfect, rigidly bonded planar interfaces, but infinite in their lateral extent. The problem has significant industrial interest. Most practical composite laminates are composed of layers of uniaxial fibers and plastic, *i.e.*, plies, whose fiber orientation directions vary from ply to ply through the thickness of the laminate. The mechanical purpose of this directional variation is to render the product stiff and strong in all in-plane directions, much as plywood is layered in cross-grain fashion. Almost no practical composite would be fabricated as a uniaxial product, because of the low bending strength normal to the fiber direction. Instead, various types of layering have been devised to give either tailored stiffness for a specific purpose or approximate in-plane isotropy, also known colloquially as a "quasi-isotropic" laminate. In fact, the approximate isotropy is achieved only in the plane of the plies, because the out-of-plane direction still has significant and unavoidable stiffness differences, since it contains no fibers.

The scale of the layering is also important. When the laminations are fine, *i.e.*, when each directional lamina is no thicker than an individual ply as we go through the thickness, only acoustic waves of relatively short wavelength will be able to discern the effect of the layering. At longer wavelengths, the laminate may behave more like an effective medium, still anisotropic, but with averaged elastic properties. On the other hand, if each lamina contains multiple numbers of individual 1/8-mm plies, then the frequency at which an acoustic wavelength approaches the layer thickness

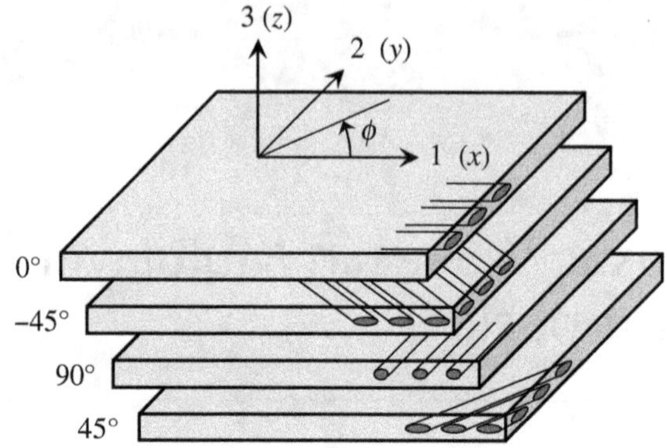

Figure 6.1. Geometry of a composite cell, showing different fiber directions in each ply (lamina). In this case the layup sequence is $[0°/-45°/90°/45°]_{2S}$. The composite laminate, composed of the unidirectional lamina, is symmetric about the midplane. Also shown is the reference coordinate system (after Rokhlin and Wang [1]).

will be proportionately lower. This is an important distinction, because it suggests the point at which the layering must be treated as a discrete substructure in order to develop an accurate description of waves in a layered medium. The situation is illustrated schematically in Fig. 6.1. The figure illustrates laminations for a quasi-isotropic composite. Other practical layup sequences can include multiple plies in a single direction, or ply group, such as $[0°_3, 90°_2]_S$, meaning that there is first a ply group of three 0° plies, followed by a group of two 90° plies, and the entire laminate is symmetric about the midplane. For this hypothetical laminate that would mean from top to bottom, three 0° plies, four 90° plies, and three 0° plies, with the total for a symmetric laminate always being an even number. Results from the well-researched area of static modeling of composite laminate mechanical properties [2,3] are applicable to the dynamic circumstances of elastic wave propagation in the limit of very long wavelength (or low frequency) only.

When preparing to consider sound propagation in such a laminate, we first postulate some coupling medium, typically water, but also possibly air, to be in contact with the laminate at its top and bottom. So, sound waves begin in the fluid, enter the solid laminate and interact with it, then exit at the opposite surface, reappearing in the semi-infinite fluid below the laminate. Alternatively, the lower fluid medium may be absent, in which case the sound reappears in the upper fluid half-space. As we are usually interested only in the resultant sound wave amplitude and phase, in either the upper or lower fluid media, the details of wave propagation and reverberation in the constituent laminate layers are irrelevant. (Only with the occurrence of defects might the details of internal field variables such as displacement and stress at locations within the laminate become important.)

The first of these kinds of calculations, in which the details of sound propagation in the layered medium are ignored in favor of the exiting sound wave field at both the entry and opposite interfaces, was proposed by Thomson [4] and Haskell [5] for the field of seismic waves. The problem is identical, except for the layer spatial dimensions and the acoustic wavelengths. In ultrasonics, the seismic dimensions and wavelengths are scaled by many orders of magnitude. Otherwise, the problem is unchanged. If the point is to find only the complex wave fields in the fluid, then calculation of all the stresses and particle displacements in each of the constituent layers is unnecessary and can be skirted, according to the *transfer matrix method* of Thomson and Haskell. A problem of much-reduced dimensionality is sufficient if the objective is to find the waves in the fluid. The plan is to write a vector of displacements and stresses equal to the product of a matrix of propagator terms times a vector of complex wave fields, essentially just a statement of Helmholtz's theorem in a layer. Combining this expression with a similar one for the adjacent layer, a new relation can be derived that connects one layer's behavior with that of the adjacent one. The resulting product of matrices is redefined to be the so-called transfer matrix. The process can then be continued recursively through the entire medium, all the way from the top of the first layer to the bottom of the last one. This is the relation sought earlier, connecting behavior at the top and bottom of the laminate. In performing the calculation, no matrix larger than 4×4 for isotropic media, or 6×6 for anisotropic media, would have to be calculated. The consequence is that the internal laminate displacements and stresses are still unknown. But, this is exactly the advantage of the method—these largely uninteresting quantities would never have to be computed. If, on the other hand, the brute force approach had been taken, then even for isotropic media in plane strain, there would be two displacements and two stresses for each of the N layers, resulting in a linear system with a total of $4N$ unknowns.

The tidy package of the transfer matrix method has a critical flaw, however, which was discovered early in its application. Owing to refraction within one or more layers of the laminate (or stack), some of the plane waves can be internally reflected, meaning that their partial waves will be evanescent within that layer. That is, the propagation constant becomes imaginary, leading from sinusoidal wave functions of the form $\exp[ikz]$ to decaying exponentials, such as $\exp[\pm k''z]$. Depending on the value of k'' and the thickness of the layer, these real-valued functions can become very large or very small. Theoretically, there is no problem, because the terms that are generated in this way will cancel mathematically in the construction of the total transfer matrix. The problem arises when we consider that to compute these quantities, we must use machines that carry only a finite number of significant digits. In the addition $X + x$, once x is smaller than X by more than the machine precision, the sum loses meaning and the calculation is compromised. As the exponential function with real arguments rises or falls so quickly, numerical instabilities arise frequently. There has been a number of attempts over the years to fix this computational flaw, but none addresses directly the fundamental problem of the resulting singular matrices. There does exist an entirely separate approach, in which the equations for all layers are assembled into a single matrix, the *global matrix* [6]. When properly implemented, this method is unconditionally stable [7], at the expense of greater complexity, but still simpler than that of solving for all variables in all layers. The extension of this method to anisotropic media has been done by Mal [8]. As there exist a number excellent reviews of this field, we will not attempt a further historical analysis here. The interested reader may consult the review article by Lowe [9], or the book by Kennett [10].

By contrast, our theoretical treatment of ultrasonic wave interaction with multidirectional composites is based on the stable stiffness matrix method, which allows us to present solutions in a concise matrix form. Some familiarity with matrix manipulations will assist the reader for a better appreciation of the theoretical sections of this chapter. The matrix formulation of ultrasonic wave solutions for multilayered composites leads conveniently to relatively simple computer coding. An additional discussion of ultrasonic wave interaction with multilayered composites is presented in the following chapter. The treatment here follows the approach of Wang and Rokhlin [11] and Rokhlin and Wang [12].

6.2 Transfer Matrix

To appreciate what is happening when partial waves in the layers become evanescent and how this event compromises the accuracy of the transfer matrix, it is instructive to analyze the problem first using the transfer matrix method. Later, we will contrast the transfer matrix with the *recursive stiffness matrix* approach that eliminates the numerical instabilities altogether. Let us begin with a multilayered plate composed of N layers, each having anisotropic elastic properties. This is a model of a multiaxial composite laminate. The geometry is shown in Fig. 6.2 The thickness of the jth layer is $h_j = z_{j-1} - z_j$, and the coordinate value at the top of the jth layer is z_{j-1}. The particle displacement vector in the jth layer can be expressed as

$$\mathbf{u}_j = \sum_{n=1}^{3} \left(a^{+n}\mathbf{p}^{+n} \exp[ik_z^{+n}(z - z_j)] + a^{-n}\mathbf{p}^{-n} \exp[ik_z^{-n}(z - z_{j-1})]\right)_j$$

$$\exp[i(k_x x - \omega t)], \tag{6.1}$$

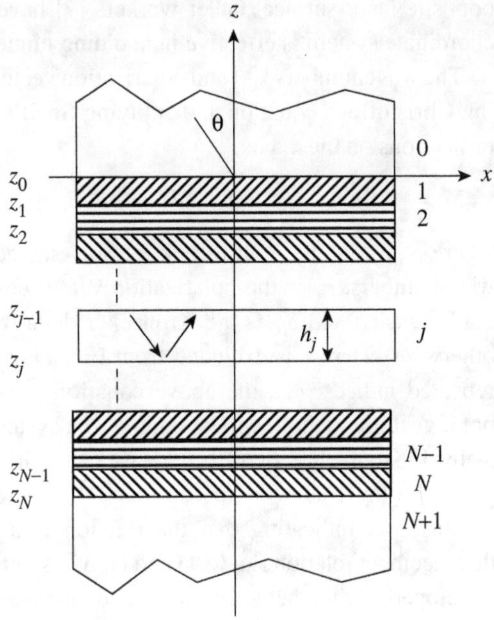

Figure 6.2. Coordinate system of a layered medium composed of N anisotropic layers, where z_{j-1} is the coordinate at the top of the jth layer, and the thickness of the jth layer is $h_j = z_{j-1} - z_j$ (after Wang and Rokhlin [11]).

where $\mathbf{u}_j$ is the vector displacement, $n = 1, 2, 3$ represents the nth partial wave, $a^{\pm n}$ are the amplitudes of each partial wave, and the superscripts + or – denote waves with wavevector components along the positive or negative z directions. The polarization directions of the nth + and –, or positive- and negative-going, partial waves are given by $\mathbf{p}^{\pm n}$, corresponding to wavevector components $k_z^{\pm n}$. The explicit expressions for the displacement vectors in the jth layer are $\mathbf{u}_j = (u_x, u_y, u_z)_j^{\mathrm{T}}$, superscript T is a transpose; the unit polarization vectors are $\mathbf{p}^{\pm n} = (p_x^{\pm n}, p_y^{\pm n}, p_z^{\pm n})^{\mathrm{T}}$. Here, as everywhere else in this book, vectors and higher-order tensors are in bold, and the same symbol in italic accompanied with a subscript or subscripts denotes the given component of the vector or higher-order tensor. Consequently, bold characters with subscripts are not components of tensors since the subscript is not a free index but an identifier (*e.g.*, serial number of a layer).

There is no y dependence in Eq. (6.1) because the coordinate system has been chosen so that the incident acoustic plane coincides with the (x, z) plane. A local coordinate system origin is selected to be at the *top* of the layer for the – going partial waves and at the layer *bottom* for + going partial waves. This choice aids in the suppression of singularities when partial waves become evanescent. With this choice of local coordinate system, the spatially dependent exponential term of each evanescent wave equals unity at the interface where they are initiated, decaying as they propagate to the opposite layer surface. Other workers [7] have also shown that this local coordinate system is effective in avoiding numerical overflow.

The wavenumbers $k_z^{\pm n}$ and polarization vectors $\mathbf{p}^{\pm n}$ can be found by solving Christoffel's equation and applying Snell's law to all the wavenumber projections on the x axis,

$$(c_{ij\ell n}k_j k_n - \rho\omega^2\delta_{i\ell})p_\ell = 0, \tag{6.2}$$

where $c_{ij\ell n}$ is the fourth-rank elastic stiffness tensor, k_n and k_j are wavenumbers, p_ℓ is the polarization vector component, ρ is the laminate mass density, and $\delta_{i\ell}$ is the Kronecker delta ($\delta_{i\ell} = 1$ for $i = \ell$, and zero otherwise). Here ℓ is excluded from Einstein's summation convention for repeated indices; *i.e.*, the above equation is not a single scalar equation but a vector equation, corresponding to a system of three scalar equations. Referring to the first two chapters, we can write the stress traction vector $\boldsymbol{\sigma}_j$ in the jth layer as $\boldsymbol{\sigma}_j = (\sigma_{31}, \sigma_{32}, \sigma_{33})_j^{\mathrm{T}}$, where the superscript T denotes a transpose, indicating that the traction is a column vector. Using the displacement relation Eq. (6.1) and Hooke's law $\sigma_{ij} = c_{ij\ell n}\epsilon_{\ell n}$ and relations developed earlier between strain and displacement, we can express the traction vector in terms of the wave fields,

$$\boldsymbol{\sigma}_j = \sum_{n=1}^{3} \left(a^{+n}\mathbf{d}^{+n}\exp[ik_z^{+n}(z - z_j)] + a^{-n}\mathbf{d}^{-n}\exp[ik_z^{-n}(z - z_{j-1})]\right)_j \exp[i(k_x x - \omega t)]. \tag{6.3}$$

The expression $\exp[i(k_x x - \omega t)]$ common to every partial wave for any layer is just another way to state Snell's law. Traction vectors for different partial waves are $\mathbf{d}^{\pm n} = (d_1^{\pm n}, d_2^{\pm n}, d_3^{\pm n})^{\mathrm{T}}$, where indices 1, 2, 3 denote components of the vectors. We will use a running index $q(= 1, 2, 3)$ to specify those individual components. Recall that we have already obtained in Chapter 4 the relations between components of stress and displacement on the interface between two anisotropic solids for use in boundary conditions, Eqs. (4.11) and (4.13). Comparing Eqs. (4.11), (4.13), and (6.3) we have for components of $\mathbf{d}^{\pm n}$ in the jth layer

$$(d_q^{\pm n})_j = i(c_{q3tm}k_m^{\pm n}p_t^{\pm n})_j. \tag{6.4}$$

The number 3 in the second position of the stiffness tensor subscript acknowledges that we are interested only in the tractions on ply interfaces normal to the z or 3 direction. The transfer matrix method connects displacements and traction stresses from one side of the layer to the other because these are precisely the quantities that are affected by layer interface conditions.

The transfer matrix method proceeds by expressing both the displacements and stresses (tractions) on the upper surface of the jth layer in terms of the amplitudes, polarizations, wavenumbers, and propagators. The construction of the transfer matrix here is far from obvious, however, because we are dealing with anisotropic elastic media. Thomson's [4] original treatment was limited to isotropic materials. The extension to anisotropic media became essential only when the problem of wave propagation in layered anisotropic composites first appeared [13, 14].

Our objective is to relate displacement and traction vectors on the top surface of the layer $\mathbf{u}_j, \boldsymbol{\sigma}_j$ to those, $\mathbf{u}_{j+1}, \boldsymbol{\sigma}_{j+1}$, at the bottom surface. They will be connected by a square matrix which is called the *transfer matrix*. In preparation for the next relation, let us introduce several new quantities: the matrix $\mathbf{P}^{\pm}(3 \times 3) = [\mathbf{p}^{\pm 1}, \mathbf{p}^{\pm 2}, \mathbf{p}^{\pm 3}]$ is composed of polarization unit vectors $\mathbf{p}^{\pm n}$ for each partial wave n, $\mathbf{D}^{\pm}(3 \times 3) = [\mathbf{d}^{\pm 1}, \mathbf{d}^{\pm 2}, \mathbf{d}^{\pm 3}]$ is likewise a matrix of $\mathbf{d}^{\pm n}$ traction vectors, and a further matrix $\mathbf{H}^{\pm}$ $(3 \times 3) = \mathrm{Diag}[e^{ik_z^{\pm 1} h_j}, e^{ik_z^{+2} h_j}, e^{ik_z^{+3} h_j}]$ is formed by distributing the propagators $\exp(ik_z^{\pm n} h_j)$ down the diagonal of a 3×3 matrix, whose other elements are zero. Here, $h_j = z_{j-1} - z_j$ is the thickness of the jth layer. The complete expression is quite complicated; we seek here instead a shorthand terminology we can use to calculate efficiently. Using the notations outlined above, Eq. (6.1) and (6.3) for the top of the layer ($z = z_{j-1}$), can be cast in the following way

$$\begin{bmatrix} \mathbf{u}_{j-1} \\ \boldsymbol{\sigma}_{j-1} \end{bmatrix} = \begin{bmatrix} \mathbf{P}^- & \mathbf{P}^+\mathbf{H}^+ \\ \mathbf{D}^- & \mathbf{D}^+\mathbf{H}^+ \end{bmatrix}_j \begin{bmatrix} \mathbf{A}_j^- \\ \mathbf{A}_j^+ \end{bmatrix}, \tag{6.5}$$

where $\mathbf{A}_j^{\pm} = [a^{\pm 1}, a^{\pm 2}, a^{\pm 3}]_j^{\mathrm{T}}$ is the vector of partial wave amplitudes. Likewise, at the bottom of the jth layer ($z = z_j$) the displacements and tractions are

$$\begin{bmatrix} \mathbf{u}_j \\ \boldsymbol{\sigma}_j \end{bmatrix} = \begin{bmatrix} \mathbf{P}^-\mathbf{H}^- & \mathbf{P}^+ \\ \mathbf{D}^-\mathbf{H}^- & \mathbf{D}^+ \end{bmatrix}_j \begin{bmatrix} \mathbf{A}_j^- \\ \mathbf{A}_j^+ \end{bmatrix}. \tag{6.6}$$

Eliminating the amplitude vectors $\mathbf{A}^{\pm}$ from Eqs. (6.5) and (6.6), we find a relationship between the displacements and stresses at the top and at the bottom of the jth layer,

$$\begin{bmatrix} \mathbf{u}_j \\ \boldsymbol{\sigma}_j \end{bmatrix} = \begin{bmatrix} \mathbf{P}^-\mathbf{H}^- & \mathbf{P}^+ \\ \mathbf{D}^-\mathbf{H}^- & \mathbf{D}^+ \end{bmatrix}_j \begin{bmatrix} \mathbf{P}^- & \mathbf{P}^+\mathbf{H}^+ \\ \mathbf{D}^- & \mathbf{D}^+\mathbf{H}^+ \end{bmatrix}_j^{-1} \begin{bmatrix} \mathbf{u}_{j-1} \\ \boldsymbol{\sigma}_{j-1} \end{bmatrix} = \mathbf{B}_j \begin{bmatrix} \mathbf{u}_{j-1} \\ \boldsymbol{\sigma}_{j-1} \end{bmatrix}. \tag{6.7}$$

The layer transfer matrix $\mathbf{B}(6 \times 6)$ is constructed from four submatrices $\mathbf{B}_{ik}(3 \times 3)$, where $i, k = 1, 2$,

$$\mathbf{B}_j(6 \times 6) = \begin{bmatrix} \mathbf{B}_{11} & \mathbf{B}_{12} \\ \mathbf{B}_{21} & \mathbf{B}_{22} \end{bmatrix}_j. \tag{6.8}$$

The process by which this matrix has been constructed can be repeated recursively for each successive layer until all N layer transfer matrices in the laminate have been computed. Now, we wish to add an additional layer j on the bottom of the sandwich-like system of $J - 1$ layers; its transfer matrix $\mathbf{B}_j$ multiplies the *total* transfer matrix (denoted with a tilde accent $[\tilde{\,}]$) $\tilde{\mathbf{B}}_{J-1} = \prod_{i=1}^{J-1} \mathbf{B}_i$ from the left, where $\tilde{\mathbf{B}}_1 = \mathbf{B}_1$. Continuing, we have for the total transfer matrix $\tilde{\mathbf{B}}_J$ of all J layers

$$\tilde{\mathbf{B}}_J = \mathbf{B}_j \tilde{\mathbf{B}}_{J-1}. \tag{6.9}$$

Upper case indices J and $J - 1$ denote the total number of layers considered, while lower case index j indicates the jth layer only. The advantage of the recursive transfer matrix computation is that only two transfer matrices need to be stored: the total $\tilde{\mathbf{B}}_{J-1}$ and the current $\mathbf{B}_j$ matrix for layer j. The total (global) transfer matrix for the layered composite structure relates the displacements and stresses at the top ($z = z_0$) and bottom ($z = z_n$) surfaces of the structure. The total matrix yields the relationship

$$\begin{bmatrix} \mathbf{u}_N \\ \boldsymbol{\sigma}_N \end{bmatrix} = \tilde{\mathbf{B}} \begin{bmatrix} \mathbf{u}_0 \\ \boldsymbol{\sigma}_0 \end{bmatrix}. \tag{6.10}$$

While the total transfer matrix is computed recursively using Eq. (6.9), it is instructive to write it as a direct product of individual matrices

$$\tilde{\mathbf{B}} = \prod_{j=1}^{N} \mathbf{B}_j = (\mathbf{B}_N(\mathbf{B}_{N-1} \cdots (\mathbf{B}_3(\mathbf{B}_2\mathbf{B}_1) \cdot \cdot)) = \tilde{\mathbf{B}}_N, \tag{6.11}$$

illustrating the cascading nature of the matrix product. The transfer matrices in the above equation are, of course, not commutative, so the order of the product is important. The grouping of terms to the right of the second equals sign in Eq. (6.10) indicates that the products are formed recursively; first, $\mathbf{B}_1$ is multiplied from the left by $\mathbf{B}_2$ and then this product is multiplied from the left by $\mathbf{B}_3$. The sequence continues to $\mathbf{B}_{N-1}$ and $\mathbf{B}_N$, completing the total transfer matrix. The ellipsis dots before $\mathbf{B}_{N-1}$ stand for the omitted $\mathbf{B}_j$ and left-hand parentheses, and those on the right of $\mathbf{B}_1$ stand for the absent right-hand parentheses.

As we know, however, the total transfer matrix is, under some conditions, not numerically stable. At the root of the difficulty lies the choice made in the variable matrix. In Eq. (6.7), the displacement and traction vectors at the upper layer surface $(\mathbf{u}_{j-1}, \boldsymbol{\sigma}_{j-1})$ occupy the column matrix on the right side, while these same vectors evaluated at the lower layer surface $(\mathbf{u}_j, \boldsymbol{\sigma}_j)$ occupy another column matrix on the left side side of the equation. Although this choice for the distribution of variables is clearly intuitive, it accounts for all the problems, and changing the choice can suppress the instabilities.

The matrix product in Eq. (6.7) contains the seeds of the problem. At large enough incident angles, a critical angle is reached in some layer or layers. At that point, the wavenumbers become imaginary, and the corresponding partial waves are evanescent. For the **H** matrix this means that the only nonzero terms, the ones on the diagonal, will become small, all the more so if the layer is thick and/or if the frequency is high, *i.e.*, many sound wavelengths subtend the layer. From the first line in Eq. (6.7), we see that the **H** matrix multiplies elements in an entire column of both matrices in the product. With increasing layer thickness or increasing frequency, these matrices will develop columns of zeroes and become singular, and the inverse of the right-hand matrix will be undefined. This condition is the origin of the numerical instabilities in the conventional transfer matrix method. No simple extension of the precision of the calculation can overcome the exponential functions with real arguments. They quickly overwhelm any half measures. Instead, what is needed is a complete reformulation of the problem to eliminate the singular matrices in a fundamental way. This is the theme of the next section.

6.3 Stiffness Matrix

6.3.1 *General formulation of the stiffness matrix for an anisotropic layer*

Writing a matrix composed of the displacement and traction vectors (as the transfer matrix method does) is a very natural thing to do because these are just the quantities that must be conserved across the interface using welded contact boundary conditions. Moreover, using the transfer matrix to relate these same quantities from one interface to the next also has natural appeal. For this approach to function, however, there must exist an actual "transfer" of wave information from one layer boundary (the input) to another (the output). When partial waves are evanescent this transfer is interrupted, and the communication between the boundaries (or input and output ports) is weak. The transfer function formulation requires the determination of all the unknown parameters at the input port from all the parameters found at the output port. This situation is acceptable only as long as the numerical calculation can tolerate the demands thus placed on its level of precision. When the coupling becomes weak enough, numerical instabilities ensue. The choice of dynamical variables and the need for the existence of a transfer of wave information ultimately lead to the numerical problems encountered in evaluating the transfer matrix.

Fixing the problem requires a rethinking of these natural choices for matrix elements and the recursive procedure to construct the total matrix. In a purely practical sense the goal is to isolate the **H** matrices in parts of the calculation where they do not lead to singular matrices. One choice that leads to a different and desirable outcome is a reorganization of Eqs. (6.5–6.6) to place the displacements, at both the top and bottom of the jth layer, in a single column matrix, and then to collect the tractions in another column matrix. Assembling separate displacement and stress component vectors we have, instead of Eqs. (6.5) and (6.6),

$$\begin{bmatrix} \mathbf{u}_{j-1} \\ \mathbf{u}_j \end{bmatrix} = \begin{bmatrix} \mathbf{P}^- & \mathbf{P}^+\mathbf{H}^+ \\ \mathbf{P}^-\mathbf{H}^- & \mathbf{P}^+ \end{bmatrix}_j \begin{bmatrix} \mathbf{A}_j^- \\ \mathbf{A}_j^+ \end{bmatrix} = \mathbf{E}_j^u \mathbf{A}_j \tag{6.12}$$

$$\begin{bmatrix} \boldsymbol{\sigma}_{j-1} \\ \boldsymbol{\sigma}_j \end{bmatrix} = \begin{bmatrix} \mathbf{D}^- & \mathbf{D}^+\mathbf{H}^+ \\ \mathbf{D}^-\mathbf{H}^- & \mathbf{D}^+ \end{bmatrix}_j \begin{bmatrix} \mathbf{A}_j^- \\ \mathbf{A}_j^+ \end{bmatrix} = \mathbf{E}_j^\sigma \mathbf{A}_j \tag{6.13}$$

Eliminating the common amplitude "vector" $\mathbf{A}_j$ (whose components $\mathbf{A}_j^\pm$ are true vectors) in Eqs. (6.12–6.13), we have

$$\begin{aligned} \begin{bmatrix} \boldsymbol{\sigma}_{j-1} \\ \boldsymbol{\sigma}_j \end{bmatrix} &= \begin{bmatrix} \mathbf{D}^- & \mathbf{D}^+\mathbf{H}^+ \\ \mathbf{D}^-\mathbf{H}^- & \mathbf{D}^+ \end{bmatrix}_j \begin{bmatrix} \mathbf{P}^- & \mathbf{P}^+\mathbf{H}^+ \\ \mathbf{P}^-\mathbf{H}^- & \mathbf{P}^+ \end{bmatrix}_j^{-1} \begin{bmatrix} \mathbf{u}_{j-1} \\ \mathbf{u}_j \end{bmatrix} \\ &= \mathbf{K}^j \begin{bmatrix} \mathbf{u}_{j-1} \\ \mathbf{u}_j \end{bmatrix}, \end{aligned} \tag{6.14}$$

where the matrix $\mathbf{K}^j$ is now called the *stiffness matrix* because it relates stresses only to displacements only directly through the layer properties. In this chapter, *stiffness* and *compliance matrices* refer to the relationships among tractions and displacements at interfaces, which are very different from the 6×6 stiffness and compliance matrices used in other parts of this book to relate stress and strain tensor elements at a given point.

The stiffness $\mathbf{K}$ and compliance $\mathbf{S} = \mathbf{K}^{-1}$ matrices for a layer j are defined as

$$\mathbf{K}^j = \begin{bmatrix} \mathbf{D}^- & \mathbf{D}^+\mathbf{H}^+ \\ \mathbf{D}^-\mathbf{H}^- & \mathbf{D}^+ \end{bmatrix}_j \begin{bmatrix} \mathbf{P}^- & \mathbf{P}^+\mathbf{H}^+ \\ \mathbf{P}^-\mathbf{H}^- & \mathbf{P}^+ \end{bmatrix}_j^{-1} = \begin{bmatrix} \mathbf{K}_{11}^j & \mathbf{K}_{12}^j \\ \mathbf{K}_{21}^j & \mathbf{K}_{22}^j \end{bmatrix} \tag{6.15}$$

and

$$\mathbf{S}^j = \begin{bmatrix} \mathbf{P}^- & \mathbf{P}^+\mathbf{H}^+ \\ \mathbf{P}^-\mathbf{H}^- & \mathbf{P}^+ \end{bmatrix}_j \begin{bmatrix} \mathbf{D}^- & \mathbf{D}^+\mathbf{H}^+ \\ \mathbf{D}^-\mathbf{H}^- & \mathbf{D}^+ \end{bmatrix}_j^{-1} = \begin{bmatrix} \mathbf{S}_{11}^j & \mathbf{S}_{12}^j \\ \mathbf{S}_{21}^j & \mathbf{S}_{22}^j \end{bmatrix}. \tag{6.16}$$

Notice that the $\mathbf{H}$ submatrices are no longer located in a single column, nor along the diagonal. This is exactly the robust rearrangement we had sought, and we shall provide further commentary later on. In the next several paragraphs, however, let us examine some details of the calculation and consequences of the choices leading to Eqs. (6.15–6.16). For example, in obtaining the compliance $\mathbf{S}$ matrix, we recall that $(\mathbf{AB})^{-1} = \mathbf{B}^{-1}\mathbf{A}^{-1}$. As we will see in the following sections for fluid-loaded composites and for problems with stress-free boundary conditions, more compact formulations can be obtained using the compliance matrix. The following useful relations exist among the submatrix elements of the compliance matrix $\mathbf{S}_{ij}$ and elements of the stiffness matrix $\mathbf{K}_{ij}$

$$\begin{aligned} \mathbf{S}_{11} &= (\mathbf{K}_{11} - \mathbf{K}_{12}\mathbf{K}_{22}^{-1}\mathbf{K}_{21})^{-1} \\ \mathbf{S}_{12} &= -\mathbf{K}_{11}^{-1}\mathbf{K}_{12}\mathbf{S}_{22} \\ \mathbf{S}_{21} &= -\mathbf{K}_{22}^{-1}\mathbf{K}_{21}\mathbf{S}_{11} \\ \mathbf{S}_{22} &= (\mathbf{K}_{22} - \mathbf{K}_{21}\mathbf{K}_{11}^{-1}\mathbf{K}_{12})^{-1}, \end{aligned} \tag{6.17}$$

These relations will be useful when we will deal with guided waves.

Consider the stiffness matrix for the lower semi-space with the coordinate origin at the interface ($z = z_N$) in Fig. 6.2. In this case, only three waves ($\mathbf{A}^-$) propagate in the $-z$ direction from the surface to infinity. The displacements and stresses in the semi-space $(N + 1)$ can be obtained from the displacements and tractions on the interface $z = z_N$ (see Eq. (6.12) and Eq. (6.13)). These quantities are $\mathbf{u}_N = \mathbf{P}^-\mathbf{A}^-$ and $\boldsymbol{\sigma}_N = \mathbf{D}^-\mathbf{A}^-$. We define the stiffness matrix for the lower semi-space as the surface stiffness matrix $\mathbf{K}_S^+$ which is given by

$$\boldsymbol{\sigma}_N = \mathbf{D}^-(\mathbf{P}^-)^{-1}\mathbf{u}_N = \mathbf{K}_S^+\mathbf{u}_N, \tag{6.18}$$

where $\mathbf{K}_S^+ = \mathbf{D}^-(\mathbf{P}^-)^{-1}$ is the 3×3 stiffness matrix for the lower semi-space. The reason for the apparently contradictory definition for the lower semi-space surface matrix $\mathbf{K}_S^+$, with the superscript $(+)$, is that it is defined for the *upper* surface of the lower semi-space, and for the upper semi-space the surface stiffness matrix is defined on its *lower* surface, emphasized by the superscript $(-)$ in $\mathbf{K}_S^-$,

$$\boldsymbol{\sigma}_0 = \mathbf{D}^+(\mathbf{P}^+)^{-1}\mathbf{u}_0 = \mathbf{K}_S^-\mathbf{u}_0, \tag{6.19}$$

where $\mathbf{K}_S^- = \mathbf{D}^+(\mathbf{P}^+)^{-1}$ is the surface stiffness matrix for the upper semi-space. As is customary, the $(+)$ and $(-)$ signs indicate positive- and negative-outgoing normals, respectively, at the corresponding surfaces. Likewise, we define the surface compliance matrix $\mathbf{S}_S = \mathbf{K}_S^{-1}$.

To introduce attenuation into the composite lamina we will assume elastic constants c_{q3tm} in Eq. (6.4) to be complex with a small negative imaginary part (this can be estimated from experiment or chosen empirically). From a computational standpoint, a small attenuation can be useful because it precludes the appearance of unphysical singularities. Choosing a negative imaginary part in the elastic constants produces positive wave attenuation for propagation in both the positive and negative z directions. For example, for propagation in the positive z direction, $\exp(ik_z z) = \exp[-k_z'' z]\exp[ik_z' z]$. In this case, the normalized attenuation coefficient, or the attenuation per wavelength, is frequency independent, while the attenuation coefficient, or the attenuation per unit distance, is proportional to frequency. In principle, the imaginary parts of the elastic constants can be selected so that the attenuation coefficient exhibits a frequency dependence different from linear. In comparisons of experimental and modeling results shown below, however, the normalized attenuation is taken independent of frequency (linear attenuation coefficient).

The stiffness matrix **K** becomes singular if the determinant of the matrix $\mathbf{E}_j^u$ in Eq. (6.12) is zero (owing to the fact that its inverse is used in computing **K**). This condition occurs at resonance frequencies of a layer under constrained boundary conditions ($\mathbf{u} = 0$) at both top and bottom layer surfaces. Therefore, at a given value of k_x, there will be a number of angular frequencies ω_r, where the stiffness matrix is singular. The compliance matrix, moreover, becomes singular at resonant frequencies corresponding to free boundary conditions ($\boldsymbol{\sigma} = 0$), *i.e.*, when the determinant of the matrix $\mathbf{E}_j^\sigma$ in Eq. (6.13) is zero. (The inverse of $\mathbf{E}_j^\sigma$ is used in calculating **S**, as seen in Eq. (6.16).) If one includes material attenuation, then all the resonance frequencies become complex, and therefore **K** and **S** will always be regular for real ω.

Note on matrices

We have cast the (6×6) matrices in our matrix equations in the form of (2×2) matrices, where each of the four matrix elements is a submatrix of dimension (3×3). This representation simplifies the matrix manipulations. Matrix multiplication with (2×2) square matrices is done for submatrices in the same way as with scalar matrix elements (the only proviso being that matrices generally do not commute). Moreover, the use of submatrices simplifies the calculation of the matrix inverse.

We return now to the question of how the stiffness matrix suppresses the numerical instabilities in the transfer matrix approach. An examination of the first line in Eq. (6.14) reveals that the **H** matrices are no longer located on a single column, nor along the diagonal, both risky places to have vanishing matrix elements. Instead, the **H** matrices are distributed away from the diagonal and never all in one column. Plus, the diagonal elements of the matrices in Eq. (6.14) are independent of layer thickness. This rearrangement indeed satisfies the criteria we postulated for avoiding the numerical instabilities of the transfer matrix method.

Why does this reformulation work? Under weak coupling conditions (*i.e.*, the occurrence of evanescent partial waves) the stiffness matrix approach naturally de-couples the input and output ports. That is, the stiffness seen at one port becomes independent from the stiffness loading the layer at the other port. The stiffness matrix approach is always stable because its elements are combinations of traction-free and clamped stiffnesses that always exist. The fact is that the transfer matrix approach is stable only when there actually *is* transfer; but as we have seen, this is not always the case.

Our job, however, is far from done. Due to the rearrangement of elements in the matrix on the right-hand side of Eq. (6.14), the task of constructing the layer stiffness matrix and, beyond that, the total stiffness matrix is not as simple as before. To be general, let us denote the layer stiffness matrix for the first layer ($j = 1$) by $\mathbf{K}^A$ and the one for the second layer ($j = 2$) by $\mathbf{K}^B$. Then, according to Eq. (6.14)

$$\begin{bmatrix} \boldsymbol{\sigma}_0 \\ \boldsymbol{\sigma}_1 \end{bmatrix} = \begin{bmatrix} \mathbf{K}^A_{11} & \mathbf{K}^A_{12} \\ \mathbf{K}^A_{21} & \mathbf{K}^A_{22} \end{bmatrix} \begin{bmatrix} \mathbf{u}_0 \\ \mathbf{u}_1 \end{bmatrix}, \qquad \begin{bmatrix} \boldsymbol{\sigma}_1 \\ \boldsymbol{\sigma}_2 \end{bmatrix} = \begin{bmatrix} \mathbf{K}^B_{11} & \mathbf{K}^B_{12} \\ \mathbf{K}^B_{21} & \mathbf{K}^B_{22} \end{bmatrix} \begin{bmatrix} \mathbf{u}_1 \\ \mathbf{u}_2 \end{bmatrix}. \tag{6.20}$$

By solving for $\boldsymbol{\sigma}_1$ and $\mathbf{u}_1$ in the relations above, these terms can be eliminated, so that $\boldsymbol{\sigma}_0$ and $\mathbf{u}_0$ can be related directly to $\boldsymbol{\sigma}_2$ and $\mathbf{u}_2$ through elements of both stiffness matrices,

$$\begin{bmatrix} \boldsymbol{\sigma}_0 \\ \boldsymbol{\sigma}_2 \end{bmatrix} = \begin{bmatrix} \mathbf{K}^A_{11} + \mathbf{K}^A_{12}(\mathbf{K}^B_{11} - \mathbf{K}^A_{22})^{-1}\mathbf{K}^A_{21} & -\mathbf{K}^A_{12}(\mathbf{K}^B_{11} - \mathbf{K}^A_{22})^{-1}\mathbf{K}^B_{12} \\ \mathbf{K}^B_{21}(\mathbf{K}^B_{11} - \mathbf{K}^A_{22})^{-1}\mathbf{K}^A_{21} & \mathbf{K}^B_{22} - \mathbf{K}^B_{21}(\mathbf{K}^B_{11} - \mathbf{K}^A_{22})^{-1}\mathbf{K}^B_{12} \end{bmatrix} \begin{bmatrix} \mathbf{u}_0 \\ \mathbf{u}_2 \end{bmatrix}. \tag{6.21}$$

This process can be repeated iteratively across the entire laminate. If we rename the stiffness matrix in Eq. (6.21) by $\mathbf{K}^A$ and layer stiffness matrix for the next two layers ($j = 3$) by $\mathbf{K}^B$, the calculation leading from Eq. (6.20) to Eq. (6.21) can be repeated, eventually encompassing the entire layer stack. Restated as an algorithm, where $\tilde{\mathbf{K}}^J$ is the total stiffness matrix for the upper J layers and $\mathbf{K}^j$ is the layer stiffness matrix for the jth layer, we have

$$\tilde{\mathbf{K}}^J = \begin{bmatrix} \tilde{\mathbf{K}}^{J-1}_{11} + \tilde{\mathbf{K}}^{J-1}_{12}(\mathbf{K}^j_{11} - \tilde{\mathbf{K}}^{J-1}_{22})^{-1}\tilde{\mathbf{K}}^{J-1}_{21} & -\tilde{\mathbf{K}}^{J-1}_{12}(\mathbf{K}^j_{11} - \tilde{\mathbf{K}}^{J-1}_{22})^{-1}\mathbf{K}^j_{12} \\ \mathbf{K}^j_{21}(\mathbf{K}^j_{11} - \tilde{\mathbf{K}}^{J-1}_{22})^{-1}\tilde{\mathbf{K}}^{J-1}_{21} & \mathbf{K}^j_{22} - \mathbf{K}^j_{21}(\mathbf{K}^j_{11} - \tilde{\mathbf{K}}^{J-1}_{22})^{-1}\mathbf{K}^j_{12} \end{bmatrix}. \tag{6.22}$$

(It is critically important to distinguish between $\mathbf{K}^j$, the matrix for the jth layer alone, and $\tilde{\mathbf{K}}^J$, the *total* stiffness matrix for all layers from 1 through J.) The only mathematical concern here is the matrix inversion, but because the submatrices $\mathbf{K}^j_{11}$ and $\mathbf{K}^j_{22}$ contain no real exponential functions, they will be well behaved as layer thickness increases (or acoustic wavelength decreases). The inverse matrix $(\mathbf{K}^j_{11} - \tilde{\mathbf{K}}^{J-1}_{22})^{-1}$ in Eq. (6.22) will always be regular.

We now consider some important symmetry properties of the submatrices assuming that the elastic stiffnesses are real, *i.e.*, there is no attenuation in the layers. (The symmetry properties apply only for lossless layers.) In what follows, the designation **M** stands for either the **S** matrix or the **K** matrix. We subdivide the generalized stiffness/compliance matrix $\mathbf{M}(6 \times 6)$ into four (3×3) submatrices, denoted $\mathbf{M}_{11}$, $\mathbf{M}_{12}$, $\mathbf{M}_{21}$, and $\mathbf{M}_{22}$. For a lossless layer of general anisotropy, the following stiffness/compliance matrix symmetries exist:

$$\mathbf{M}_{11}^{\mathrm{T}} = -\mathbf{M}_{22}, \quad \mathbf{M}_{22}^{\mathrm{T}} = -\mathbf{M}_{11}, \quad \mathbf{M}_{12}^{\mathrm{T}} = \mathbf{M}_{12}, \quad \mathbf{M}_{21}^{\mathrm{T}} = \mathbf{M}_{21} \tag{6.23}$$

$$\mathbf{M}_{11}^{\mathrm{H}} = \mathbf{M}_{11}, \quad \mathbf{M}_{22}^{\mathrm{H}} = \mathbf{M}_{22}, \quad \mathbf{M}_{12}^{\mathrm{H}} = -\mathbf{M}_{21}, \quad \mathbf{M}_{21}^{\mathrm{H}} = -\mathbf{M}_{12}, \tag{6.24}$$

where superscript T indicates transpose, H means the Hermitian conjugate (*i.e.*, the transpose of its complex conjugate $\mathbf{M}_{ij}^{\mathrm{H}} = (\mathbf{M}_{ij}^{*})^{\mathrm{T}}$, and (*) denotes a complex conjugate). Using the definition of Hermitian conjugate and combining Eqs. (6.23) and (6.24), one obtains useful properties for the submatrices

$$\mathbf{M}_{11}^{*} = \mathbf{M}_{11}^{\mathrm{T}}, \quad \mathbf{M}_{22}^{*} = \mathbf{M}_{22}^{\mathrm{T}}, \quad \mathbf{M}_{12}^{*} = -\mathbf{M}_{21}, \quad \mathbf{M}_{21}^{*} = -\mathbf{M}_{12} \tag{6.25}$$

It follows from Eq. (6.25) that all diagonal elements of submatrices $\mathbf{M}_{11}$ and $\mathbf{M}_{22}$ are real.

A laminated composite will still contain a single plane of mirror symmetry, as long as each lamina is rotated only about the axis normal to its fibers. This is the case, for example, of a natively orthotropic material rotated about its out-of-plane direction (the axis labeled z or 3 in Fig. 6.1). The effective average elastic properties become monoclinic, but a mirror symmetry plane (the (x, y) or (1,2) plane) still exists. The existence of such a mirror symmetry plane implies these further symmetry relations among the stiffness/compliance submatrix elements,

$$\mathbf{M}_{11} = \begin{bmatrix} M^{11} & M^{12} & M^{13} \\ M^{12} & M^{22} & M^{23} \\ -M^{13} & -M^{23} & M^{33} \end{bmatrix}, \quad \mathbf{M}_{12} = \begin{bmatrix} M^{14} & M^{15} & M^{16} \\ M^{15} & M^{25} & M^{26} \\ M^{16} & M^{26} & M^{36} \end{bmatrix}$$

$$\mathbf{M}_{21} = \begin{bmatrix} -M^{14} & -M^{15} & M^{16} \\ -M^{15} & -M^{25} & M^{26} \\ M^{16} & M^{26} & -M^{36} \end{bmatrix}, \quad \mathbf{M}_{22} = \begin{bmatrix} -M^{11} & -M^{12} & M^{13} \\ -M^{12} & -M^{22} & M^{23} \\ -M^{13} & -M^{23} & -M^{33} \end{bmatrix}. \tag{6.26}$$

Owing to symmetry, only 12 elements M^{ij} of the matrix **M** are independent. The combination of properties in Eq. (6.25) and Eq. (6.26) implies that elements M^{13}, and M^{23} of submatrix $\mathbf{M}_{11}$, and elements M^{16} and M^{26} of submatrix $\mathbf{M}_{12}$ are purely imaginary, and the remaining elements of matrix **M** are purely real in a lossless medium.

6.3.2 Global stiffness matrix

The total or global matrix for the multilayered anisotropic structure is

$$\begin{bmatrix} \sigma_0 \\ \sigma_N \end{bmatrix} = \begin{bmatrix} \mathbf{K}_{11} & \mathbf{K}_{12} \\ \mathbf{K}_{21} & \mathbf{K}_{22} \end{bmatrix} \begin{bmatrix} \mathbf{u}_0 \\ \mathbf{u}_N \end{bmatrix}, \tag{6.27}$$

where the submatrices $\mathbf{K}_{ij}$ of the total stiffness matrix are obtained using the recursive algorithm shown in Eq. (6.22) as discussed earlier in this chapter. We call the total stiffness, or compliance, matrix for the layered structure the *global matrix*. This designation must be distinguished from the similar term "Global matrix method" [6] noted in the Introduction, which describes the high-order matrix assembled from all layers in the structure. In this book, the term global stiffness or compliance matrix only relates to the matrices introduced above and obtained by the recursive scheme Eq. (6.22).

6.3.3 Relation between transfer and stiffness matrices

Since both stiffness and transfer matrices relate stresses and displacements on surfaces of the layer, the two kinds of matrices are related to each other. While these relations are not suitable for stable computations in composite layers, they would be important in the treatment of very thin layers. The relation between the transfer matrix **B** and the stiffness matrix **K** is given by [12],

$$\mathbf{B}_{6\times 6} = \begin{bmatrix} -(\mathbf{K}_{12})^{-1}\mathbf{K}_{11} & (\mathbf{K}_{12})^{-1} \\ \mathbf{K}_{21} - \mathbf{K}_{22}(\mathbf{K}_{12})^{-1}\mathbf{K}_{11} & \mathbf{K}_{22}(\mathbf{K}_{12})^{-1} \end{bmatrix}. \tag{6.28}$$

The stiffness matrix can also be represented through the transfer matrix elements

$$\mathbf{K}_{6\times 6} = \begin{bmatrix} -(\mathbf{B}_{12})^{-1}\mathbf{B}_{11} & (\mathbf{B}_{12})^{-1} \\ \mathbf{B}_{21} - \mathbf{B}_{22}(\mathbf{B}_{12})^{-1}\mathbf{B}_{11} & \mathbf{B}_{22}(\mathbf{B}_{12})^{-1} \end{bmatrix}. \tag{6.29}$$

Equations Eqs. (6.28) and (6.29) are exact analytical relations. The layer stiffness matrix cannot, however, be calculated directly using Eq. (6.29) for composite layers in the range outside of the transfer matrix stable computation, *i.e.*, when composite layers are not very thin or wavelengths are long. To obtain the layer stiffness matrix, computations must in general be based on Eq. (6.15). Likewise, computation of **B** using Eq. (6.28) is also unstable outside of the computational stability range of the **B** matrix. The relations in Eqs. (6.28) and (6.29) are the same for **S** matrix.

Analytically, Eqs. (6.28) and (6.29) are also applicable in the computation of the global stiffness/transfer matrices **K** and **B**. They can be used computationally in the range of stability of global matrix **B**. In this chapter, we have developed the stiffness matrix for cascading layers from the top to the bottom of the layered structure (see Fig. 6.2). If the stacking is inverted, so that it runs from the bottom to the top, instead, the relations in Eqs. (6.28) and (6.29) and the recursive relation Eq. (6.22) should be correspondingly modified. It is a useful exercise to compare equations in these cases and re-derive some of them.

6.4 Scattering Coefficients for a Fluid-Loaded Composite Laminate

In this section, we discuss ultrasonic wave interactions with a multilayered composite laminate immersed in a fluid. We have addressed in Chapter 4 the scattering of ultrasonic waves on an interface between a fluid and a unidirectional composite half-space. The reflection and transmission coefficients were obtained from the solutions of linear systems of equations representing boundary or interfacial conditions between fluid and anisotropic solid. For a composite laminate, we couple the reflected and transmitted fields using the total stiffness matrix of the multilayered composite structure, in essence utilizing boundary conditions on both composite fluid interfaces. In Chapter 9, when we discuss composite characterization using ultrasonic transducers in air, our analysis will be applicable to this case also, if the immersion fluid takes on the properties of the air.

To find the reflected and transmitted fields, we utilize the total compliance matrix $\mathbf{S}$ for the multidirectional composite structure. As in Eq. (6.27) we obtain $\mathbf{S}$ for a stack of N layers,

$$\begin{bmatrix} \mathbf{u}_0 \\ \mathbf{u}_N \end{bmatrix} = \begin{bmatrix} \mathbf{S}_{11} & \mathbf{S}_{12} \\ \mathbf{S}_{21} & \mathbf{S}_{22} \end{bmatrix} \begin{bmatrix} \boldsymbol{\sigma}_0 \\ \boldsymbol{\sigma}_N \end{bmatrix}, \tag{6.30}$$

or obtain $\mathbf{S}$ by matrix inversion,

$$\mathbf{S} = \mathbf{K}^{-1} = \begin{bmatrix} \mathbf{S}_{11} & \mathbf{S}_{12} \\ \mathbf{S}_{21} & \mathbf{S}_{22} \end{bmatrix}. \tag{6.31}$$

As before $\mathbf{u}_0$ and $\boldsymbol{\sigma}_0$ are vectors of displacement and traction respectively on the composite top surface, and $\mathbf{u}_N$ and $\boldsymbol{\sigma}_N$ is the same pair on the bottom surface.

Before we continue, the reader should be reminded that for the sake of simplicity our notation is necessarily contracted. When considered as z-dependent field distributions, $\mathbf{u}_j$ or $\boldsymbol{\sigma}_j$ are displacement and traction vectors in the jth layer (see in Eqs. (6.1) and (6.3)). When they are independent of z, however, they represent specialized values at the jth interface, *i.e.*, at the bottom of the jth layer (see Eqs. (6.5) and later). Continuing this notation for the upper and lower fluid half-spaces, let us assume that an ultrasonic wave is incident on an arbitrarily oriented multilayered composite structure in a fluid from the top fluid half-space as shown in Fig. 6.2. We seek a solution for the displacements in the fluid directly. Let us denote the wave solution in the upper fluid half-space with a subscript "0" ($z > 0$ in Fig. 6.2) as

$$\mathbf{u}_0 = a_{\text{in}}(\mathbf{p}_{\text{in}}e^{-ik_zz} + R\mathbf{p}_{\text{r}}e^{ik_zz})e^{i(k_xx-\omega t)}, \tag{6.32}$$

and for the transmitted wave in the lower fluid half-space, the $(N+1)$th medium,

$$\mathbf{u}_{N+1} = a_{\text{in}}T\mathbf{p}_{\text{t}}e^{-ik_zz}e^{i(k_xx-\omega t)}, \tag{6.33}$$

where a_{in} is the incident amplitude, $R = a_{\text{r}}/a_{\text{in}}$ and $T = a_{\text{t}}/a_{\text{in}}$ are reflection and transmission coefficients, respectively, and a_{r} and a_{t} are reflected and transmitted amplitudes. The polarization vector is $\mathbf{p}_\alpha$, which equals the wave normal $\mathbf{n}_\alpha$ (α stands for "in" (incident), "r" (reflected), or "t" (transmitted) waves). The normal stress component $\sigma_{zz} = -p$, where p is the pressure, is found as

$$\sigma_{zz} = \rho_f V_f^2 \left(\frac{\partial u_x}{\partial x} + \frac{\partial u_z}{\partial z} \right), \tag{6.34}$$

where ρ_f is the fluid density, and V_f is the fluid wavespeed. The normal components of displacements and stresses on the top ($z = z_0$) and the bottom ($z = z_N$) contact interfaces between the fluid and composite are, suppressing the common term $\exp[i(k_xx - \omega t)]$,

$$u_z^0 = a_{\text{in}}(R-1)\cos\theta_f, \qquad \sigma_{zz}^0 = a_{\text{in}}i\omega\rho_f V_f(R+1), \tag{6.35}$$

$$u_z^N = -a_{\text{in}}T\cos\theta_f, \qquad \sigma_{zz}^N = a_{\text{in}}i\omega\rho_f V_f T, \tag{6.36}$$

where the continuity of the normal stress and displacement components on the interface is accounted for. The vanishing of the tangential stresses at the fluid–solid interfaces does not have to be enforced separately because continuity of the tangential tractions is inherently included in the model and there are no shear stresses in the fluid. The wavevector angle θ_f of a plane wave in the fluid is measured from the normal. The coordinate z is taken to be zero on both surfaces because the phase delay in the plate is accounted for by the stiffness matrix and the selection of the local coordinate system for each layer. To simplify the notation, we have used in Eqs. (6.35) and (6.36) superscripts instead of subscripts to identify the displacement and stress components at the zeroth and Nth interfaces, *i.e.*, at the top and the bottom of the composite plate.

The reflected and transmitted field in the fluid must satisfy the boundary conditions on the top and bottom surfaces of the composite structure. On the composite–fluid interfaces we have continuity of the normal stress and displacement components and the vanishing of the shear stress components ($\sigma_{zx}^0 = \sigma_{zy}^0 = \sigma_{zx}^N = \sigma_{zy}^N = 0$). Displacements and stresses on the surfaces of this structure are given by the total compliance matrix of the structure, Eq. (6.30). Therefore, the two equations from the compliance matrix relation

Eq. (6.30) between displacement and stress can be particularized for the normal displacement components (u_z^0, u_z^N) only as

$$u_z^0 = S_{11}^{33}\sigma_{zz}^0 + S_{12}^{33}\sigma_{zz}^N \tag{6.37}$$

and

$$u_z^N = S_{21}^{33}\sigma_{zz}^0 + S_{22}^{33}\sigma_{zz}^N, \tag{6.38}$$

where the terms S_{mn}^{33} are the (3,3) elements in the $\mathbf{S}_{mn}(3 \times 3)$ submatrices that comprise the 6×6 total compliance matrix $\mathbf{S}$ in Eq. (6.31). Tangential displacement components on the surfaces of the structure do not participate in the boundary conditions for loading with an ideal (inviscid) fluid.

To find the unknown reflection R and transmission T coefficients in the fluid, we now substitute Eqs. (6.35–6.36) for the normal displacements and stresses $(u_z^0, u_z^N, \sigma_{zz}^0, \sigma_{zz}^N)$ into Eqs. (6.37–6.38) relating R and T to elements of the total stiffness matrix of the composite multilayered structure

$$\Lambda(1 - R) = (1 + R)S_{11}^{33} + TS_{12}^{33}, \tag{6.39}$$

and

$$\Lambda T = (1 + R)S_{21}^{33} + TS_{22}^{33}, \tag{6.40}$$

where $\Lambda = -\cos\theta_f/(i\omega\rho_f V_f)$. Then, for the transmission and reflection coefficients, we have

$$T = -\frac{2\Lambda S_{21}^{33}}{(S_{11}^{33} + \Lambda)(S_{22}^{33} - \Lambda) - S_{21}^{33}S_{12}^{33}} \tag{6.41}$$

and

$$R = -\frac{(S_{11}^{33} - \Lambda)(S_{22}^{33} - \Lambda) - S_{21}^{33}S_{12}^{33}}{(S_{11}^{33} + \Lambda)(S_{22}^{33} - \Lambda) - S_{21}^{33}S_{12}^{33}}. \tag{6.42}$$

If the bottom fluid half-space is absent and the bottom surface of the composite is free, the normal stress on the lower surface of the composite structure is zero, $\sigma_{zz}^N = 0$. In that case, we obtain for the reflection coefficient

$$R = -\frac{S_{11}^{33} - \Lambda}{S_{11}^{33} + \Lambda}. \tag{6.43}$$

As we noted with respect to Eq. (6.25) S_{11}^{33} is real, while Λ is purely imaginary, and thus for a lossless composite $|R| = 1$. This problem can be of practical significance in the case of ultrasonic wave interaction with a composite vessel of small curvature immersed in a fluid, but containing air inside.

When the material description requires only two partial waves, such as orthotropic media along a symmetry direction or isotropic media, then the

3×3 submatrices are replaced by 2×2 submatrices, and all terms of the form S^{33}_{mn} in Eqs. (6.41–6.42) are replaced by S^{22}_{mn}.

An example of a numerical calculation of wave reflection and transmission, using the transfer matrix, for a $[0°/90°]_{8S}$ composite laminate with a plane wave incident at 45° and a fiber orientation angle ϕ of 30°(see Fig. 6.1) is shown in Fig. 6.3. At this relatively high incident angle, the transfer matrix begins to show critical instabilities at a frequency of only 700 kHz. Beyond this point in frequency the calculation is completely useless, owing to the effects of the exponentials with real arguments in layers where the partial waves are evanescent. By contrast, the stiffness matrix approach renders the entire spectrum with no instability at all, as seen in Fig. 6.4. The frequency range of the calculation has been expanded in Fig. 6.4 to reach out past 8 MHz. Clearly, well beyond the point where the transfer matrix method can no longer be used, the stiffness matrix approach is unconditionally stable. To drive this point home, the region from 990 to 1000 MHz is calculated and plotted at the far right-hand side of the graph in Fig. 6.4. Although quite an unrealistic frequency range for ultrasonic tests of a 16-ply composite laminate, the calculation demonstrates the complete stability of the stiffness

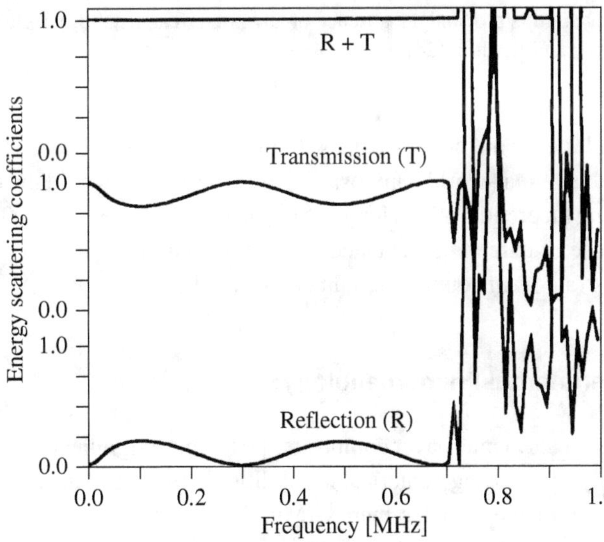

Figure 6.3. Energy reflection and transmission coefficients for a $[0°, 90°]_{8S}$ composite layer stack calculated by the *transfer matrix method* with an incident plane wave at 45°. In-plane fiber orientation angle is 30°. Also recorded is the sum of the two coefficients, demonstrating that they indeed add to unity. Note the instabilities that develop after 700 kHz (after Wang and Rokhlin [24]).

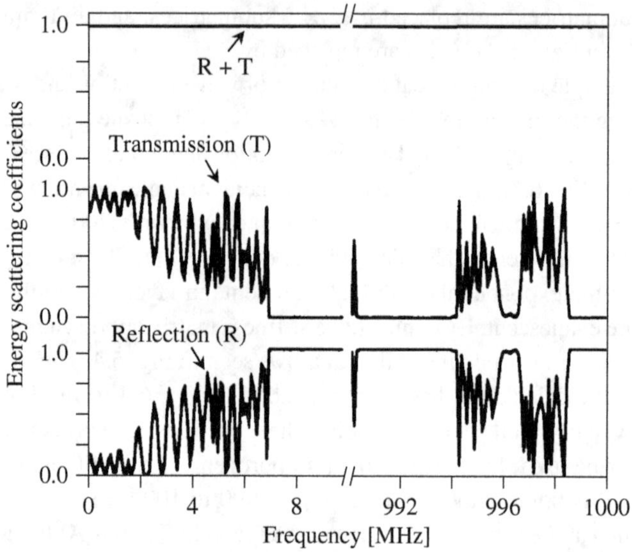

Figure 6.4. Energy reflection and transmission coefficients for the same composite layer stack as in Fig. 6.3 calculated by the *stiffness matrix method* for an incident plane wave at 45°. In-plane fiber orientation angle is the same at 30°. Again, the sum of the two coefficients adds everywhere to unity. Note the extended frequency range in this plot and the absence of any numerical instability (after Wang and Rokhlin [24]).

matrix under extreme circumstances. In practical ultrasonic tests, the non-ideal aspects of real-world composite sample fabrication would overwhelm any existing theoretical model long before the frequency reached 1000 MHz. The purpose of extending the calculation to such a high frequency is only to demonstrate its unconditional numerical stability.

6.5 Experimental Phenomenology

Modeling waves in multilayer composites poses more significant challenges than simply suppressing numerical instabilities in the scattering coefficients calculated with the transfer matrix. Multilayer composite laminates are complicated structures that have granularity on multiple scales. The ply dimension is typically 1/8 mm, or about eight plies to the millimeter. The individual plies, however, have granularity at other dimensions as well. Of course, there are the individual high-modulus fibers, whose diameters are in the range of 3–5 μm. By contrast, ultrasonic wavelengths in an individual composite ply at 50 MHz (higher than the frequency of most practical tests) are at least 50 μm, far longer than the fiber dimension. For most considerations of ultrasonic testing, the fiber dimension can generally be ignored and the ply considered as an anisotropic continuum. Even this apparently valid assumption is not always justified. Although the individual fibers are very small compared to an acoustic wavelength, substantial local variations in fiber density can easily be at the scale of an acoustic wavelength in the ply [16]. These variations can be the size of moderate frequency ultrasonic wavelengths in the composite, and they constitute yet another scale dimension of a fabricated composite laminate. A close examination of Fig. 6.5 shows both the fibers and the variation in their density throughout a ply, as well as variations in ply thickness. This level of variability is typical of multiaxial composites. The fiber density variations lead to spatially varying local stiffnesses. These variations in the local stiffness occurring on the scale of the ultrasonic wavelength can markedly influence the backscattering behavior of the laminate. Uniformly distributed reinforcement would appear to be macroscopically homogeneous and would exhibit very small ultrasonic backscattering. Real fabricated composites, on the other hand, can have a strong backscattered signal that can even be exploited to assess mis-orientations in the ply layup scheme, to detect matrix

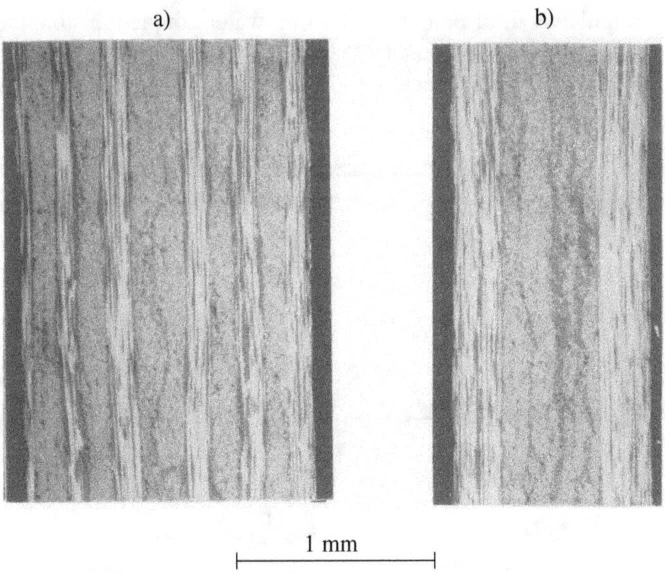

Figure 6.5. Optical micrographs of typical composite laminates. Frame a) shows a $[0°/90°]_{3S}$ layup, and frame b) is a polished cross-section of a $[0°_2/90°_2]_S$ layup. The lighter regions with long stringy black lines are the plies with fibers running top to bottom in the plane of the figure; the darker regions with point-like black areas are plies with fibers running normal to the figure plane (after Chimenti and Nayfeh [15]).

cracking [17, 18], to reveal porosity [19], or to induce the propagation of Lamb waves [20].

Idealized mathematical modeling of ultrasound in composite laminates makes some natural, but not always justified, assumptions. For one, a multiaxial laminate is assumed to be composed of layers of uniform (and often equal) thickness. From Fig. 6.5, it is clear that even with careful fabrication the composite can have plies of variable thickness. Another quite conventional assumption is that each ply can be modeled as an anisotropic, but homogeneous, continuum. Again, from the figure above there is little doubt that this assumption is a gross idealization. Yet, under some conditions and for some kinds of wave propagation and scattering, the typical modeling idealizations of real composite laminates can be reasonably well justified. Certainly, to abandon these convenient assumptions would imply either detailed structural knowledge of a laminate (in practice this seldom exists), or valid statistics on the spatial variations in fiber volume fraction and ply thickness, as well as the occurrence factors such as porosity, wavy fibers, incomplete fiber wetting, and others. It makes more sense to assume a perfect laminate for computational simplicity and then argue that deviations from behavior predicted on that basis may be caused by these non-ideal elements.

For impulsive ultrasonic waves from water-coupled piston radiator transducers reflecting from a quasi-isotropic composite laminate having a layup sequence of $[0/-45°/90°/+45°]_{2S}$, the time-domain data are shown in Fig. 6.6. The signal in Fig. 6.6(a) shows clearly the reflections from laminate boundaries, where the smaller variations between the large

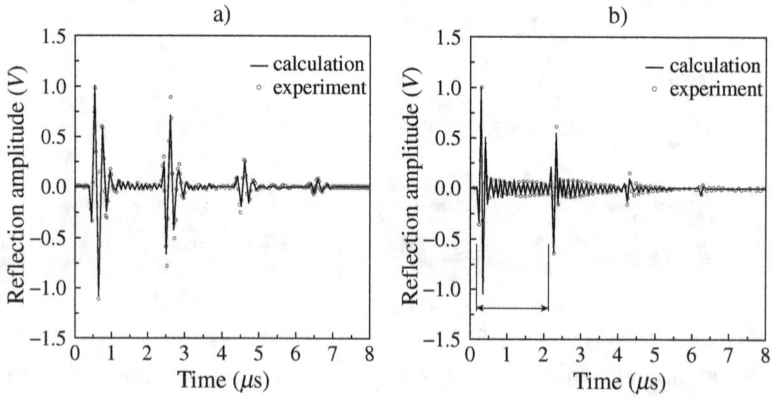

Figure 6.6. Reflected ultrasonic signals at normal incidence from a $[0°/-45°/90°/+45°]_{2S}$ laminate. Frame (a) reflected amplitude with pulse having 5 MHz center frequency; frame (b) same, but with 10-MHz pulse. At higher frequency, reverberations between major reflections are now visible (after Wang and Rokhlin [21]).

reflections are related to the ultrasonic behavior or reflections between or within a ply. At the higher frequency of 10 MHz, the acoustic information between the laminate boundary reflections becomes spatially more coherent, and we can discern reverberations caused by wave reflection at the ply boundaries. At normal incidence we might expect there to be little or no mechanism for the wave to reflect, because there is nominally no change in the acoustic impedance from ply to ply. When composite laminates are fabricated, however, they are often cured under pressure, forcing excess resin from the pre-impregnated fiber bundles (pre-preg) comprising a single ply. At the outer laminate surfaces, so-called bleeder cloths are positioned to absorb this excess resin during cure. Within the laminate, however, the excess resin tends to accumulate at the ply internal boundaries, leaving a thin resin-rich region, whose acoustic impedance is indeed quite different from the average value for a typical ply. It is from this micron-thick resin-rich zone that the reverberations between the boundary reflections shown in Fig. 6.6(b) occur. In the calculations presented in this figure, this resin-rich zone has been modeled by a 6-μm thick layer of epoxy resin only. The good agreement between the experimental data and the model calculation at the higher frequency in Fig. 6.6(b) confirms the existence of the layer and its approximate thickness. In frame (a) at lower frequency (or longer acoustic wavelength), the ultrasonic wave cannot sample the details of the thin resin-rich zone, and its properties are simply averaged with the rest of the composite, so that the only remaining signals are the laminate boundary reflections.

A brief review of the geometry of such experiments is useful at this stage. There is a variety of geometrical schemas in which measurements of wave propagation or wave reflection and transmission in composites can be made. As composite laminates present themselves to experimenters primarily in the form of plate-like shapes (at least on a local scale), the study of guided waves in composites is a natural extension of wave scattering. Moreover, because composites are so often inspected nondestructively using water coupling, the concept of leaky guided waves often takes center stage. Two different, but related, geometries for examining composites ultrasonically are shown in Fig. 6.7. In each of these cases the wave traverses the sample and is detected by either the same, or a different (usually identical), transducer. Although the details of the geometry may be a bit different (in these and other possible geometries), the fundamental nature of the ultrasonic interaction is basically the same.

When the composite laminate of Fig. 6.6 is insonified at various incident angles θ and orientation angles ϕ, the results are shown in Fig. 6.8. In Fig. 6.8 we see that the transmission amplitude drops quickly as the incident angle changes from 0° to 10°. This decrease in transmitted amplitude is nearly independent of incident plane orientation angle ϕ. The effect can

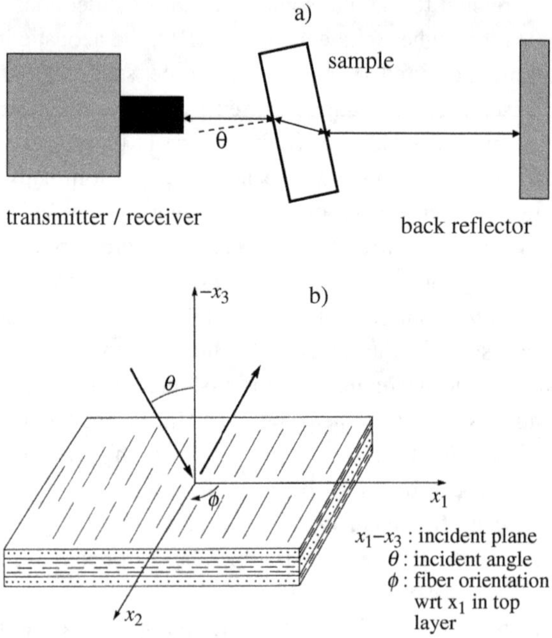

Figure 6.7. Experimental geometry for measuring wave reflection, transmission, or guided wave generation in composite laminates. In frame (a) the wave passes through the sample and reflects from the back surface reflector, passing a second time through the sample. In frame (b) the wave is either received in reflection after two traversals of the sample, or in transmission after a single traversal. The ply orientation angle ϕ is shown in frame (b) (after Wang and Rokhlin [21], and Chimenti and Nayfeh [15]).

be understood by considering the large difference between the in-plane and out-of-plane lamina elastic anisotropy (see Table 3.4). At incident angles between 20° and 40°, there is almost no transmitted energy. Then a very obvious transmission peak emerges around 50° and its maximum position and amplitude changes for different orientation angles. At the 25° orientation angle, there is another transmission peak near an incident angle of 15°.

Time-domain signals from the same composite laminate as sampled in Fig. 6.8 can be seen in Fig. 6.9, where the geometry of Fig. 6.7(a) has been used. The agreement between the data and a model calculation, accounting for the details of the laminate's resin-rich regions and the ultrasound beam geometry is impressively good. At an incident angle of 30° the amplitude falls nearly to zero; an examination of the data in Fig. 6.8 shows that not only at an orientation angle of 25°, but in fact at all the other angles displayed in that figure, the transmitted amplitude is also quite low. The trace at normal incidence (0°) shows echoes that are quite well separated

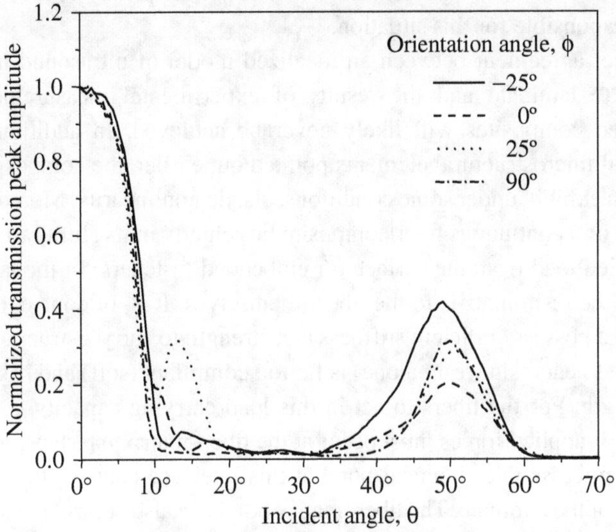

Figure 6.8. Normalized transmitted ultrasonic peak signals versus incident angle at various orientation angles measured on a $[0°/-45°/90°/+45°]_{2S}$ laminate. Each trace shows the transmitted amplitude for a 2.25-MHz center frequency pulse at the orientation angle shown in the legend (after Wang and Rokhlin [21]).

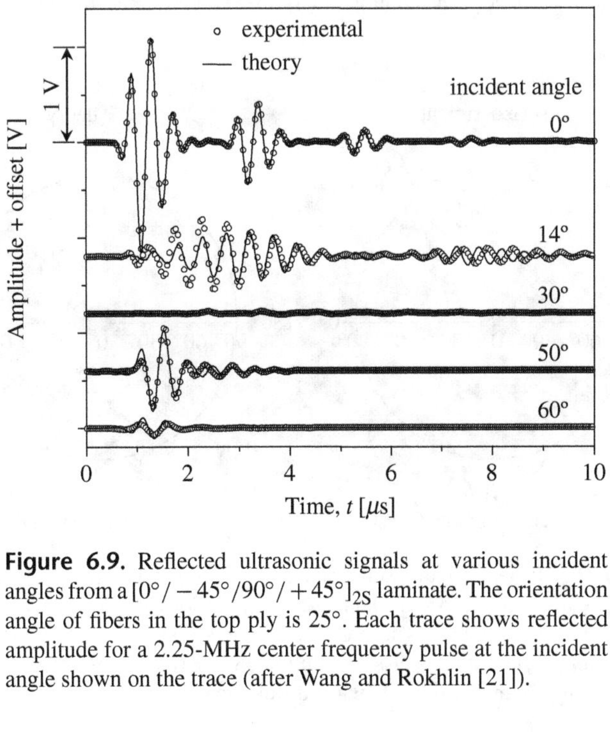

Figure 6.9. Reflected ultrasonic signals at various incident angles from a $[0°/-45°/90°/+45°]_{2S}$ laminate. The orientation angle of fibers in the top ply is 25°. Each trace shows reflected amplitude for a 2.25-MHz center frequency pulse at the incident angle shown on the trace (after Wang and Rokhlin [21]).

with no reverberation between them. The lower frequency of 2.25 MHz is likely responsible for this situation.

Perfect agreement between an idealized model of ultrasound in a real composite laminate and the results of experimental measurements on fabricated composites will likely never be achieved. In addition to the non-ideal microstructural elements pointed out earlier, the composite resin itself can exhibit, under some conditions, elastic nonlinearity. Moreover, the strength of a continuous fiber composite lies chiefly in its ability to transfer the applied load from the surface (or embedded fasteners) of the laminate, through the resin matrix, to the fibers themselves. It is, of course, only the fibers that possess sufficient stiffness and strength to carry a structural load. The cured neat resin (resin alone) is far too compliant (soft) and has almost no strength. For the fibers to act in this load-carrying capacity, a reliable transfer of applied forces must occur at the fiber–matrix interface, meaning that the microscopic chemical bond at this interface controls the ultimate strength of the laminate. The fiber–matrix bond can also control the laminate stiffness, and degraded bond interfaces have been revealed using ultrasonic nondestructive evaluation, in which only the small-strain elastic parameters are evaluated.

A measure of the full spectrum of transmission phenomena can be seen in Fig. 6.10, which has been done in double through-transmission using the geometry of Fig. 6.7(a). The transmission properties of the composite have been measured at orientation angles from 0° to 180° with increments of 1°.

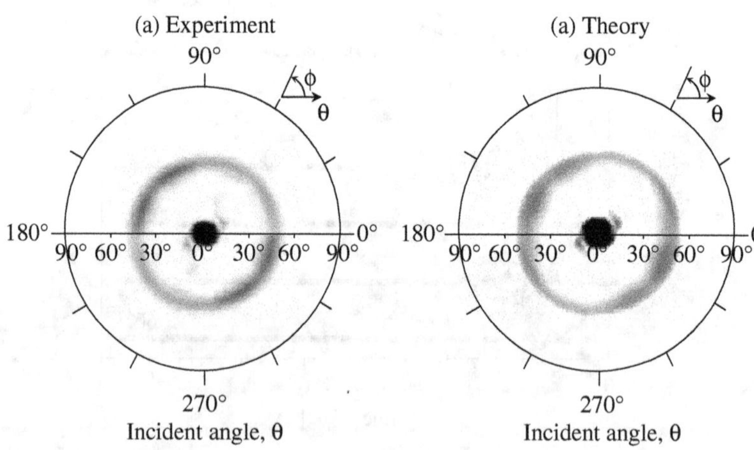

Figure 6.10. Behavior of ultrasonic double through-transmission peak amplitude in a $[0°/-45°/90°/+45°]_{2S}$ laminate for variations in the orientation angle ϕ and the incident angle θ. The transducer center frequency is 2.25 MHz. The gray level represents the transmission amplitude value (black is the highest). Frame (a) experiment, (b) theory (after Wang and Rokhlin [21]).

The transmission peak amplitude is represented as an image of Fig. 6.10(a). In this gray scale, the radial direction represents the incident angle θ and the circumferential direction represents the orientation angle ϕ. The gray level represents the amplitude; the darker the gray level, the higher the amplitude. Figure 6.10(b) shows the corresponding theoretical transmission peak amplitude distribution calculated for the composite laminate. In both the calculation and experimental results, one can see that the amplitude has two-fold rotation symmetry about the horizontal axis. The figure has, however, no reflection symmetry, and at incident angles close to normal incidence the multidirectional composite has behavior similar to an isotropic layer.

Results of a different kind of experiment are shown in Fig. 6.11. This experiment has been performed with a pitch–catch setup, as illustrated in Fig. 6.7(b). Data are acquired in these experiments by exciting a broadband ultrasonic transducer with low-level rf tone bursts from a function generator. The 20-μs bursts propagate through a water path long enough to place the sample in the transducer's far-field at all frequencies. The reflected acoustic beam is detected by a second, matched transducer positioned at the same angle and distance to the sample. A highly reflective tungsten foil is used to find the origin of coordinates, where the two beams intersect at the sample surface. Then, the receiver is displaced away from the intersection by about the transducer radius, partially suppressing sidelobe interference, as we will

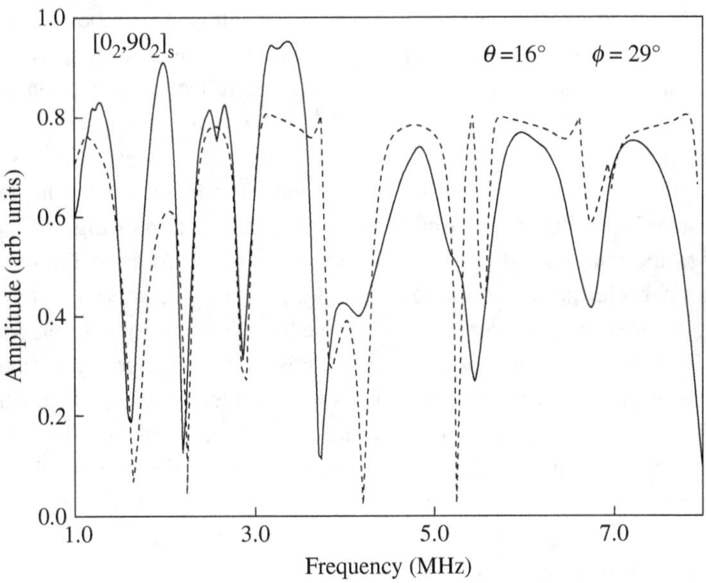

Figure 6.11. Two-transducer reflection spectrum of a $[0°_2/90°_2]_S$ laminate with $\theta = 16°$ and $\phi = 29°$. Experiment is the solid curve, while theory is dashed. Amplitude is in arbitrary units (after Nayfeh and Chimenti [13]).

demonstrate in Chapter 8 [23]. Data are accumulated by a stepwise scanning of the frequency, while the reflected signal is recorded after processing with a broadband receiver/amplifier and envelope detector, sampled by a gated integrating amplifier, and digitized for computer processing and plotting. The transducer spectral characteristics are acquired in a separate experiment with a frequency-independent reflector (such as the tungsten foil) and used to normalize the experimental data. Further experimental details can be found in [22].

The incident angle is θ, which defines only the central axis of the transducer beam, and the ply orientation angle here is denoted by ϕ, referenced to the fiber direction in the topmost ply of the laminate. The general trends of the data are reproduced in the calculations, but it would not be accurate to characterize this as an excellent fit. The difference arises from the way the prediction is calculated. In Fig. 6.11, the reflected signal amplitude is computed from the plane wave reflection coefficient. In the experiment, however, the incident ultrasound is a collimated transducer beam composed of a spectrum of plane waves. So, the measured transducer voltage on the receiver contains the combined effects of the reflection coefficient of the sample, as well as the directivity functions of the transmitter and receiver. By neglecting the wave diffraction phenomena and estimating the received signal with only the nominal reflection coefficient, the comparison between theory and measurement is not as good as it could be. To help minimize the strongest effects of wave diffraction, the data of Fig. 6.11 have been acquired with the ultrasonic probes deployed as described above, somewhat suppressing the finite-beam effects. In particular, the rapid variations predicted in the reflection coefficient near 4 MHz and 5 MHz and not reproduced in the measurement are likely a consequence of finite-beam diffraction. We will explore a deeper analysis of this relationship between receiver voltage and wave diffraction in Chapter 8.

There is another factor complicating the interpretation of Fig. 6.11, and that is the coupling of the incident ultrasonic wave from the fluid to in-plane particle motion in the laminate. Ideally, a plane wave (or a beam from a line source) incident on an isotropic layer or orthotropic layer, where the incident plane contains a symmetry axis, could be expected to couple to out-of-plane particle motion only. No mechanism in that case exists to facilitate coupling of wave motion out of the incident plane. This makes sense: the wave motion from a line source in the fluid is polarized in the vertical plane, and, once in the solid, this same polarization should be seen, unless the solid is anisotropic and the incident plane contains no symmetry direction. Multiaxial composites, where the incident plane does not contain a symmetry axis, will couple wave motion in the vertical plane to motion out of this plane in the solid by mechanisms discussed in Chapter 3. Exactly this situation occurs in Fig. 6.11. At an orientation angle $\phi = 29°$, the incident plane contains no symmetry direction, and all allowed partial waves will, in general, be excited: the quasilongitudinal and both quasishear waves. This fact alone gives rise to a more complicated reflection coefficient than for reflection along a symmetry axis.

Another example of this behavior is seen in Fig. 6.12, where the fiber axis in the top ply lies in the incident plane. So, here $\theta = 12°$ and $\phi = 0°$, implying that the incident plane contains a symmetry axis of the multilayered composite. In this figure, we find much closer agreement between a prediction based on the plane wave reflection coefficient and the result of measurements made with piston radiator transducers. Moreover, this conclusion holds over a wide frequency range from 1 to 12 MHz. Only the depth of the minima and some of the sharpest features of the prediction are not reproduced in the experimental data. This lack of perfect agreement can surely be ascribed to the smoothing effect of the finite-beam transducers and possibly to non-ideal elements of the composite laminate itself.

Away from a symmetry direction, the behavior is again quite complicated, as shown in Fig. 6.13, where the incident angle is $\theta = 12°$, as in Fig. 6.12, and the orientation angle is $\phi = 60°$. The less-regular appearance of reflection minima, corresponding to excitation of guided modes with particle motion

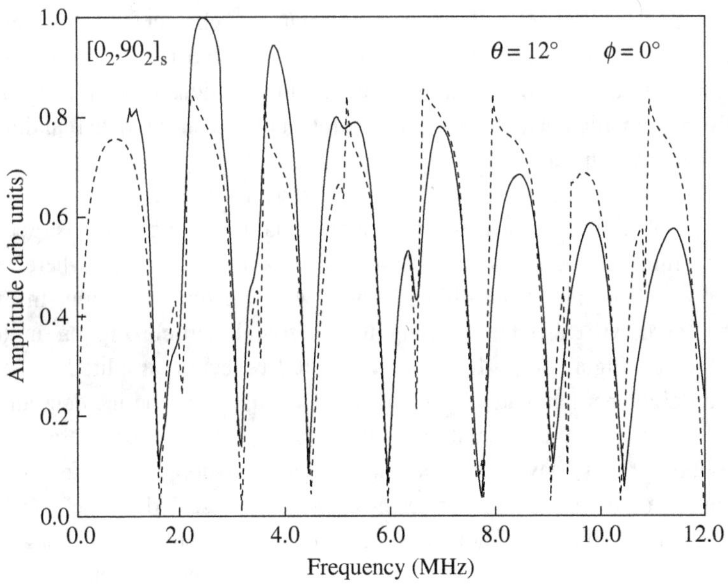

Figure 6.12. Two-transducer reflection spectrum of a $[0°_2/90°_2]_S$ laminate with $\theta = 12°$ and $\phi = 0°$. Fibers in the topmost layer are aligned in the incident plane. Experiment is the solid curve, while theory is dashed (after Nayfeh and Chimenti [13]).

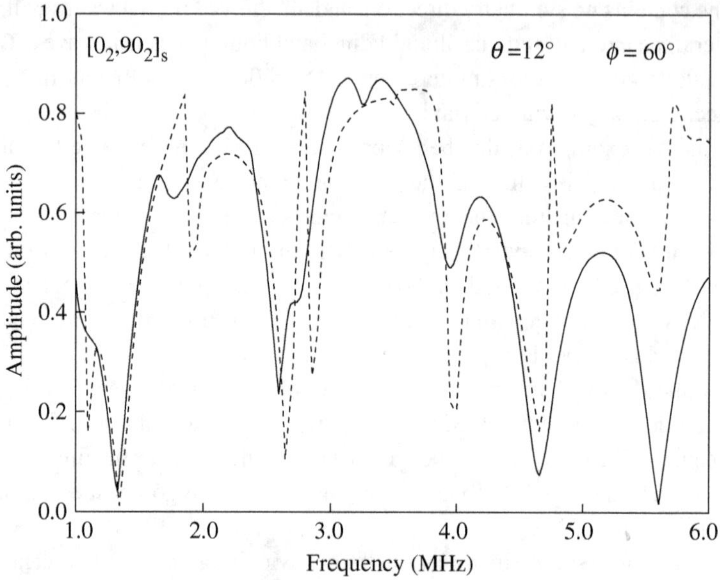

Figure 6.13. Two-transducer reflection spectrum of a $[0°_2/90°_2]_S$ laminate with $\theta = 12°$ and $\phi = 60°$. Experiment is the solid curve, while theory is the dashed curve (after Nayfeh and Chimenti [13]).

components polarized in the horizontal plane, is characteristic of waves incident in planes that contain no symmetry direction. Here again, the depth of the minima and the smoothing of the rapid variations in the plane wave reflection coefficient are likely to be the result of measuring with real finite-diameter transducers.

Although this chapter deals with multilayered composites, implying varying fiber direction with depth through the laminate plies, it is instructive to examine the ultrasonic behavior of uniaxial laminates, where the incident plane makes an arbitrary angle with the fiber direction. In that case there will also be coupling to horizontal particle motion in the composite. Figure 6.14 shows the measured reflection amplitude results for an eight-ply laminate of thickness 0.92 mm, where the incident angle is $\theta = 12°$ and the orientation angle is $\phi = 30°$. The data here, as in previous graphs, have been only normalized to suppress the frequency dependence of the transducer sensitivity and then scaled at one point to the height of the prediction. Comparison with measurements along the fiber direction at this same incident angle suggests that the sharp, shallow minima superimposed over the deeper zeroes measured near 1.5, 3.1, 4.5, and 6.2 MHz are weaker resonances possibly caused by the coupling to horizontal particle motion. (They are largely absent in reflection data from samples measured along a symmetry direction, with *e.g.*, $\phi = 0°$.

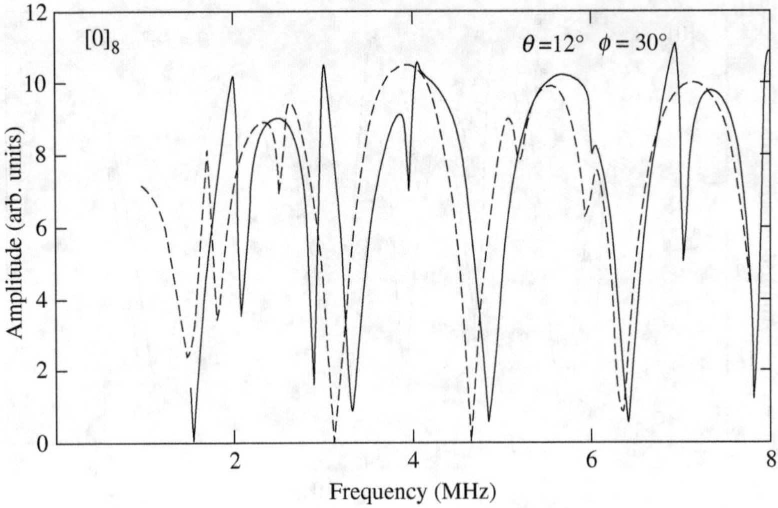

Figure 6.14. Reflected amplitude spectrum of a uniaxial $[0°]_8$ laminate with $\theta = 12°$ and $\phi = 30°$. Theory is the solid curve, while experiment is dashed. Deep minima correspond to guided wave resonances with particle motion polarized principally in the vertical plane; smaller, shallower minima are related to coupling to modes with horizontally polarized wave motion (after Chimenti and Nayfeh [22]).

This observation can be confirmed by an examination of Fig. 6.12). We can see the same phenomena in another data set in Fig. 6.15, where vertically polarized guided wave resonances are indicated by deep minima in both calculation and measured data. Smaller, shallow resonances closer to the reflection maxima are likely related to plate-wave resonances induced by coupling between modes containing vertical and horizontal particle motion. Some of the shallower resonances predicted in the model calculation are masked in the experiment by the spatial averaging effects of the piston radiators [23]. These, and other outstanding issues relating to finite beams and diffraction, will be examined more thoroughly in Chapter 7.

Collecting data on reflection minima from experiments performed at many different incident angles, such as the ones illustrated above, and plotting the data on a single graph of reflection minima, together with a model calculation, yields a result shown in Fig. 6.16. The discrete open circles are the experimental data, while the solid curves are the result of a theoretical calculation accounting only for the anisotropy of the laminate. The data are assembled by collecting reflection minima, but the zeroes of the reflection coefficient are similar to the guided wave mode spectrum in the fluid-loaded plate. The curves here are quite complicated because of the strong coupling

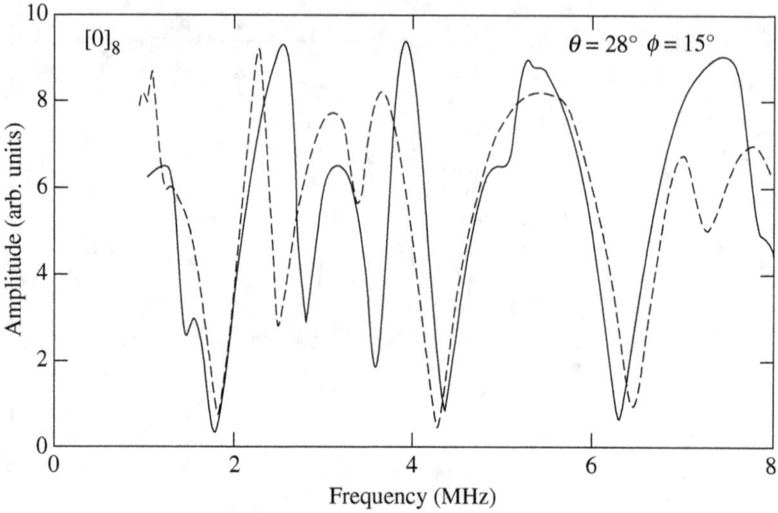

Figure 6.15. Reflected amplitude spectrum of a uniaxial $[0°]_8$ laminate with $\theta = 28°$ and $\phi = 15°$. Theory is the solid curve, while experiment is dashed. Deep guided wave resonances are well predicted by the calculation (after Nayfeh and Chimenti [22]).

between in-plane and out-of-plane particle motion, owing to the orientation of the laminate. Along a symmetry axis, this coupling would not exist, and the curves measured in that case are much simpler and not multiply connected, as we see in this circumstance with $\phi = 60°$.

The relationship here between minima in the reflection coefficient and the occurrence of guided waves, which will be the subject of the next chapter, has been mentioned several times. Guided waves are normal modes of the traction-free plate. In the plate they can be straight-crested waves whose partial wave superposition, longitudinal and shear in isotropic media, or quasilongitudinal and two quasishear waves in anisotropic media, exactly satisfy the conditions of vanishing surface tractions on both parallel surfaces of the uniform plate. (Layered plates have much more complicated guided wave mode structure.) In that case the partial waves within the plate will be in resonance in the transverse, or thickness, direction only. In the in-plane direction, the wave propagates. The transverse resonance condition implies that the tangential particle motion at the plate surfaces must be identical and either in phase, leading to symmetric modes, or out of phase, leading to antisymmetric modes (the normal particle motion exhibits opposite symmetry). If the fluid surrounding the plate can be considered tenuous (not too great a perturbation), then surface displacements caused by the guided wave propagation will not be disturbed very much by the fluid's presence. So, this in-phase or out-of-phase motion will persist.

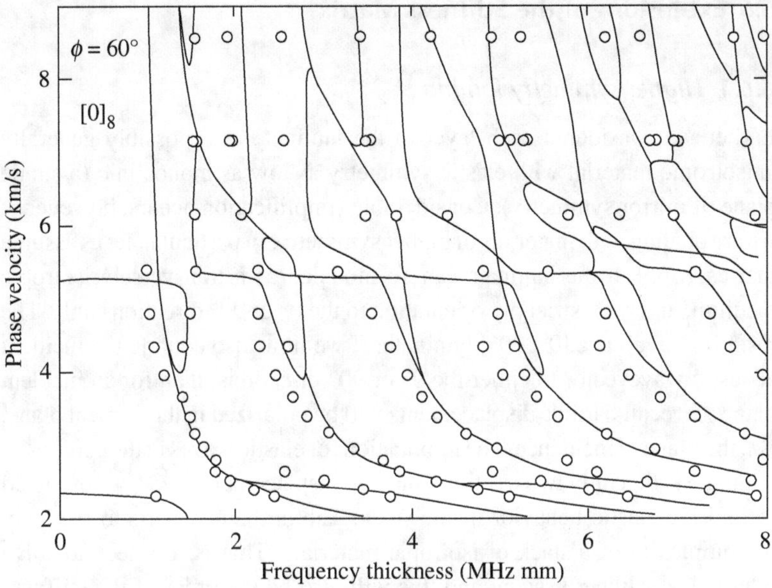

Figure 6.16. Loci of reflection coefficient amplitude minima plotted versus frequency–thickness product for a uniaxial composite laminate, where the fiber direction makes a 60° angle with the incident plane (see Fig. 6.7); experimental data are shown as discrete open circles, and the theoretical calculation is represented by the solid curves. This plot is equivalent to the spectrum of guided waves in the fluid-loaded plate. The complexity of the curves and their connections arise because of wave mode coupling between horizontal and vertical particle motion. Data taken along a symmetry axis (where the coupling is absent) would appear much simpler (after Nayfeh and Chimenti [22]).

If we now imagine an incident plane wave at the plate surface and ask under what conditions the reflection amplitude would approach zero, those conditions must be when the plate effectively "disappears." That unusual circumstance occurs in an approximate sense when the normal particle displacements at the plate surfaces are identical; in that case a particle in the fluid just at the top surface executes the *same motion* (either in phase or out of phase) as a fluid particle just at the bottom surface and in the same vertical position. But this is just the condition for transverse resonance, or the existence of a guided wave mode in the plate when the fluid loading is negligible. For this reason, the occurrence of guided waves coincides with reflection minima (and by extension, transmission maxima). A more rigorous demonstration of this point can be made using mathematical analysis in the complex plane and considering poles and zeroes of the reflection coefficient, but we will save this deeper technical discussion for Chapter 8.

6.6 Extensions of the Stiffness Matrix

6.6.1 *Higher symmetry lamina*

Earlier we considered each layer in the laminate as a possibly generally anisotropic material with elastic symmetry as low as monoclinic (a single plane of mirror symmetry). Considerable simplification occurs, however, if we are dealing with materials of higher symmetry. In particular, let us assume that each ply in the laminate can be modeled as a transversely isotropic medium, and we restrict ply orientation to the 0° or 90° directions only. The resulting layup is a [0°, 90°] laminate. If we now also confine the incident acoustic wavevector to either the 0° or 90° directions, then for an incident plane wave, all particle displacements will be polarized in the vertical plane, *i.e.*, the plane of incidence. The implication for elastic waves in the composite laminate is that only two tractions and two displacements need be computed to find the elastic behavior of the stack. The problem reverts to the level of complexity of a stack of isotropic materials. That is, we need to solve only a 4 × 4 linear system, and this will also be the order of the stiffness matrix. In the 90° direction the ply indeed has isotropic symmetry, and in the 0° direction, although anisotropic, there is no coupling to waves outside the incident plane, as there would be for an arbitrary ply orientation. These simplifications permit an analytical, although complicated, solution for the stiffness matrix.

When waves are polarized in the vertical plane, the Christoffel equation reverts back to two distinct solutions. (The third is a degenerate shear wave.) The solutions for the z components of the longitudinal and shear wavevectors from Christoffel's equation are given by

$$k_z^{\ell,s} = \left[\frac{-B \mp \sqrt{B^2 - 4AC}}{2A}\right]^{1/2}, \tag{6.44}$$

where the minus sign belongs to k_z^{ℓ}, and

$$\begin{aligned} A &= C_{33}C_{55} \\ B &= (C_{11}k_x^2 - \rho\omega^2)C_{33} + (C_{55}k_x^2 - \rho\omega^2)C_{55} - (C_{13} + C_{55})^2 k_x^2 \\ C &= (C_{11}k_x^2 - \rho\omega^2)(C_{55}k_x^2 - \rho\omega^2). \end{aligned} \tag{6.45}$$

The terms C_{ij} are the elastic stiffnesses of the ply. In the case of elastic isotropy, $C_{11} = C_{33} = C_{13} + 2C_{55}$. The resulting formulas for the z projections of the wavevectors are especially simple,

$$k_z^{\ell} = \sqrt{\frac{\rho\omega^2}{C_{13} + 2C_{55}} - k_x^2}, \qquad k_z^{s} = \sqrt{\frac{\rho\omega^2}{C_{55}} - k_x^2}. \tag{6.46}$$

Finally, the components of the layer stiffness matrix $\mathbf{K}^j$ are

$$K_{11} = (g_s m_\ell - g_\ell m_s)(m_-(e_\ell^2 e_s^2 - 1) + m_+(e_\ell^2 - e_s^2))/\Delta$$

$$K_{12} = (-(g_\ell + g_s)m_+(e_\ell^2 + e_s^2) + (g_\ell - g_s)m_-(e_\ell^2 e_s^2 + 1)$$
$$+ 4e_\ell e_s(g_\ell m_s + g_s m_\ell))/\Delta$$

$$K_{13} = 2(g_\ell m_s - g_s m_\ell)(m_s e_\ell - m_\ell e_s + m_\ell e_\ell^2 e_s - m_s e_s^2 e_\ell)/\Delta$$

$$K_{14} = 2(g_s m_\ell - g_\ell m_s)(e_\ell - e_s + e_\ell e_s^2 - e_\ell^2 e_s)/\Delta$$

$$K_{22} = ((f_\ell - f_s)(m_-(e_\ell^2 e_s^2 - 1) - m_+(e_\ell^2 - e_s^2)))/\Delta$$

$$K_{21} = -K_{12}, \quad K_{23} = K_{14}$$

$$K_{24} = 2(f_\ell - f_s)(m_\ell e_\ell - m_s e_s + m_\ell e_\ell e_s^2 - m_s e_s e_\ell^2)/\Delta$$

$$K_{31} = -K_{13}, \qquad K_{32} = K_{14}, \qquad K_{33} = -K_{11}, \qquad K_{34} = K_{12},$$

$$K_{41} = K_{14}, \qquad K_{42} = -K_{24}, \qquad K_{43} = -K_{12}, \qquad K_{44} = -K_{22}, \tag{6.47}$$

where $m_\pm = m_\ell \pm m_s$, $f_{\ell,s} = i(C_{13}k_x + C_{33}m_{\ell,s}k_z^{\ell,s})$, $g_{\ell,s} = iC_{55}(m_{\ell,s}k_x + k_z^{\ell,s})$, and $e_{\ell,s} = \exp[ihk_z^{\ell,s}]$, where h is the layer thickness. Further terms necessary to complete the calculation are given by

$$m_{\ell,s} = \frac{\rho\omega^2 - C_{11}k_x^2 - C_{55}(k_z^{\ell,s})^2}{(C_{13} + C_{55})k_x k_z^{\ell,s}}$$

$$\Delta = m_-^2(1 + e_\ell^2 e_s^2) - m_+^2(e_\ell^2 + e_s^2) + 8m_\ell m_s e_\ell e_s. \tag{6.48}$$

There exist a number of interesting symmetry relations among the elements of the stiffness matrix, and other quantities, in this special case [11, 12], but consideration of these would take the discussion too far afield at this point.

6.6.2 Stiffness matrix for a fluid

In a common practical circumstance, a composite laminate is immersed in a coupling fluid, usually water. In that case, development of a stiffness matrix for the fluid is important. When bounded by an ideal fluid, the shear tractions on the bounding layers of the laminate will vanish. Moreover, an ideal fluid supports no shear stress; so, for the fluid, C_{55} vanishes and $C_{11} = C_{33} \to C_{13f}$, the only nonzero elastic stiffness. The z component of the wavevector in the fluid k_z^f is then

$$k_z^f = \sqrt{\frac{\rho_f\omega^2}{C_{13f}} - k_x^2}, \tag{6.49}$$

and the acoustic wavespeed in the fluid is $V_f = \sqrt{C_{13f}/\rho_f}$. Using these simplifications, the stiffness matrix for a fluid layer is

$$\mathbf{K}_f = \frac{i\omega\rho_f V_f}{\cos\theta_f(a-2)} \begin{bmatrix} 0 & 0 & 0 & 0 & 0 & 0 \\ 0 & 0 & 0 & 0 & 0 & 0 \\ 0 & 0 & a & 0 & 0 & -b \\ 0 & 0 & 0 & 0 & 0 & 0 \\ 0 & 0 & 0 & 0 & 0 & 0 \\ 0 & 0 & b & 0 & 0 & -a \end{bmatrix}, \tag{6.50}$$

where $a = 1 + \exp[2ihk_z^f]$, $b = 2\exp[ihk_z^f]$, and $\cos\theta_f = \sqrt{1-(k_x/k^f)^2}$. Here, $k^f = \omega/V_f$, ρ_f is the fluid mass density, and θ_f is the angle measured from the normal of a plane wave incident upon, reflected from, or transmitted through, the laminate from a fluid. Although the matrix in Eq. (6.50) is clearly of lower dimension than six, it is convenient to maintain the 6×6 form so that $\mathbf{K}_f$ can be combined easily with the total stiffness matrix of the stack of layers. The rows and columns of zeroes in $\mathbf{K}_f$ correspond to the two positive- and two negative-going quasishear waves that do not exist in the fluid.

6.6.3 Stiffness matrix for imperfect boundary conditions

In our previous discussion of the total stiffness matrix for multilayered composites, we have assumed continuity of both displacements and tractions across the interface between different layers, *i.e.*, the so-called "welded" boundary conditions (BC). As we noted in Chapter 4, slip boundary conditions, which assume a vanishing of shear tractions in the plane of the interface, represent a good approximation for a thin layer of fluid couplant between a solid transducer wedge and a composite. This is the method most often used for angle-beam contact inspection. Moreover, these BCs quite accurately model any mechanical lubricated contact. Examples given in Chapter 4 (Figs. 4.13 and 4.14) show that there are significant differences in ultrasonic reflection/transmission coefficients when one replaces "welded" boundary conditions by slip boundary conditions.

How can we incorporate slip boundary conditions into the stiffness matrix framework? We can reformulate the stiffness matrix in Eq. (6.50) for very small values of h/λ_f (h is the fluid layer thickness and λ_f is the ultrasonic wavelength in the fluid). In doing so, we can simply substitute $h = 0$ and obtain $a = b = 2$ for the matrix elements in Eq. (6.50). We need to be careful, however, with the term $(a-2)$ in the denominator of the prefactor, which vanishes when $a = 2$. To avoid this singularity, we replace $(a-2)$ by the first term in its Taylor expansion, obtaining $(a-2) = 2ihk_z^f$. This substitution yields for the stiffness matrix of a thin fluid layer $\mathbf{K}_{\text{tfl}}$,

$$\mathbf{K}_{\text{tfl}} = \frac{C_{13f}}{h\cos^2\theta_f}\begin{bmatrix} 0 & 0 & 0 & 0 & 0 & 0 \\ 0 & 0 & 0 & 0 & 0 & 0 \\ 0 & 0 & 1 & 0 & 0 & -1 \\ 0 & 0 & 0 & 0 & 0 & 0 \\ 0 & 0 & 0 & 0 & 0 & 0 \\ 0 & 0 & 1 & 0 & 0 & -1 \end{bmatrix}. \tag{6.51}$$

As we can see, in this case the stiffness matrix elements are inversely proportional to the thin-layer thickness. Substituting Eq. (6.51) into Eq. (6.14), we find for the normal stress–displacement relation across the interface, $\sigma_{zz}^{j-1} = \sigma_{zz}^{j} = K_{\text{n}}(u_z^{j-1} - u_z^{j})$, with $K_{\text{n}} = C_{13f}/(h\cos^2\theta_f)$. This is a very clear physical result: to satisfy the load transfer across the interface for a given normal stress, one should decrease the displacement difference $(u_z^{j-1} - u_z^{j})$ proportionally with the decrease of the thin-layer thickness. For $h = 0$ this displacement difference should be zero at finite stress, *i.e.*, continuity of normal displacements across the interface. In this way, slip boundary conditions will be satisfied at very large K_{n}. In reality, the thin-layer thickness is always finite, because it is limited by the surface roughness of two solids squeezed together. For a very thin layer, fluid viscosity may also play a role. (For a deeper discussion of this effect see Ref [10] of Chapter 4.)

Let us now generalize the above discussion to the case of imperfect or so-called "spring" boundary conditions. Interfacial imperfections, micro-defects or micro-disbonds (generally, micro-damage which can be considered as interfacial cracks) introduce additional compliance at the interface. Spring boundary conditions allow us to simulate the effect of interfacial imperfections on ultrasonic wave scattering from an imperfect interface in a composite and thus select (or optimize) ultrasonic methods for interface characterization. The stiffness matrix for the imperfect boundary conditions can be shown [12] to take the form

$$\mathbf{K}_{\text{spring}} = \begin{bmatrix} K_{\text{t1}} & 0 & 0 & -K_{\text{t1}} & 0 & 0 \\ 0 & K_{\text{t2}} & 0 & 0 & -K_{\text{t2}} & 0 \\ 0 & 0 & K_{\text{n}} & 0 & 0 & -K_{\text{n}} \\ K_{\text{t1}} & 0 & 0 & -K_{\text{t1}} & 0 & 0 \\ 0 & K_{\text{t2}} & 0 & 0 & -K_{\text{t2}} & 0 \\ 0 & 0 & K_{\text{n}} & 0 & 0 & -K_{\text{n}} \end{bmatrix}, \tag{6.52}$$

where the subscript n refers to normal and t to transverse. The stiffness matrix elements are the two shear K_{t1}, K_{t2} and normal K_{n} spring constants. The spring stiffness matrix Eq. (6.52) is equivalent to that for slip BCs, or Eq. (6.51); when the shear stiffnesses vanish $K_{\text{t1}} = K_{\text{t2}} = 0$. Usually, spring constants can be considered phenomenologically as parameters of an imperfect interface. In some cases they can be related to a specific micromechanical model of the damaged interface. For a thin imperfect interface layer of thickness h, the spring constants are given in terms of effective elastic stiffnesses of the layer [12],

$$K_{t1} = \frac{(C_{44}C_{55} - C_{45}^2)}{C_{44}h}, \qquad K_{t2} = \frac{(C_{44}C_{55} - C_{45}^2)}{C_{55}h}, \qquad K_n = \frac{C_{33}}{h}. \tag{6.53}$$

In some cases, the spring constants K_n, K_{t1}, K_{t2} in Eq. (6.52) are usefully considered to be complex, accounting phenomenologically for scattering losses as the wave interacts with interface imperfections. To be consistent with the definition of wave attenuation in Section 6.3.1, the sign of the additional imaginary part of the spring constant should be negative.

When the stiffness matrix in Eq. (6.52) is employed in the calculation, the imperfect boundary conditions are naturally included in the recursive algorithm Eq. (6.22) (or slip boundary conditions with use of Eq. (6.51)). For very stiff springs ($K_{t,n} \to \infty$) the reflected and transmitted ultrasonic amplitudes become independent of the spring stiffness, and the imperfect boundary conditions transit to the welded boundary conditions, where the range of this transition depends on the material's elastic properties. For composites, the welded boundary conditions will be satisfied at spring constants of order 10^{20} N/m^3 at typical inspection frequencies. Further discussion of this often-overlooked topic is beyond the scope of this chapter.

6.6.4 *Stiffness matrix computations for a monoclinic lamina*

To compute wave propagation in multilayer composites, the layer is represented by arbitrary rotation of an orthotropic unidirectional lamina or ply around the symmetry axis z, which is elastically equivalent to a layer with monoclinic symmetry. The constitutive relations in this case are represented by Eq. (1.33) of Chapter 1, and the matrix elements (which may be compared with Eq. (5.7) of Chapter 5 and Eq. (4.27) of Chapter 4) are given by

$$\begin{aligned} G_{11} &= C_{11}\xi^2 - \rho\omega^2 + C_{55}k_z^2, & G_{12} &= C_{16}\xi^2 + C_{45}k_z^2, \\ G_{13} &= (C_{13} + C_{55})\xi k_z, & G_{22} &= C_{66}\xi^2 - \rho\omega^2 + C_{44}k_z^2, \\ G_{23} &= (C_{36} + C_{45})\xi k_z & G_{33} &= C_{55}\xi^2 - \rho\omega^2 + C_{33}k_z^2, \end{aligned} \tag{6.54}$$

where ρ is the lamina density, and ξ takes the place of k_x. If the z axis is a material symmetry axis, then $k_z^{-n} = -k_z^{+n}$, and therefore we have a sixth-order polynomial in k_z, the six roots of which satisfy the conditions

$$k_z^{-1} = -k_z^{+1}, \quad k_z^{-2} = -k_z^{+2}, \quad k_z^{-3} = -k_z^{+3}. \tag{6.55}$$

The stress coefficient $\mathbf{d}^{\pm n}$ used in Eq. (6.3) for the (x, y) plane is obtained from the relation Eq. (6.4)

$$\mathbf{d}^{\pm n} = \begin{bmatrix} d_1^{\pm n} \\ d_2^{\pm n} \\ d_3^{\pm n} \end{bmatrix} = i \begin{bmatrix} C_{55}(p_x^{\pm n} k_z^{\pm n} + p_z^{\pm n} \xi) + C_{45} p_y^{\pm n} k_z^{\pm n} \\ C_{45}(p_x^{\pm n} k_z^{\pm n} + p_z^{\pm n} \xi) + C_{44} p_y^{\pm n} k_z^{\pm n} \\ C_{13} p_x^{\pm n} \xi + C_{33} p_z^{\pm n} k_z^{\pm n} + C_{36} p_y^{\pm n} \xi \end{bmatrix}, \tag{6.56}$$

where ξ is common for all partial waves. The polarization vectors are computed using Eqs. (4.35), (4.33), and the normalization condition Eq. (4.36) of Chapter 4 with the substitution $G_{ik} = \omega^2 T_{ik}$. After normalization of the matrix elements in Eq. (6.56) by $p_x^{\pm n}$ and ξ, it is easy to recognize an equivalency between the components of the stress coefficients $\mathbf{d}^{\pm n}$ expressed in Eq. (6.56) and those of the vectors $\mathbf{D}_n$ obtained in Eq. (5.18) of Chapter 5.

Next, the 6×6 matrices $\mathbf{P}^{\pm}$, $\mathbf{D}^{\pm}$, and $\mathbf{H}^{\pm}$ are formed. If the z axis is a material symmetry axis ((x, y) plane is a mirror symmetry plane), then we obtain

$$\mathbf{H}^- = \mathbf{H}^+$$

$$\mathbf{P}^- = -\mathbf{I}\mathbf{P}^+, \quad \text{where} \quad \mathbf{I} = \begin{bmatrix} 1 & 0 & 0 \\ 0 & 1 & 0 \\ 0 & 0 & -1 \end{bmatrix} \tag{6.57}$$

$$\mathbf{D}^- = -\mathbf{I}\mathbf{D}^+.$$

We have all necessary matrix elements to calculate the stiffness matrix $\mathbf{K}$ or the compliance matrix $\mathbf{S}$.

6.6.5 Simple asymptotic method to compute stiffness matrix

Asymptotic method description

As we have shown in this chapter, the computation of ultrasonic wave interaction with a multilayered composite structure proceeds in several steps:

1. Calculating stiffness or compliance matrices for a single anisotropic layer (lamina in composites);
2. Combining, by the recursive algorithm, stiffness matrices for different layers of the structure as a total (global) stiffness matrix and finally;
3. Using the total stiffness or compliance matrices for finding reflection and transmission amplitudes of ultrasonic waves interacting with the structure or solving other problems to be addressed.

Step 1 is the most difficult for computer implementation. The objective of this section is to briefly outline a simple and accurate method [25–27] to compute a stiffness matrix for a composite lamina (step 1), which in general can be of arbitrary symmetry or even be piezoelectric. In this section, however, we provide a description for a monoclinic lamina, which is the most relevant for composites. The method does not require obtaining wave solutions inside each layer. It allows one to calculate the global stiffness matrix for the composite structure using only matrix multiplication and inversion, and the recursive stiffness matrix algorithm described earlier. We call this the *Asymptotic Stiffness Matrix Method* because originally it was applied [28] to obtain asymptotic boundary conditions between two anisotropic solids.

The method is built on the foundation of Stroh's [29–32] formalism developed to study wave propagation in anisotropic media. While it is very different from the Christoffel method described in detail in the previous chapters, the results obtained are identical, as must be true for these two equivalent approaches. The idea of Stroh's formalism is to reduce the equations of motion for anisotropic media to a system of first-order differential equations for a state vector **Q** composed of the displacement and traction vectors, which we have already come across in different situations,

$$\mathbf{Q} = \begin{bmatrix} \mathbf{u} \\ \boldsymbol{\sigma} \end{bmatrix} \tag{6.58}$$

The solution of the system of differential equations is elegantly presented in the form of a transfer matrix exponential **B** (matrix propagator)

$$\mathbf{B} = \exp[-i\mathbf{A}h], \tag{6.59}$$

where h is the layer thickness and **A** is the fundamental acoustic matrix, which is explicitly given as a function of the material properties, wavenumber, and frequency. The construction of the transfer matrix exponential in Eq. (6.59) requires some care; for more information on the correct procedure, see [32]. For a generally anisotropic material the **A** matrix is 6×6. The principal idea of the asymptotic stiffness matrix method lies in the utilization of the asymptotic representation of the transfer matrix Eq. (6.59) as a second-order approximation [28], suitable for a thin anisotropic layer,

$$\mathbf{B}_{\mathrm{II}} = (\mathbf{I} + i\frac{h}{2}\mathbf{A})^{-1}(\mathbf{I} - i\frac{h}{2}\mathbf{A}), \tag{6.60}$$

where **I** is the 6×6 identity matrix and i is the imaginary unit. The series built on this approximation asymptotically converges to the exact solution [25]. The solution depends only on the layer thickness h and the acoustic matrix **A**, for which a very simple explicit representation for a monoclinic medium [33] can be obtained.

$$\mathbf{A} = \begin{bmatrix} \mathbf{A}_{11} & \mathbf{A}_{12} \\ \mathbf{A}_{21} & \mathbf{A}_{22} \end{bmatrix}, \tag{6.61}$$

where

$$\mathbf{A}_{11} = -k_x \begin{bmatrix} 0 & 0 & 1 \\ 0 & 0 & 0 \\ C_{13}/C_{33} & C_{36}/C_{33} & 0 \end{bmatrix} \quad \mathbf{A}_{12} = (-i) \begin{bmatrix} S_{55} & -S_{45} & 0 \\ -S_{45} & S_{44} & 0 \\ 0 & 0 & 1/C_{33} \end{bmatrix} \tag{6.62}$$

$$\mathbf{A}_{21} = -i \begin{bmatrix} k_x^2 Q_{11} - \rho\omega^2 & k_x^2 Q_{16} & 0 \\ k_x^2 Q_{16} & Q_{66}k_x^2 - \rho\omega^2 & 0 \\ 0 & 0 & -\rho\omega^2 \end{bmatrix} \qquad \mathbf{A}_{22} = \mathbf{A}_{11}^{\mathrm{T}}. \tag{6.63}$$

The elements of matrix **A** are related to the elements of the matrix of elastic stiffnesses by

$$S_{44} = C_{55}/(C_{44}C_{55} - C_{45}^2) \qquad S_{55} = C_{44}/(C_{44}C_{55} - C_{45}^2) \tag{6.64}$$

$$S_{45} = C_{45}/(C_{44}C_{55} - C_{45}^2) \qquad Q_{ij} = C_{ij} - C_{i3}C_{j3}/C_{33} \tag{6.65}$$

In our coordinate system the matrix of elastic stiffnesses for lamina with monoclinic symmetry is given by Eq. (1.33) of Chapter 1,

$$\begin{bmatrix} \sigma_1 \\ \sigma_2 \\ \sigma_3 \\ \sigma_4 \\ \sigma_5 \\ \sigma_6 \end{bmatrix} = \begin{bmatrix} C_{11} & C_{12} & C_{13} & 0 & 0 & C_{16} \\ C_{12} & C_{22} & C_{23} & 0 & 0 & C_{26} \\ C_{13} & C_{23} & C_{33} & 0 & 0 & C_{36} \\ 0 & 0 & 0 & C_{44} & C_{45} & 0 \\ 0 & 0 & 0 & C_{45} & C_{55} & 0 \\ C_{16} & C_{26} & C_{36} & 0 & 0 & C_{66} \end{bmatrix} \begin{bmatrix} \epsilon_1 \\ \epsilon_2 \\ \epsilon_3 \\ \epsilon_4 \\ \epsilon_5 \\ \epsilon_6 \end{bmatrix}. \tag{6.66}$$

At normal incidence, when the wave propagates along the z axis, $k_x = 0$ and $\mathbf{A}_{11} = \mathbf{A}_{22} = 0$; also, $\mathbf{A}_{21} = i\rho\omega^2\mathbf{I}$, where $\mathbf{I}$ is the 3×3 identity matrix, and $\mathbf{A}_{12}$ does not change.

The second-order asymptotic stiffness matrix $\mathbf{K}_{\mathrm{II}}$ is [25]

$$\mathbf{K}_{\mathrm{II}} = \begin{bmatrix} -(2\mathbf{I} - ih\mathbf{A}_{12}) & (2\mathbf{I} - ih\mathbf{A}_{12}) \\ -(2\mathbf{I} - ih\mathbf{A}_{22}) & (2\mathbf{I} - ih\mathbf{A}_{22}) \end{bmatrix}^{-1} \begin{bmatrix} (2\mathbf{I} - ih\mathbf{A}_{11}) & -(2\mathbf{I} - ih\mathbf{A}_{11}) \\ (2\mathbf{I} - ih\mathbf{A}_{21}) & -(2\mathbf{I} - ih\mathbf{A}_{21}) \end{bmatrix}. \tag{6.67}$$

For calculations at higher material symmetries (including isotropic), the only change required is to modify the elastic stiffness matrix Eq. (6.66), accordingly. No changes to the algorithm itself are needed. For arbitrary anisotropy and piezoelectricity, the matrix **A** can be found in [25]. High-order approximations for the stiffness matrix are found in [25, 26], and in reference [26] the relation of the method to a general theory of geometric integrators is established.

Stiffness matrix computation for a thick anisotropic layer

To find the stiffness matrix of a thicker layer of thickness H, the layer thickness is subdivided into $N = 2^n$ thin layers according to $h = H/N$, or $H = 2^n h$, and the total stiffness matrix of the thick layer is obtained from the thin layer stiffness matrix using the recursive algorithm given in Eq. (6.22), and repeated here for clarity,

$$\tilde{\mathbf{K}}^J = \begin{bmatrix} \tilde{\mathbf{K}}_{11}^{J-1} + \tilde{\mathbf{K}}_{12}^{J-1}(\mathbf{K}_{11}^j - \tilde{\mathbf{K}}_{22}^{J-1})^{-1}\tilde{\mathbf{K}}_{21}^{J-1} & -\tilde{\mathbf{K}}_{12}^{J-1}(\mathbf{K}_{11}^j - \tilde{\mathbf{K}}_{22}^{J-1})^{-1}\mathbf{K}_{12}^j \\ \mathbf{K}_{21}^j(\mathbf{K}_{11}^j - \tilde{\mathbf{K}}_{22}^{J-1})^{-1}\tilde{\mathbf{K}}_{21}^{J-1} & \mathbf{K}_{22}^j - \mathbf{K}_{21}^j(\mathbf{K}_{11}^j - \tilde{\mathbf{K}}_{22}^{J-1})^{-1}\mathbf{K}_{12}^j \end{bmatrix}. \tag{6.68}$$

To obtain the total stiffness matrix for $N = 2^n$ subdivisions, only n recursive operations are required because in each operation the layer thickness is doubled. The error analysis performed in [25] allows us to obtain an estimate of the optimal number of composite structure subdivisions $N_{\text{opt}} = 2^{n_{\text{opt}}}$, where n_{opt} is the optimal number of recursive operations to compute the total stiffness matrix. For a second-order approximation it is

$$n_{\text{opt}} = \frac{1}{4}\log_2 \frac{\alpha_0}{e_0}, \tag{6.69}$$

where e_0 is the minimum round-off error of the computer (estimated to be 10^{-16} for double precision computations); the term α_0 is given by $\alpha_0 = \frac{1}{12}(k_s H)^2$, and k_s is the slow shear wavenumber. At this number of optimal subdivisions, the error for a second-order approximation is essentially determined by the round-off error of recursive operations, or about 10^{-8}. For high-order approximations, the error is several orders of magnitude smaller, and the number of optimal subdivisions is likewise smaller [25]. The recursive algorithm converges to the exact stiffness matrix and is computationally stable for an arbitrary frequency–thickness fH product.

Summary of computational steps for an arbitrary multilayered composite

For a layered structure, the computational algorithm to obtain the total stiffness matrix recursively is separated into two major steps, and several minor ones, summarized below:

1. Obtain the cell stiffness

    a) For each lamina rotate the coordinate system in the propagation plane and compute the matrix **A**.

b) Subdivide each homogeneous layer H_m of the layered structure into $N = 2^n$ thin layers.
c) Find the optimal number of subdivisions (taking the number of recursive operations n between 10 and 16 will be usually sufficient).
d) Calculate the stiffness matrix for the thin layer.
e) Using the recursive process of Eq. (6.68) a total of n times, compute the total stiffness matrix of the lamina.
f) Repeat the process for each lamina in the cell.

2. Compute the total stiffness/compliance matrix of the composite structure applying cell structure and symmetry using the recursive procedure of Eq. (6.22), or Eq. (6.68).

Bibliography

1. S. I. Rokhlin and L. Wang, "Ultrasonic waves in layered anisotropic media: characterization of multidirectional composites," Int. J. Solids Struct. **39**, 4133–4149 (2002).
2. R. M. Christensen, *Mechanics of Composite Materials* (John Wiley & Sons, New York, 1979).
3. S. W. Tsai and H. T. Hahn, *Introduction to Composite Materials* (Technomic Publishing, Lancaster, 1980).
4. W. T. Thomson, "Transmission of elastic waves through a stratified solid medium," J. Appl. Phys. **21**, 89–93 (1950).
5. N. Haskell, "Dispersion of surface waves on multilayered media," Bull. Seismol. Soc. Am. **11**, 17–34 (1953).
6. L. Knopoff, "A matrix method for elastic wave problems," Bull. Seismol. Soc. Am. **54**, 431–438 (1964).
7. H. Schmidt and F. B. Jensen, "A full wave solution for propagation in multilayered viscoelastic media with application to Gaussian beam reflection at fluid–solid interfaces," J. Acoust. Soc. Am. **77**, 813–825 (1985).
8. A. K. Mal, "Wave propagation in layered composite laminates under periodic surface loads," Wave Motion **9**, 231–238 (1988).
9. M. J. S. Lowe, "Matrix techniques for modeling ultrasonic waves in multi-layered media," IEEE Trans. Ultrason. Ferroelectr. Freq. Control (UFFC) **42**, 525–542 (1995).
10. B. L. Kennett, *Seismic Wave Propagation in Stratified Media* (Cambridge University Press, New York, 1983).
11. L. Wang and S. I. Rokhlin, "Stable reformulation of the transfer matrix method for wave propagation in layered anisotropic media," Ultrasonics **39**, 413–424 (2001).
12. S. I. Rokhlin and L. Wang, "Stable recursive algorithm for elastic wave propagation in layered anisotropic media: Stiffness matrix method," J. Acoust. Soc. Am. **112**, 822–834 (2002).
13. A. H. Nayfeh and D. E. Chimenti, "Elastic wave propagation in fluid-loaded multiaxial anisotropic media," J. Acoust. Soc. Am. **89**, 542–549 (1991).

14. A. H. Nayfeh, "The general problem of elastic wave propagation in multilayered anisotropic media," J. Acoust. Soc. Am. **89**, 1521–1531 (1991).
15. D. E. Chimenti and A. H. Nayfeh, "Ultrasonic reflection and guided wave propagation in biaxially laminated composite plates," J. Acoust. Soc. Am. **87**, 1409–1415 (1990).
16. R. A. Roberts, "Porosity characterization in fiber-reinforced composites by use of ultrasonic backscatter," in *Review of Progress in Quantitative Nondestructive Evaluation*, eds. D. O. Thompson and D. E. Chimenti, vol. **6** (Plenum Press, New York, 1987), pp. 1147–1156.
17. Y. Bar-Cohen and R. L. Crane, "Acoustic backscatter imaging of sub-critical flaws in composites," Mater. Eval. **40**, 970–975 (1982).
18. V. K. Kinra, A. S. Ganpatye, and K. Maslov, "Ultrasonic ply-by-ply detection of matrix cracks in laminated composites," J. Nondestr. Eval. **25**, 39–51 (2006).
19. J. Qu and J. D. Achenbach, "Analytical treatment of polar backscattering from porous composites," in *Review of Progress in Quantitative Nondestructive Evaluation*, eds. D. O. Thompson and D. E. Chimenti, vol. **6** (Plenum Press, New York, 1987), pp. 1137–1146.
20. P. B. Nagy, A. Jungman, and L. Adler, "Measurements of backscattered leaky Lamb waves in composite plates," Mater. Eval. **46**, 95–100 (1988).
21. L. Wang and S. I. Rokhlin, "Ultrasonic wave interaction with multidirectional composites: Modeling and experiment," J. Acoust. Soc. Am. **114**, 2582–2595 (2003).
22. A. H. Nayfeh and D. E. Chimenti, "Ultrasonic wave reflection from orthotropic plates with application to fibrous composites," J. Appl. Mech. **55**, 863–870 (1988).
23. O. I. Lobkis, A. Safaeinili, and D. E. Chimenti, "Precision ultrasonic reflection studies in fluid-coupled plates," J. Acoust. Soc. Am. **99**, 2727–2736 (1996).
24. L. Wang and S. I. Rokhlin, "An efficient stable recursive algorithm for elastic wave propagation in layered anisotropic media," in *Review of Progress in Quantitative Nondestructive Evaluation*, eds. D. O. Thompson and D. E. Chimenti, vol. **21** (AIP Press, New York, 2002), pp. 115–122.
25. L. Wang and S. I. Rokhlin, "Modeling of wave propagation in layered piezoelectric media by a recursive asymptotic method," IEEE Trans. Ultrason. Ferroelect. Freq. Control (UFFC) **51**, 1060–1071 (2004).
26. L. Wang and S. I. Rokhlin, "Recursive geometric integrators for wave propagation in a functionally-graded multilayered elastic medium," J. Mech. Phys. Solids **52**, 2473–2506 (2004).
27. L. Wang and S. I. Rokhlin, "Recursive asymptotic stiffness matrix method for analysis of surface acoustic wave devices on layered piezoelectric media," Appl. Phys. Lett. **81**, 4049–4051 (2002).
28. S. I. Rokhlin and W. Huang, "Ultrasonic wave interaction with a thin anisotropic layer between two anisotropic solids. II. Second-order asymptotic boundary conditions," J. Acoust. Soc. Am. **94**, 3405–3420 (1993).
29. A. N. Stroh, "Steady state problems in anisotropic elasticity," J. Math. Phys. **41**, 77–103 (1962).
30. A. H. Fahmy and E. L. Adler, "Propagation of acoustic waves in multilayers: A matrix description," Appl. Phys. Lett. **20**, 495–497 (1973).
31. E. L. Adler, "Matrix method applied to acoustic waves in multilayers," IEEE Trans. Ultrason. Ferroelect. Freq. Control (UFFC) **37**, 485–490 (1990).

32. L. Wang and S. I. Rokhlin, "A compliance/stiffness matrix formulation of general Green's function and effective permittivity for piezoelectric multilayers," IEEE Trans. Ultrason. Ferroelectr. Freq. Control (UFFC) **51**, 453–463 (2004).
33. S. I. Rokhlin and W. Huang, "Ultrasonic wave interaction with a thin anisotropic layer between two anisotropic solids: Exact and asymptotic-boundary-condition methods," J. Acoust. Soc. Am. **92**, 1729–1742 (1992).

7

Waves in Periodically Layered Composites

Composite materials, unless they are quite thin, often include periodic layering, where laminated plates composed of alternating uniaxial plies in two or more directions result in more evenly distributed in-plane stiffness. The oriented plies can generally be reduced to a unit cell geometry which repeats throughout the laminate and is composed of sublayers each having highly directional in-plane stiffness, but identical out-of-plane properties. As the transverse isotropy of a uniaxial fibrous ply derives from the geometry of the two-phase material, composite laminates of these plies will have microscopic elastic stiffness tensors which change only in the plane of the laminate, as we saw in Chapter 1. The elastic properties normal to the laminate surface remain unchanged from ply to ply. In this chapter we take up the subject of waves in periodically layered plates.

7.1 Introduction and Background

Unusual guided wave dispersion effects have been observed experimentally in periodically layered plates. Shull *et al.* [1] found, for guided waves polarized in the vertical plane in plates of alternating aluminum and aramid–epoxy composites, that dispersion never scales with the frequency–thickness product, as it would in homogeneous isotropic, or layered transversely isotropic, plates. Instead, grouping of the mode curves has been observed.

In an attempt to understand this behavior in terms of periodic layering, Auld *et al.* [2] have analyzed the simpler case of SH wave propagation in periodically layered plates and have demonstrated that these observed phenomena can be attributed to the pass band and stop band structure caused by the periodic layering. In this section, we will show that *Floquet modes* play a critical role in the behavior of guided waves in plates that are periodically layered. To analyze the problem, we apply an extension of the stiffness matrix method of the previous chapter. Floquet modes, which are the characteristic modes for the infinite periodically layered medium, can be thought of as the analogy—in a periodically layered medium—to the quasilongitudinal and quasishear modes for the infinite homogeneous medium.

On the topic of infinite periodic media, many calculations, both approximate and exact, have been performed to model elastic wave propagation in this important class of structures [3, 4]. Some of these efforts have been directed toward the construction of low-frequency approximate wave propagation theories employing approaches such as effective media or higher-order continua. Most early models simplified the multilayered problem by homogenizing the individual lamina properties into a single layer with effective bulk properties [5]. Postma's approach approximated only the static response of the material. In an effort to extend the dynamic range of the modeling to include the dispersive effects of multilayered plates, Sun *et al.* [6,7] and Bedford and Stern [8] developed continuum models that included some microstructural effects. Still, these models were restricted to the quasi-static regime. The work of Sun *et al.* [6] included the highly dispersive problem of wave propagation perpendicular to the layering. This problem was also treated by Brekhovskikh [9], and Sve [10] furthered this work by investigating oblique incidence.

Dispersion characteristics have been studied by solving the equations of motion on the individual lamina of the unit cell, *i.e.* the basic repeating layer combination, applying continuity of the field variables at the boundaries and then applying Floquet's theorem to extend the problem to the entire medium. Delph *et al.* [11, 12] thoroughly analyzed the dispersion characteristics of a bilayered isotropic periodic medium using this method. Later, Braga [13] investigated a finite-thickness plate with isotropic layering immersed in a fluid. More recently, Braga and Herrmann [13, 14] detailed the features of the dispersion spectrum for anisotropic periodic media using the sextic formalism developed by Stroh, implying coupling of in-plane and out-of-plane particle motion. In these papers and others [4] are discussed the dispersion characteristics or Brillouin zones in terms of the propagating and evanescent solutions to the dispersion problem. A similar discussion of the analysis of the acoustical or mechanical filtering predicted for such periodic media is presented by Rousseau and Catignol [15]. More recent work on the application of Floquet analysis to periodically layered finite plates has been reported by Potel and de Belleval [16], Safaeinili and Chimenti [17], Safaeinili *et al.* [18], and Wang and Rokhlin [19–21].

In the following paragraphs, the highly restricted case of SH waves in a layered isotropic plate will be briefly presented to illustrate the problem of pass and stop bands and their effect on guided waves. Then, the full anisotropic analysis of the periodically layered plate based on the stiffness matrix approach will be discussed. Before we begin, however, let us examine some experimental results that will serve to motivate the discussion. Measurements shown in Fig. 7.1 have been performed on a bi-layered sample, consisting of alternating layers of aluminum and layers of epoxy impregnated with continuous aramid fibers, suggesting the trade name of this structure, ARALL^{tm}. This sample whose results are shown in Fig. 7.1 has 3 1/2 unit cells (outer layers are both aluminum); the terminology for this layering is X/Y, where X is the number of aluminum layers and Y is the number of aramid–epoxy layers, illustrated schematically in Fig. 7.2. As before, ϕ is the angle between the uniaxial fiber direction in the aramid composite and the plane of acoustic incidence. Due to the method of fabrication, the aramid–epoxy, although unixial, is best modeled as an orthotropic layer. The elastic properties of the ARALL composite constituents are collected in Table 7.1.

From Fig. 7.1, we discern a new kind of periodic behavior in guided wave dispersion. A single plate has the type of guided wave dispersion curves seen in Chapter 5. Of course, plate-wave dispersion curves will be periodic

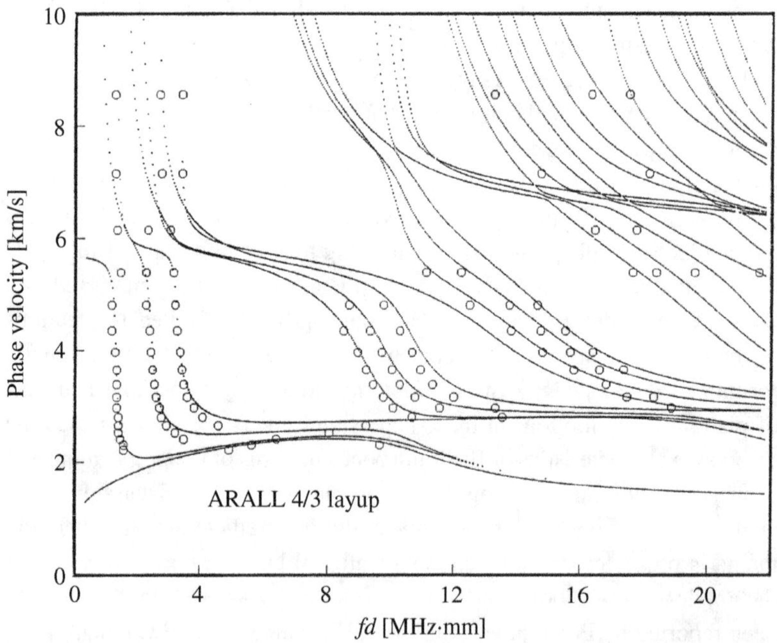

Figure 7.1. Experimental (open circles) and calculated (dotted lines) dispersion curves for a 4/3 ARALL plate, with d = 1.80 mm, and ϕ = 0° (after Shull, *et al.* [1]).

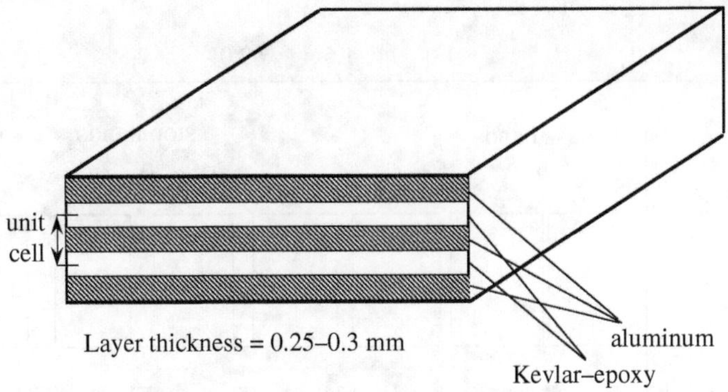

Figure 7.2. Schematic illustration of ARALL structure, showing alternating layers of aluminum with aramid (Kevlar^{tm})–epoxy. Typical dimensions are listed on the figure, which shows a 3/2 layup.

Table 7.1 Elastic properties of ARALL constituents.

Mat'l/Props	C_{11}	C_{22}	C_{33}	C_{44}	C_{55}	C_{66}	C_{23}	C_{13}	C_{12} (GPa)	ρ_s (kg/m^3)
Aluminum	110	110	110	27	27	27	56	56	56	2700
Aramid	60	5.8	8.5	1.8	2.3	2.1	5	5	1.7	1600

because of the sinusoidal nature of the dispersion relations. The curves of Fig. 7.1, however, are bunched into groups of three, and the calculated results agree quite well with the measurements. It has been discovered [1] that the same number of branches is present, but they are now collected into groups or bands. The cause of the grouping is related to the character of the partial waves comprising the guided modes. As the plate becomes dominated by periodic layering, the longitudinal and shear, or QL and two QS, partial waves into which the guided wave can be decomposed, are replaced by the bulk waves that propagate in the periodically layered medium. These normal modes of the periodic medium are called *Floquet* waves. One aspect of Floquet waves is that they propagate only in the allowed regions of the wave number solution space, in so-called pass bands. Due to destructive interference caused by reflections from the periodic layering, there are also stop bands, where no bulk wave can propagate. (Wavenumber solutions still exist, but are imaginary.)

Once there is sufficiently coherent layering to define Floquet waves, these excitations will replace the usual bulk partial waves in the guided modes, resulting in the behavior seen in Fig. 7.1. The grouping of guided modes is the result of the replacement of ordinary bulk partial waves with Floquet waves, whose existence occurs only within the pass bands of the periodic medium. The guided modes of the periodically layered plate are crowded into the

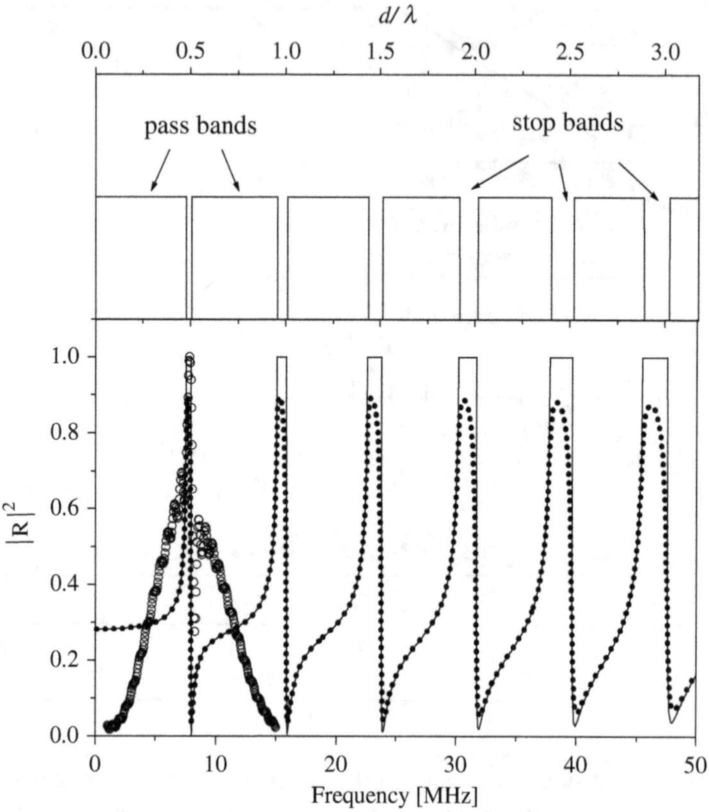

Figure 7.3. Floquet wave pass and stop bands at normal incidence for an infinite periodic medium of a [0°–45°/90°/+45°] composite layup repeated indefinitely with a unit cell formed by a lamina and an interlaminar interphase. In lower frame, reflection coefficient magnitude for a wave incident from a fluid onto a halfspace periodic medium. The solid line is calculated without attenuation, the dotted line calculated with small attenuation in the epoxy layer. The open circles are an experimental spectrum, not deconvolved to remove the transducer response (after Wang and Rokhlin [21]).

Floquet pass bands, the only portion of the solution space where they can exist. An example of this behavior for critical frequencies is seen in Fig. 7.3. The upper frame of this figure shows the existence of pass and stop bands for an infinite periodic medium of a [0°/–45°/90°/+45°] composite layup. The upper horizontal scale is plotted as d/λ or the unit cell thickness over the acoustic wavelength (the unit cell is formed by a lamina and an interlaminar interphase zone). For long wavelength (low frequency) $d/\lambda \ll 0.5$, the solid behaves like a homogenized acoustic medium. The acoustic wavelength is too long to sample the cellular structure. When $d/\lambda \geq 0.5$, however, the cell layering becomes important, and the Floquet band structure appears.

The lower frame of Fig. 7.3 shows the reflection coefficient magnitude of an acoustic wave normally incident from a fluid onto a halfspace composed of the same layered structure cited above. Here, the Floquet pass and stop bands are clearly visible as a function of frequency for both lossless media and media with small attenuation losses. With material losses, the band structure is not quite as sharply defined. The reflection coefficient result has strong implications for guided waves in periodically layered plates. Finally, the results of a measurement from 2 to 15 MHz are plotted as open circles on the predicted curves of the lower frame, showing excellent correspondence between the measured and predicted pass band (for further discussion see Section 7.4.1).

7.2 A Simple Illustration

To see how this band structure appears in a much simpler example, let us briefly consider the case of SH waves in a periodic plate composed of isotropic media. With only a single wave mode to consider, the SH wave analysis is analogous to electromagnetic, optical, or quantum mechanical problems in systems with a similarly layered structure. In practical ultrasonic composite inspections, however, the more relevant case is the propagation of guided waves polarized in the vertical plane in periodic plates, which we address directly later. Using the characteristic equation, we analyze the asymmetric unit cell of a bilayer plate with boundaries denoted **a** in Fig. 7.4. The characteristic equation can be developed using a formulation based on acoustic impedance of the layers. The layer mechanical impedance is

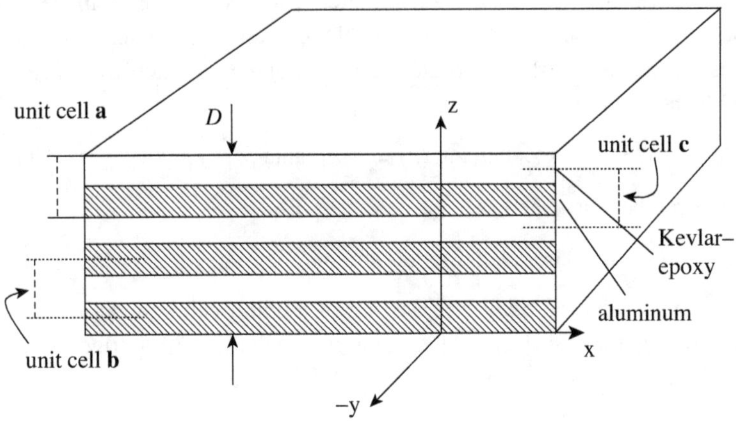

Figure 7.4. Geometry of ARALL structure, showing layering, and coordinate system. Various choices for the unit cell are shown as unit cell **a**, **b**, and **c**; unit cells **b** and **c** are symmetric. All cell choices have the same thickness d, while the total laminate thickness is D.

then incorporated into the transverse resonance condition [22]. Transverse resonance is an approach to obtain the plate-wave characteristic equation that is equivalent to satisfying interfacial continuity conditions on the tractions and displacements. Transverse resonance can be especially useful and simple when there is only a single partial wave mode, such as with SH waves. In the standing wave constructed from upward- and downward-propagating oblique plane waves in the two layers, the acoustic impedances Z^+ and Z^- looking outward from the interface toward the traction-free upper and lower boundaries, respectively, are

$$\begin{aligned} Z^+ &= iZ_2 \tan \gamma_2 d_2 \\ Z^- &= iZ_1 \tan \gamma_1 d_1 , \end{aligned} \tag{7.1}$$

where γ_i, Z_i, and related terms are given by

$$\begin{aligned} \gamma_i &= \frac{\omega}{V_{si}} R_i, \qquad i = 1, 2 \\ Z_i &= \rho_i V_{si} R_i \\ R_i &= \sqrt{1 - V_{si}^2/(\omega/\xi_p)^2} . \end{aligned} \tag{7.2}$$

Here, ξ_p is the guided wave vector in the plate, equivalent to the wavevector x-component in the geometry of Fig. 7.4, V_{si} is the shear wavespeed in the ith medium, and the subscript $i = 1, 2$ refers to the two media; R_i is a direction cosine that accounts for the projection of the partial wave vectors onto the transverse z direction. Transverse resonance occurs when the combined transverse impedance vanishes *i.e.*, $Z^+ + Z^- = 0$.

In Eq. (7.1), the two components of the bilayer have thicknesses d_1 and d_2, with $d_1 + d_2 = d$. For simplicity, let us assume from here on that $d_1 = d_2 = d/2$. Then, the interfacial continuity conditions on the shear stress σ_{yz} and on the shear particle velocity $\dot{u}_y$ at the cell interface in Fig. 7.4 yields the condition,

$$Z_2 \tan(\gamma_2 d/2) = -Z_1 \tan(\gamma_1 d/2) \tag{7.3}$$

which can be rearranged as

$$Z_1 \sin(\gamma_1 d/2) \cos(\gamma_2 d/2) + Z_2 \sin(\gamma_2 d/2) \cos(\gamma_1 d/2) = 0 , \tag{7.4}$$

for a non-symmetric bilayer consisting of the unit cell boundaries **a** with traction-free outer surfaces. After some manipulations using trigonometric identities and collecting terms in Eq. (7.4) we obtain,

$$[Z_1 + Z_2] \sin(\gamma_1 d/2 + \gamma_2 d/2) + [Z_1 - Z_2] \sin(\gamma_1 d/2 - \gamma_2 d/2) = 0 . \tag{7.5}$$

This dispersion relation is identical to the expression derived by satisfying boundary conditions on the stresses and displacements.

If we now consider a symmetric bilayer plate with traction-free boundaries **b** or **c** in Fig. 7.4, we find that wave motion solutions will occur with either symmetric or antisymmetric particle velocity distributions, corresponding to whether the shear stress σ_{yz} or the velocity $\dot{u}_y$ vanishes on the cell midplane. For *symmetric* particle displacements of the traction-free unit cell **b**, if we again write the acoustic impedances looking outward from the material boundary toward the traction-free cell boundary and the traction-free midplane, respectively, we find that Eq. (7.3) becomes

$$Z_1 \tan(\gamma_1 d/4) = -Z_2 \tan(\gamma_2 d/4) \tag{7.6}$$

The reason for this is that the cell boundary and the traction-free midplane are now only half as far from the material interface as before, as shown in Fig. 7.4. A similar calculation to the one above then leads to the dispersion relation,

$$[Z_1 + Z_2] \sin(\gamma_1 d/4 + \gamma_2 d/4) + [Z_1 - Z_2] \sin(\gamma_1 d/4 - \gamma_2 d/4) = 0\,, \tag{7.7}$$

for symmetric particle motion; this expression is identical to Eq. (7.5), but with $d/2 \to d/4$. Then, for *antisymmetric* particle displacements of the traction-free unit cell **b**, the SH particle velocity $\dot{u}_y$ must vanish at the cell symmetry plane, while the upper cell boundary remains traction-free. The upward-looking impedance is identical to the previous case, but the downward-looking impedance becomes,

$$Z^- = -iZ_2 \cot(\gamma_2 d/4)\,, \tag{7.8}$$

and Eq. (7.3) becomes

$$Z_1 \tan(\gamma_1 d/4) = Z_2 \cot(\gamma_2 d/4)\,. \tag{7.9}$$

Proceeding with the calculation yields a dispersion relation of the form,

$$[Z_1 + Z_2] \cos(\gamma_1 d/4 - \gamma_2 d/4) + [Z_1 - Z_2] \cos(\gamma_1 d/4 + \gamma_2 d/4) = 0\,. \tag{7.10}$$

For cell boundaries at **c** instead of at **b**, the subscripts 1 and 2 referred to above are interchanged. The dispersion equations obtained above are valid for a plate composed of a single unit cell. To extend this treatment to multiple cells requires the use of Floquet's theorem coupled with the transfer matrix. Instead of proceeding in that fashion, let us examine a numerical example based on the SH waves and then turn our attention back to the full anisotropic analysis using the stiffness matrix method. In the case we considered above, the shear wavespeed of aluminum is larger than the corresponding value for aramid (see Table 7.1). The pass and stop band structure for bilayer-cell isotropic periodic plate is shown in Fig. 7.5. This figure shows the normalized guided mode wavevector plotted versus the

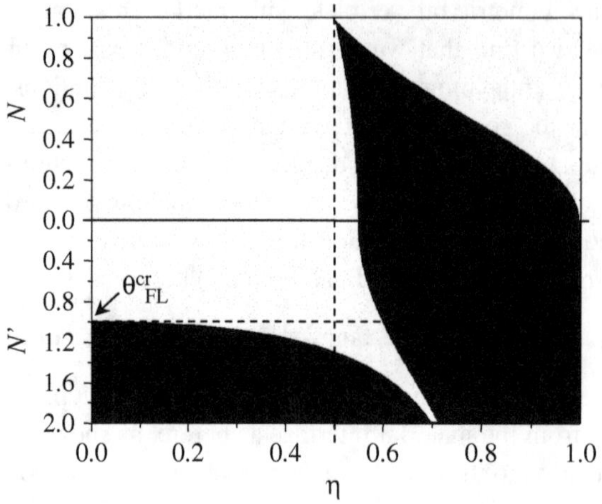

Figure 7.5. Floquet wave stop (black) and pass (white) bands for a bilayer isotropic periodic medium. The horizontal dashed line shows the critical angle $N' = 1$. The dashed vertical line indicates the upper bound of the homogenization domain $\eta = 0.5$ (after Wang and Rokhlin [21]).

normalized frequency. The parameter $N = \gamma_1/\gamma_2$, as long as γ_1 is real. Or, $N' = \Im[\gamma_1]/\gamma_2$, if γ_1 is imaginary. The frequency parameter is $\eta = \gamma_1 d/(2\pi)$, referring to the geometry of Fig. 7.4. The graph illustrates the first two Floquet pass and stop bands for the isotropic periodic medium consisting of a two-layer cell.

The dispersion curves for SH waves in a 3-cell bilayer, such as illustrated in Fig. 7.4 are shown in Fig. 7.6. Here, the elastic properties of the two layers are appropriate for aluminum and homogenized aramid-epoxy to keep the calculation as simple as possible. Guided SH waves in the isotropic bilayer exhibit all the essential dispersion behavior of the more complicated case of plate waves in fully anisotropic media, making this simplification an excellent means to study the problem. These curves, calculated as individual but highly dense points, demonstrate the pass and stop bands of a periodically layered medium, in this case a bilayer. Even with only three cells, the bulk partial waves are replaced by Floquet waves, which can propagate only within the pass bands. Due to this structural filtering, the modes of the SH guided waves are crowded into narrow bands shown in Fig. 7.6. The upper and lower band edge solutions for the second Floquet band are marked on the graph as U and L, respectively. Each successive pass band also contains four guided wave modes, although these are not labeled. The calculation in Fig. 7.6 demonstrates the nature of guided waves propagating in periodic media, where Floquet waves dominate the behavior.

7.3 Floquet Analysis for Anisotropic Periodic Plates

To complete the picture of guided wave behavior in periodic plates, let us return to the formalism developed in Chapter 6 for layered composites.

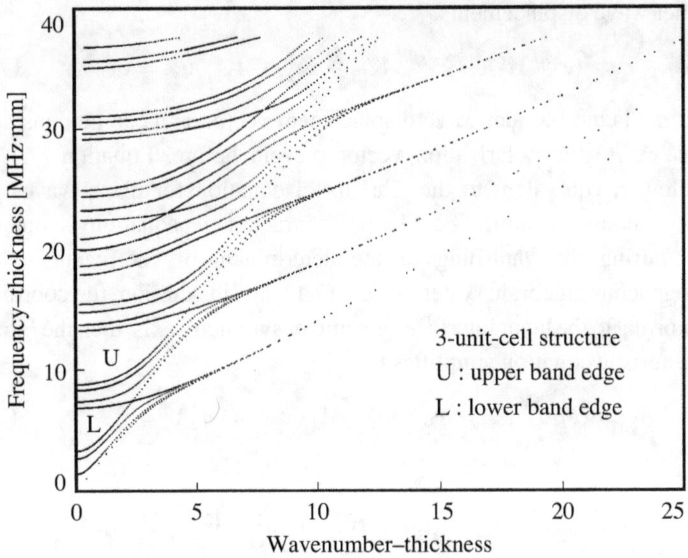

Figure 7.6. Theoretical dispersion curves of guided SH waves in a 3-cell ARALL structure, calculated as dense points. Note the bunching of modes into groups of four and comparative absence of modes between groupings. The letters U and L denote the upper and lower band edges for the second Floquet pass band (after Auld *et al.* [2]).

We seek the Floquet wave characteristic equation for waves in periodic plates. To begin, let us state the Floquet periodicity condition, also known as Floquet's theorem [23],

$$\begin{bmatrix} \mathbf{u}^+ \\ \boldsymbol{\sigma}^+ \end{bmatrix} = \mathbf{I}\exp(i\beta d)\begin{bmatrix} \mathbf{u}^- \\ \boldsymbol{\sigma}^- \end{bmatrix}, \tag{7.11}$$

where β is the Floquet wavenumber, d is the unit cell thickness, $\mathbf{u}$ and $\boldsymbol{\sigma}$ are the displacement and traction vectors at the upper $(^+)$ and lower $(^-)$ cell surfaces, and $\mathbf{I}(6 \times 6)$ is the identity matrix. At this point in the calculation, the transfer matrix is typically employed to derive the characteristic equation. Instead, let us rely on the unconditionally stable stiffness matrix approach. Consider a unit cell composed of n arbitrarily oriented anisotropic layers; then, the cell stiffness matrix $\mathbf{K}^c$,

$$\begin{bmatrix} \boldsymbol{\sigma}^+ \\ \boldsymbol{\sigma}^- \end{bmatrix} = \begin{bmatrix} \mathbf{K}^c_{11} & \mathbf{K}^c_{12} \\ \mathbf{K}^c_{21} & \mathbf{K}^c_{22} \end{bmatrix}\begin{bmatrix} \mathbf{u}^+ \\ \mathbf{u}^- \end{bmatrix}, \tag{7.12}$$

relating the stresses at the top and bottom cell surfaces, is obtained from the layer stiffness matrix $\mathbf{K}_j$ by use of the recursive algorithm (Eq. (6.22)) discussed in Chapter 6 [19]. Using the cell stiffness matrix of Eq. (7.12) and the Floquet theorem of Eq. (7.11), we obtain a system of equations for the

Floquet wave displacements,

$$(e^{i\beta d}\mathbf{K}_{21}^{c} - e^{-i\beta d}\mathbf{K}_{12}^{c} + \mathbf{K}_{22}^{c} - \mathbf{K}_{11}^{c})\mathbf{u}^{-} = 0. \tag{7.13}$$

Here, $\mathbf{u}^{-}$ is the Floquet wave displacement vector, and can be considered equivalent to the polarization vector $\mathbf{p}$ from before. Equation (7.13) is the Floquet equivalent to the Christoffel equation for plane waves in a homogeneous medium. The Floquet characteristic equation is obtained by requiring the vanishing of the determinant of coefficients of the homogeneous algebraic system in Eq. (7.13). If, in addition, the coordinate axis normal to the layer interfaces is a mirror symmetry axis, then the Floquet characteristic equation simplifies to

$$A_3\cos(3\beta d) + A_2\cos(2\beta d) + A_1\cos(\beta d) + A_0 = 0\,, \tag{7.14}$$

where

$$\begin{aligned}
A_3 &= |\mathbf{K}_{21}^{c}|\,, \qquad A_2 = (|\mathbf{M} + \mathbf{K}_{21}^{c}| + |\mathbf{M} - \mathbf{K}_{21}^{c}|)/2 - |\mathbf{M}| \\
A_1 &= (|\mathbf{K}_{12}^{c} + \mathbf{K}_{21}^{c}| + |\mathbf{K}_{12}^{c} - \mathbf{K}_{21}^{c}| - |\mathbf{K}_{12}^{c} + \mathbf{M}| - |\mathbf{K}_{12}^{c} - \mathbf{M}|)/2 \\
&\quad + |\mathbf{K}_{12}^{c}| - |\mathbf{K}_{21}^{c}| \\
A_0 &= 2|\mathbf{M}| + (|\mathbf{M} + \mathbf{K}_{21}^{c} - \mathbf{K}_{12}^{c}| + |\mathbf{M} + \mathbf{K}_{12}^{c} - \mathbf{K}_{21}^{c}| \\
&\quad + |\mathbf{K}_{12}^{c} + \mathbf{M}| - |\mathbf{K}_{12}^{c} - \mathbf{M}| + |\mathbf{K}_{21}^{c} + \mathbf{M}| - |\mathbf{K}_{21}^{c} - \mathbf{M}|)/2\,.
\end{aligned} \tag{7.15}$$

In the above, the matrix $\mathbf{M} = \mathbf{K}_{22}^{c} - \mathbf{K}_{11}^{c}$, and $|\mathbf{M}|$ denotes the determinant of the matrix $\mathbf{M}$. Equation (7.14) has a total of six independent solutions for the Floquet wavenumber β, where $-\pi < \beta d < \pi$. Due to the existence of a plane of mirror symmetry normal to the layering, we also have that $\beta_1 = -\beta_2$, $\beta_3 = -\beta_4$, and $\beta_5 = -\beta_6$. With this Floquet characteristic equation for the periodic plate, we have accomplished our limited objective.

To see how these ideas affect wave propagation in layered composites, let us examine the calculated result in Fig. 7.7. The left-hand frame shows the result for a [0°/90°] composite, and the right-hand frame is the calculation for a [0°/45°/90°/−45°] composite laminate, plotted as phase-match or incident angle (related to wavenumber $\xi_p = \omega\sin\theta/V_f$) versus frequency. At low frequency, the effect of the layering can be negligible. The upper bound of the homogenization frequency is indicated by the dashed curves labeled "Homogenization domain" in each plot. To the left of those curves, the wavelengths of the Floquet waves are long enough so that the details of the composite microstructure are not significant. Only the averaged, or homogenized, elastic properties of the layering will control wave propagation in this regime. Even here, however, just as in an anisotropic but homogeneous material, there will be phase-match angles for which the partial waves in the layered medium become evanescent with complex wavenumbers. These regions are illustrated in Fig. 7.7 by a gray

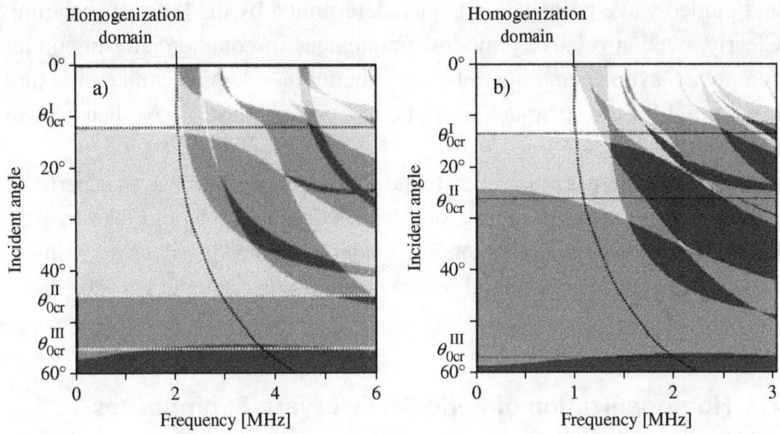

Figure 7.7. Calculated pass and stop bands of the three Floquet partial waves in wavenumber (shown as reverse incident angle) and frequency domain for (a) a [0°/90°] composite, and (b) a [0°/45°/90°/−45°] composite laminate. The propagating or incident plane is oriented along 0° top-ply direction. The gray scale coding is interpreted as follows: white—three propagating Floquet partial waves; light gray—two propagating Floquet waves; dark gray—one propagating wave; black—stop band, no propagating waves (after Wang and Rokhlin [19]).

scale coding scheme. Look at the left-hand side of each graph. Where the dispersion spectrum is white, all three Floquet waves (QL, QT1, QT2) are propagating. Light gray denotes that two of the Floquet waves propagate; dark gray, only one; and black indicates that none of the Floquet waves has a real wavenumber in that regime. In both graphs, at low incident (or phase-match) angle, all waves are propagating. Then, near 12° a critical angle occurs, removing one Floquet wave from propagating status. The left-hand zones of the two graphs are quite different from this point on, but both show the occurrence of additional critical angles, leading to dark gray and finally black regions near 60°, where no Floquet wave is propagating. At higher frequency—above about 2 MHz in frame (a) and 1 MHz in frame (b)—much more complicated behavior is observed. In these regions, to the right of the dashed *homogenization domain* curves, the detailed microstructure of the laminate begins to dominate, and this factor then determines the pass and stop band occurrence in wavenumber and frequency, both at low and high phase-match angle. The stop and pass band behavior of Floquet waves will certainly influence the propagation of guided waves in plates composed of finite numbers of unit cells whose Floquet characteristics are depicted in Fig. 7.7. When the frequency is high enough for the composite layup microstructure to control waves in the laminate, as shown on the right-hand sides of Fig. 7.7(a) and (b), the Floquet waves replace the plane partial waves of an anisotropic homogeneous medium in plates of finite thickness [1, 2], and guided wave propagation is then determined by the Floquet spectrum. Clearly, even at relatively modest frequencies in commercially important composite layup geometries, a full consideration of laminate microstructure is essential for the accurate description of guided elastic wave behavior in these materials.

Of course, there is much more to the phenomenology of waves in periodic plates, but to extend the discussion here significantly would take us well beyond the scope of this chapter. The reader interested in a deeper treatment may refer to the many original research articles on the topic [13–20].

7.4 Homogenization of Periodically Layered Composites

7.4.1 Floquet wave spectrum and signal distortion

We have already discussed in Section 7.3 the stop and pass bands of the Floquet wave spectrum shown in Fig. 7.7 for incidence from a fluid. At frequencies above the first stop band, the Floquet wave behavior is very complicated. Owing to the occurrence of stop and pass bands, this behavior influences the spectral characteristics of the transmitted and reflected ultrasonic signals. Interesting examples of measured and simulated time-domain signals transmitted through a multidirectional periodic composite plate in water at a central frequency of 2.25 MHz are shown in Fig. 6.9 of Chapter 6. We can see that at 14° incidence on a $[0°/45°/90°/-45°]_{2S}$ composite laminate, the transmitted signal is extremely distorted. This fact forecloses the possibility of composite characterization at this angle and frequency. At these conditions, we are above the first stop band. As the incident angle increases, the first stop band moves toward higher frequency. As a result, as the incident angle increases to 50°, the signal distortion is significantly reduced because at this incident angle and frequency the slow transverse wave propagates below the first stop band. At normal incidence (0° in Fig. 6.9) the signal is transmitted without distortion at 2.25 MHz. Why are the distortions not occurring at this frequency, as would be expected from Fig. 7.7 at an incident angle of 0° (at the top of the figure)?

The reason is that the Floquet wave stop and pass band spectrum shown in Fig. 7.7 is for oblique incidence, where transverse and longitudinal waves, sensitive to lamina orientation in the cell, are excited, and the slow transverse wave is affecting this spectrum. In this figure, at low incident angles the lowest frequency stop band occurs for the shear wave. At normal incidence on a fluid-immersed composite, only the longitudinal wave is excited. For a longitudinal wave in the z (or normal) direction, all lamina are assumed to have identical properties. Therefore, the ideal composite will look like a homogeneous medium, and such a material will not display band behavior. For manufactured graphite-epoxy laminates, however, there is a very thin residual interfacial epoxy layer between each ply (containing no fibers), and therefore the properties of this thin layer are different from those of the lamina. The presence of these thin residual epoxy layers at each ply interface transforms the structure into a periodic layered medium in the normal z direction (the residual epoxy layer has little effect at oblique incidence). The unit cell of this periodic structure is formed by two layers: the ply itself and a thin epoxy interfacial layer. This cell thickness is one fourth the thickness of the $[0°/45°/90°/-45°]_{2S}$ cell used for Fig. 7.7(b), and one half the cell thickness of the [0°/90°] layup used for Fig. 7.7(a). With a longitudinal wavelength about twice as large, this scales the frequency axis of the stop and pass bands spectrum shown in Fig. 7.3 for normally incident longitudinal waves. The first stop band appears at a frequency of 7.9 MHz ($\lambda_\ell = 2d_c$, where d_c is the combined thickness of the lamina and of the thin interfacial layer, and λ_ℓ is the longitudinal wavelength). Now we can see why the normally incident transmitted signal at 2.25 MHz in Fig. 6.9 is not distorted.

At frequencies below the first stop band, the Floquet behavior is similar to that of a plane wave propagating in an anisotropic dispersive homogeneous medium. Thus, it is important to have a simple way to estimate the homogenization domain.

7.4.2 Homogenization approach

It is understandable that a multilayered composite structure behaves effectively at low frequencies as a homogeneous anisotropic plate. We have discussed in the section above that ultrasonic wave interaction with such a structure and ultrasonic signal distortion are significantly affected by the frequency and incident angle range. Therefore, it is important to quantify this behavior and, in particular, to understand when the composite medium can be considered effectively homogeneous, *i.e.*, to be homogenized.

At low frequencies, below the first stop band, the Floquet waves propagate like plane waves in a homogeneous anisotropic medium. This fact forms the basis for the homogenization approach [19]. First, let us discuss the replacement of a complicated multilayered periodic medium by an equivalent homogeneous anisotropic medium. To achieve the replacement, we define the effective-medium cell as an anisotropic homogeneous cell for which the stiffness $\mathbf{K}_{\text{eff}}$ and transfer $\mathbf{B}_{\text{eff}}$ matrices are equal to the stiffness and transfer matrices of the actual cell

$$\mathbf{K}_c = \mathbf{K}_{\text{eff}}, \qquad \mathbf{B}_c = \mathbf{B}_{\text{eff}}. \tag{7.16}$$

As all effective cells are homogeneous and identical, the whole periodic anisotropic medium transforms into an anisotropic homogeneous medium. This exchange does not change the Floquet wave velocities or other properties because, owing to the equality of the stiffness matrices, the stresses and displacements on the cell boundaries do not change when we substitute the actual cell for a homogeneous effective cell.

Let us first demonstrate that for an effective homogeneous medium, the Floquet wavenumbers (calculated in Eq. (7.14)) are equal to the wavenumbers of plane waves in the effective medium (calculated by Christoffel's equation Eq. (2.51)). We will use the transfer matrix approach to demonstrate this result. Let us consider an infinite anisotropic homogeneous medium in which we select a layer (cell) and form a fictitious periodic medium. The transfer matrix **B** for the homogeneous cell is given by the transfer matrix for a single anisotropic layer. Equating determinants of the transfer matrices Eq. (6.7) and Eq. (7.11), we obtain [19]

$$\det\begin{bmatrix} \mathbf{H}^- - \mathbf{I}\exp(i\beta d) & 0 \\ 0 & (\mathbf{H}^+)^{-1} - \mathbf{I}\exp(i\beta d) \end{bmatrix} = 0, \tag{7.17}$$

where, as defined in Chapter 6, **I** is the 3×3 identity matrix, and d is the cell thickness. The 3×3 matrix $\mathbf{H}^{\pm}$ is given by

$$\mathbf{H}^{\pm} = \begin{bmatrix} \exp(ik_z^{\pm 1} d) & 0 & 0 \\ 0 & \exp(ik_z^{\pm 2} d) & 0 \\ 0 & 0 & \exp(ik_z^{\pm 3} d) \end{bmatrix}, \tag{7.18}$$

$k_z^{\pm n}$ is the z projection of the wavenumber for the nth partial wave.

Now we see that if the Floquet wavenumber β equals one of the wavenumbers k_z^i, then Eq. (7.17) is satisfied. This procedure is applied to each of the three Floquet waves. Therefore, the z component of the Floquet wavenumbers and partial wavenumbers in an effective homogeneous medium are identical. The x components will also be equal, according to Snell's law. Thus, for a uniform layer, the Floquet wave velocities are equal to the effective wave velocities.

We have just shown that an effective homogeneous anisotropic medium can be constructed, such that for an arbitrary frequency and wave propagation direction the Floquet wavenumbers calculated by Eq. (7.14) are equal to the wavenumbers of plane waves in the same medium (calculated from Christoffel's equation). For a periodic medium, however, whose cell consists of n lamina, the homogenization requirement Eq. (7.16) and the equality of Floquet wavenumbers in the periodic medium with the wavenumbers in the effective homogeneous anisotropic medium is only applicable exactly at a specific frequency and specific propagation direction in the actual anisotropic periodic medium. Even at a single frequency, different propagation directions imply different effective homogeneous anisotropic media. To produce realistic effective anisotropic media, we must require a common Christoffel tensor for all directions of wave propagation at the given frequency. This can only be done at frequencies below the first stop band, *i.e.*, in the homogenization domain. Such a homogenized medium has frequency-dependent effective elastic constants, *i.e.*, wave dispersion. Frequency dependence of effective elastic constants for composites of different layups has been studied in [19]. Further discussion of this topic is beyond our scope.

7.4.3 Homogenization domain estimation for an anisotropic cell

As discussed above, the composite can be considered as an effective homogeneous medium at frequencies below the first stop band. For specific applications, it is important to estimate the upper frequency bound of this domain. Of course, one can use the Floquet spectral zones calculated using Eq. (7.14) as shown in Fig. 7.7. A simple estimation method based on physical considerations is a better choice.

We can generalize the condition for the first stop band of an SH wave, as shown in Fig. 7.5. In anisotropic media the situation is more complicated, however, and we have three plane waves, and therefore we have to consider three wavelengths. For a selected orientation of the propagation plane, we compute the z components of wavevectors for three waves in all lamina, selecting the shortest wavelength. For example, for the [0°/90°] cell, the first stop band will be determined by the smallest wavelength (lowest ultrasonic velocity) of the partial wave at a given Floquet wave incident angle θ. It will be smaller than the actual smallest Floquet wave velocity, because the velocity of the partial wave varies from ply to ply in the given propagation plane. We obtain the approximate upper frequency f_{MAX} bound of the periodic composite medium [19,21]

$$f_{\mathrm{MAX}} < \frac{0.5V_{\min}(\theta_i)}{d\cos\theta_i}, \tag{7.19}$$

where d is the cell thickness, θ_i is the incident angle from the fluid, related to the Floquet wave propagation direction by Snell's law. Then, $V_{\min}(\theta_i)$ is the minimum velocity among all bulk waves in each ply of the cell excited by a wave incident from the fluid at an angle of θ_i. An average of the lowest velocities for different lamina in the cell can also be used in Eq. (7.19). The inverse dependence of the frequency bound in Eq. (7.19) on $\cos\theta_i$ relates to the association of the stop bands with the z projections of the wavenumbers. The estimated homogenization domain using Eq. (7.19) for [0°/90°] and [0°/45°/90°/−45°] composites is shown in Fig. 7.7 by the dotted curves starting at the first stop band. One can see that Eq. (7.19) gives a conservative estimate of the upper frequency bound of the homogenization domain.

Comparing homogenization ranges in Fig. 7.7, we note that the periodic cell [0°/45°/90°/−45°] (four lamina) is twice as thick as the [0°/90°] (two lamina) and thus the homogenization frequency range for those composites is scaled by a factor of two. Both estimation and exact calculation show that the upper bound of the homogenization frequency increases as the incident angle increases.

7.5 Lamina Moduli Measurement by Floquet waves

7.5.1 Problem statement

Ultrasonic characterization of elastic properties of multi-ply composites has been a longstanding challenge for experimentalists. One method of measuring the elastic constants is to use low-frequency ultrasonic waves. In this case, the inhomogeneous composite may be considered as an effective homogeneous medium. Therefore we can use the same procedure developed for a homogeneous layer to reconstruct the effective elastic constants from wave velocity measurements. Examples of this approach have been given by Rose *et al.* [24], Hosten [25, 26], and Lobkis *et al.* [27]. The effective elastic constants are assumed to be equal to the static constants, and the lamina properties are determined from them using inversion of a static micromechanical model for multi-ply composites [25]. Owing to structural inhomogeneity, however, the ultrasonically measured effective elastic constants of a multi-ply composite are frequency dependent [19] and therefore cannot be precisely related to the static elastic constants. Chu and Rokhlin [28] have proposed a method to determine lamina elastic moduli for [0°/90°] metal matrix composites without the effective media approach. Their method neglects the effect of wave interference within a single ply in the laminate on the time-delay measurement. Such an assumption is valid only for composites with thick plies or for composites with a small impedance mismatch between the 90° and 0° lamina, as in the metal-matrix composites that were considered. Therefore, the method cannot be used for graphite/epoxy multi-ply composites.

Direct measurement of the unidirectional ply elastic constants in multilayered composites may give significant advantages because the lamina moduli can be considered frequency independent over a much broader frequency range and will thus be equal to the static values. (Graphite fibers have 5–10 μm diameters and therefore the lamina can be considered nondispersive at frequencies below 10 MHz.) Elastic moduli of lamina in composites can be measured by low-frequency acoustic microscopy [29]. This approach is more complicated than that described next. The focus of this section is to extend the experimental through-transmission method, described in Chapter 3 for unidirectional composites, to the case of lamina and effective elastic property measurements in multidirectional composites. For this, we use the Floquet wave concept as described by Wang and Rokhlin [20].

7.5.2 Floquet wave lamina moduli determination

Multi-ply graphite-epoxy composites are extremely inhomogeneous owing to the strong ply anisotropy, and there is significant wave velocity variability from ply to ply. The ply thickness is usually between 100 and 150 μm and is less than the ultrasonic spatial pulse length. The ultrasonic signals interfere within the lamina, and we cannot calculate the total time delay by simply adding the time delays for different plies. Therefore, the conventional meaning of wave velocity is not applicable. The velocity and elastic property measurements in such materials require a conceptual step forward by accounting for the inhomogeneity of the structure and the ply-wave interference.

Owing to the periodicity of the composite structure, the Floquet wave approach seems well suited for characterization of the medium. Formation of Floquet waves in a periodic medium results naturally from wave interference in the structure, and their wavespeed has a clear physical meaning. In through-transmission ultrasonic measurements the spatial length of the ultrasonic pulse is less than the composite plate thickness, and thus propagation of Floquet waves can be treated as if in an unbounded periodic medium. The Floquet wave method for lamina moduli measurement is based on a method of time-delay data processing different from that obtained by the self-reference, double through-transmission method described in Chapter 3 (Sections 3.1.5, 3.4). Recall that for homogeneous media, reconstruction of elastic moduli requires calculating the time delay iteratively using the phase velocity determined from the Christoffel equation and then minimizing the residual difference between measured and calculated values. To calculate the time delay for a periodic medium, we use the Floquet wave phase velocity obtained from Eq. (7.14). In this method, the time delays are measured for a set of incident angles in one or several orientations of the incident plane. The unknown lamina moduli of the composite are found by minimizing the sum of the squares of the deviations between the experimental and calculated time delays, using the Floquet theory, where the elastic constants are considered as variables in a multi-dimensional space.

Briefly, with reference to Chapter 3, the time delay difference $\Delta t_{\theta_i} = t_{\theta_i} - t_n$ between the time of flight t_{θ_i} at angle θ_i and the reference time of flight t_n at normal incidence is obtained versus incidence angle θ_i. To improve the precision in the Δt_{θ_i} measurement, the phase correction [30] is used for transverse waves in the composite after the first critical angle. Adopting Eqs. (3.16) and (3.20) for Δt_{θ_i} in the self-reference, double through-transmission method, we obtain

$$\Delta t_{\theta_i} = \frac{2H}{1/V_f - 1/V_n} - \frac{H\cos(\theta_r - \theta_i)}{V_f \cos\theta_r} + \frac{H}{V_F(\theta_r)\cos\theta_r}, \tag{7.20}$$

where V_n is the velocity in the sample measured at normal incidence, and V_f is the sound speed in the immersion fluid. The term H is the thickness of the sample, $V_F(\theta_r)$ is the Floquet wave phase velocity in the composite at angle θ_r, and θ_r is the refraction angle determined from Snell's law.

$$\theta_r = \arcsin\{\sin(\theta_i)V_F(\theta_r)/V_f\}. \tag{7.21}$$

The phase velocity $V_F(\theta_r)$ at refraction angle θ_r is calculated from Eqs. (7.20) and (7.21) using the measured phase velocity at normal incidence V_n and the time-delay change Δt_{θ_i} at the incident angle θ_i.

Figure 7.8 shows results of time delay measurements (circles) for two composites, $[0°/45°/90°/-45°]_{12S}$ (graphite/PEEK thermoplastic matrix composite) and $[0°/90°]_{4S}$ (graphite/epoxy composite). The thickness of the $[0°, 90°]_{4S}$ sample is 2.5 mm, and the lamina thickness is 0.156 mm. The thickness of the $[0°/45°/90°/-45°]_{12S}$ sample is 14.6 mm, and the lamina thickness is 0.152 mm. The density for both samples is taken to be 1600 kg/m^3. Measurements with signals of center frequency 1 MHz for the 12S laminate and 2 MHz for the 4S laminate were used, respectively. (The central frequency is just below the first stop band for each composite.) In both cases, the lamina are considered transversely isotropic (five independent elastic constants). This limitation is not essential and can easily be extended

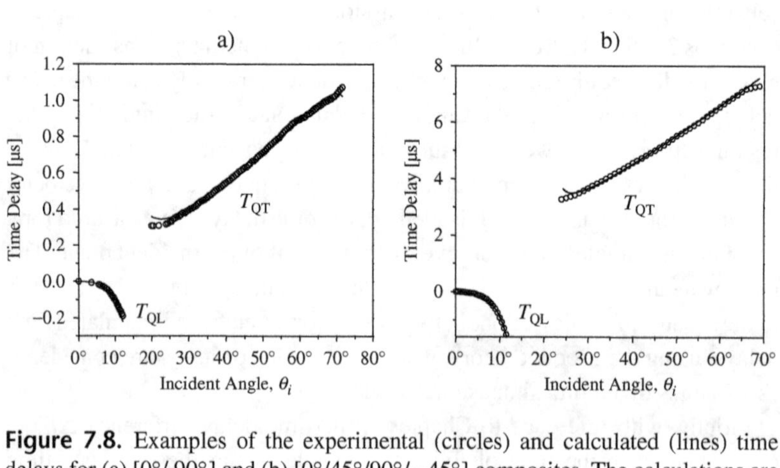

Figure 7.8. Examples of the experimental (circles) and calculated (lines) time delays for (a) [0°/ 90°] and (b) [0°/45°/90°/–45°] composites. The calculations are done based on reconstructed lamina moduli for each composite (after Wang and Rokhlin [20]).

to orthotropic symmetry. The lamina elastic constant C_{33} ($C_{33} = C_{22} = \rho V_n^2$, ρ is the sample density) is determined at normal incidence by the pulse-echo method. The four other elastic constants (C_{11}, C_{12}, C_{23}, C_{55}) are determined using the Floquet-wave-based method described above based on the measured time delays Δt_{θ_i}. It is important to note that to determine four elastic constants by this method, measurements in only one plane of incidence are required. (For an orthotropic lamina, measurements should be performed in two incident planes.) The calculated time delays in those composites based on reconstructed lamina moduli are shown in Fig. 7.8 by solid lines. Lamina moduli obtained for the 4S composite are given in the first row of Table 7.2. The reader is referred to [20] for additional detail and discussion of the elastic moduli.

7.5.3 Effective property determination: Static limit

From the practical point of view, not only are lamina moduli needed for composite design, but the static effective elastic properties are also of great importance, as a goal of the final composite design and as a quality check of the manufacturing process. Usually those static effective moduli are determined by destructive tests, which are quite expensive and cannot be performed on serviceable manufactured parts. Nondestructive ultrasonic measurements are practically the only alternative.

In the past, ultrasonically measured effective moduli of multilayered composites were assumed to be equal to the composite static effective moduli. As we will see, in some cases, one may introduce significant error using this assumption because of the frequency dependence of the laminated composite effective moduli. The Floquet wave method allows one to obtain the lamina properties directly and thus compute correctly the effective static elastic properties of multidirectional composites.

After the lamina properties are measured, we can determine the effective dynamic elastic properties of the homogenized multidirectional composite. Within a homogenization frequency domain, we transform the composite using the Floquet wave method to a dispersive homogeneous anisotropic

Table 7.2 Reconstructed lamina and effective homogeneous elastic moduli in GPa for [0°/90°] composite at 2 MHz and calculated from lamina static effective moduli (after Wang and Rokhlin [20]).

	C_{11}	C_{22}	C_{12}	C_{23}	C_{55}
Measured lamina	121	12.1	4.2	8.6	9.3
Effective Floquet wave (2 MHz)	57.8	12.1	6.16	6.68	3.7
Floquet wave (static)	66.4	12.1	4.59	6.35	3.0
Effective plane wave (2 MHz)	55.3	12.1	—	7.06	3.5

medium. One can see from the Floquet wave spectrum (see Fig. 7.7) that in the homogenization domain, which is at frequencies below the first stop band, the Floquet wave behavior in the periodic medium is equivalent to that of a plane wave in an effective anisotropic medium (three waves may propagate and their critical angles exist). As we have mentioned in the previous section, at a given frequency we can calculate the slowness surface for Floquet waves and match it to the slowness surface for bulk waves in homogenized anisotropic media [19]. The Floquet slowness surface is calculated by Eq. (7.14) using the known composite layout and lamina properties. The match is performed using least squares minimization, searching for the unknown effective elastic constants which are determined uniquely from the plane wave slowness surface.

As an example, in Fig. 7.9, the effective elastic constant C_{11} is plotted as a function of frequency for the $[0°/90°]_{4S}$ composite. As one can see, the dynamic elastic constant C_{11} at 2 MHz is 20% less than the value at 0.1 MHz (essentially static), which is a significant difference. The effective elastic moduli obtained from the lamina properties by the Floquet wave homogenization at 2 MHz are given for the $[0°/90°]_{4S}$ composite in the second row of Table 7.2. In the same way, calculated effective static moduli are given in the third row.

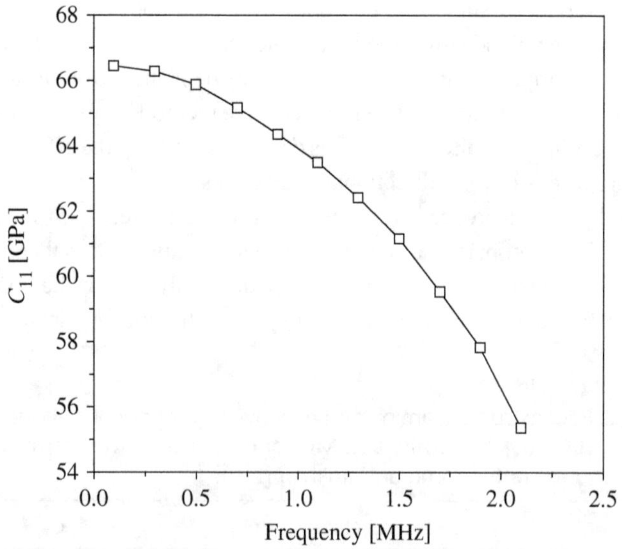

Figure 7.9. Frequency-dependent effective elastic constant C_{11} calculated from the measured lamina elastic constants using dynamic homogenization method for $[0°/90°]_{4S}$ composite (after Wang and Rokhlin [20]).

As the measurements are performed within the homogenization domain of the multidirectional composite, one can use the same time delay experimental data and reconstruct the effective elastic constants of the composite by an inversion of the Christoffel equation, as was done in Chapter 3. The effective moduli obtained this way are given in the last row in Table 7.2. These effective moduli are reasonably close to those shown in the second row, which are computed from the lamina properties by the Floquet wave homogenization method (relative deviation is about 5%). Ideally, in the homogenization domain, both methods should provide identical results. Any discrepancies indicate a homogenization error from pulse bandwidth overlap with the first stop band. Moreover, the plane wave method cannot recover C_{12} from 0° incident plane orientation measurements. To determine this constant would require experimental data in an additional incident plane. The most important deficiency is that the plane wave method alone cannot yield the effective static moduli.

7.6 Computation of the Guided Wave Spectrum

Earlier in this chapter we showed some examples of guided waves in multilayered periodic composites. Here, in this final section, we briefly outline how one can stably compute dispersion characteristics of the guided wave both in periodically layered composites and in general multidirectional composites of arbitrary layup and anisotropy using the stiffness matrix method of Chapter 6.

The characteristic equation for guided modes is determined from the total stiffness matrix of the structure Eq. (6.27). If the top and bottom surfaces are free, then

$$\begin{bmatrix} \boldsymbol{\sigma}_0 \\ \boldsymbol{\sigma}_N \end{bmatrix} = 0, \quad \text{and} \quad \mathbf{K} \begin{bmatrix} \mathbf{u}_0 \\ \mathbf{u}_N \end{bmatrix} = 0, \tag{7.22}$$

and the guided wave dispersion equation is

$$\det[\mathbf{K}] = 0. \tag{7.23}$$

Using properties of determinants of block matrices, Eq. (7.23) can be rewritten as

$$\begin{aligned} \det[\mathbf{K}] &= \det[\mathbf{K}_{11}]\det[\mathbf{K}_{22} - \mathbf{K}_{21}\mathbf{K}_{11}^{-1}\mathbf{K}_{12}] \\ &= \det[\mathbf{K}_{22}]\det[\mathbf{K}_{11} - \mathbf{K}_{12}\mathbf{K}_{22}^{-1}\mathbf{K}_{21}] = 0. \end{aligned} \tag{7.24}$$

Now using Eq. (6.17) for sub-matrices S_{ij} of the compliance matrix we have

$$\det[\mathbf{K}] = \det[\mathbf{K}_{11}]\{\det[\mathbf{S}_{22}]\}^{-1} = \det[\mathbf{K}_{22}]\{\det[\mathbf{S}_{11}]\}^{-1} = 0, \tag{7.25}$$

where the property $\det[\mathbf{S}_{11,22}^{-1}] = \{\det[\mathbf{S}_{11,22}]\}^{-1}$ was used. This representation reduces the order of determinants to be computed.

For a fluid-loaded layered anisotropic structure, the dispersion equation for leaky guided modes is determined from the reflection and transmission coefficients (Eqs. (6.41) and (6.42)) by equating the denominator to zero,

$$(S_{11}^{33} + \Lambda)(S_{22}^{33} - \Lambda) - S_{21}^{33}S_{12}^{33} = 0. \tag{7.26}$$

For vanishing zero fluid density, $\Lambda \to \infty$, so Eq. (7.26) cannot be solved directly. It can, however, be demonstrated that Eq. (7.26) asymptotically approaches Eq. (7.23) in this situation [31] and eventually becomes

$$\{S_{11}^{33}\}^{-1} = 0. \tag{7.27}$$

Bibliography

1. P. J. Shull, D. E. Chimenti, and S. K. Datta, "Elastic guided waves and the Floquet concept in periodically layered plates,"J. Acoust. Soc. Am. **95**, 99–108 (1994).
2. B. A. Auld, D. E. Chimenti, and P. J. Shull, "Shear horizontal wave propagation in periodically layered composites," IEEE Trans. Ultrason. Ferroelect. Freq. Contr. **43**, 319–325 (1996).
3. L. Brillouin, *Wave Propagation in Periodic Structures* (Dover, New York, 1946).
4. T. C. T. Ting and P. Chadwick, "Harmonic waves in periodically layered anisotropic elastic composites," in *Wave Propagation in Structural Composites*, AMD-vol. 90, edited by A. K. Mal and T. C. Ting, (ASME, New York, 1988), pp. 69–79.
5. G. W. Postma, "Wave propagation in a stratified medium," Geophys. **20**, 780–806 (1955).
6. C. T. Sun, J. D. Achenbach, and G. Herrmann, "Time-harmonic waves in a stratified medium propagating in the direction of the layering," J. Appl. Mech. **35**, 408–411 (1968).
7. C. T. Sun, J. D. Achenbach, and G. Herrmann, "Continuum theory for a laminated medium," J. Appl. Mech. **35**, 467–475 (1968).
8. A. Bedford and M. Stern, "Towards a diffusing continuum theory of composite materials," J. Appl. Mech. **38**, 8–14 (1971).
9. L. M. Brekhovskikh, *Waves in Layered Media* (Academic, New York, 1980).
10. C. Sve, "Time-harmonic waves traveling obliquely in a periodically laminated medium," J. Appl. Mech. **71**, 477–482 (1971).
11. T. J. Delph, G. Herrmann, and R. K. Kaul, "Harmonic wave propagation in a periodically layered, infinite elastic body: Antiplane strain," J. Appl. Mech. **45**, 343–349 (1978).
12. T. J. Delph, G. Herrmann, and R. K. Kaul, "Harmonic wave propagation in a periodically layered, infinite elastic body: Plane strain, analytical results," J. Appl. Mech. **46**, 113–119 (1979).
13. A. M. B. Braga, "Wave propagation in anisotropic layered composites," PhD Thesis, Stanford University, 1990.
14. A. M. B. Braga and G. Herrmann, "Floquet waves in anisotropic periodically layered composites," J. Acoust. Soc. Am. **91**, 1211–1227 (1992).
15. M. Rousseau and Ph. Gatignol, "Acoustic propagation in a periodically stratified medium," (in French), Acustica **64**, 188–194 (1987).

16. C. Potel and J-F. de Belleval, "Propagation in an anisotropic periodically multilayered medium," J. Acoust. Soc. Am. **93**, 2669–2677 (1993).
17. A. Safaeinili and D. E. Chimenti, "Floquet analysis of guided waves in periodically layered composites," J. Acoust. Soc. Am. **98**, 2336–2342 (1995).
18. A. Safaeinili, D. E. Chimenti, B. A. Auld, and S. K. Datta, "Floquet analysis of guided waves propagating in periodically layered composites," Compos. Eng. **5**, 1471–1476 (1995).
19. L. Wang and S. I. Rokhlin, "Floquet wave homogenization of periodic anisotropic media," J. Acoust. Soc. Am. **112**, 38–45 (2002).
20. L. Wang and S. I. Rokhlin, "Floquet wave ultrasonic method for determination of single ply moduli in multidirection composites," J. Acoust. Soc. Am. **112**, 916–924 (2002).
21. L. Wang and S. I. Rokhlin, "Ultrasonic wave interaction with multidirectional composites: Modeling and experiment," J. Acoust. Soc. Am. **114**, 2582–2595 (2003).
22. B. A. Auld, *Acoustic Fields and Waves in Solids*, 2nd ed (Robert E. Krieger, Malabar, 1989).
23. E. I. Ince, *Ordinary Differential Equations* (Dover, New York, 1956).
24. W. R. Rose, S. I. Rokhlin, and L. Adler, "Evaluation of anisotropic properties of graphite-epoxy composite plate using Lamb waves," in *Review of Progress in Quantitative Nondestructive Evaluation*, eds. D. O. Thompson and D. E. Chimenti, vol. **6** (Plenum Press, New York, 1987), pp. 1111–1118.
25. B. Hosten, "Stiffness matrix invariants to validate the characterization of composite materials with ultrasonic methods," Ultrasonics **30**, 365–371 (1992).
26. B. Hosten, "Ultrasonic through-transmission method for measuring the complex stiffness moduli of composite materials," in *Handbook of Elastic Properties of Solids, Liquids and Gases*, eds. A. G. Every and W. Sachse, vol. **1** (Academic Press, San Diego, 2001), pp. 67–86.
27. O. I. Lobkis, D. E. Chimenti, and H. Zhang, "In-plane elastic property characterization in composite plates," J. Acoust. Soc. Am. **107**, 1852–1858 (2000).
28. Y. C. Chu and S. I. Rokhlin, "A method for determination of elastic constants of a unidirectional lamina from ultrasonic bulk velocity measurements on [0°/90°] cross-ply composites," J. Acoust. Soc. Am. **96**, 342–354 (1994).
29. L. Wang and S. I. Rokhlin, "Time resolved line focus acoustic microscopy of layered anisotropic media: application to composites," IEEE Trans. Ultrason. Ferroelect. Freq. Control (UFFC) **49**, 1231–1244 (2001).
30. R. L. Wang, A. I. Lavrentyev, and S. I. Rokhlin, "Beam and phase effects in angle-beam through-transmission method of ultrasonic velocity measurement," J. Acoust. Soc. Am. **113**, 1551–1559 (2003).
31. S. I. Rokhlin and L. Wang, "Stable recursive algorithm for elastic wave propagation in layered anisotropic media: Stiffness matrix method," J. Acoust. Soc. Am. **112**, 822–834 (2002).

8

Measurement of Scattering Coefficients

8.1 Introduction

In the previous chapters, we saw how waves in composites behaved under various circumstances, depending on material anisotropy and wave propagation direction. The most important function that describes guided wave propagation, and the plate elastic behavior on which propagation depends, is the reflection coefficient (RC) or transmission coefficient (TC). More generally, we can call either one simply, the *scattering coefficient* (SC). It is clear that the elastic properties of the composite are closely tied to the SC, and in turn the scattering coefficient determines the dispersion spectrum of the composite plate. Measuring the SC provides a route to the inference of the elastic properties. To measure the SC, we need only observe the reflected or transmitted ultrasonic field of the incident acoustic energy. In doing so, however, the scattered ultrasonic field is influenced by several factors, both intrinsic and extrinsic.

Clearly, the scattered ultrasonic field of an incident acoustic beam falling on the plate from a surrounding or contacting fluid will be strongly influenced by the RC or TC of the plate material. The scattering coefficients are in turn dependent on the plate elastic properties and structural composition: fiber and matrix properties, fiber volume fraction, layup geometry, and perhaps other factors. These elements are not, however, the only ones to determine the amplitude and spatial distribution of energy in the scattered ultrasonic field. Extrinsic factors such as the finite transmitting and receiving transducers, their focal lengths, and their placement with respect to the sample under study can make contributions to the signal as important as the SC itself. Therefore, a systematic study of the role of the transducer is essential for a complete understanding and correct interpretation of acoustic signals in the scattered field. The interpretation of these signals leads ultimately to the inference of composite elastic properties.

As we pointed out in Chapter 5, the near coincidence under some conditions of guided plate wave modes with the zeroes of the reflection coefficient (or peaks in the transmission coefficient) has been exploited many times to reveal the plate's guided wave mode spectrum [1–4]. The simultaneous occurrence of the guided waves with the reflection zeroes was first mentioned by Cremer [5] and further analyzed by Schoch [6] in his more comprehensive and quantitative treatment of this subject. In that paper, Schoch expresses the reflected field as a spectral integral over the incident field weighted by the reflection coefficient and by a propagator term that accounts for diffraction in the plane of incidence.

It was this reflection integral that Bertoni and Tamir [7] later evaluated approximately for an incident Gaussian beam by contour integration in their analysis of leaky Rayleigh waves. An analytical formulation for reflection from plates, following the lines of the Bertoni and Tamir result, was given by Pitts *et al.* [8]. Comparisons between numerical or analytical evaluations of this integral formulation of the reflected field and experimental measurements have been made by several authors for both plates and half-spaces [9–12]. In most cases, the evaluations have been limited to the 2-D formulation suggested by Schoch, and included an approximate representation of the incident beam. The success of these comparisons has generally varied, depending on the experimental geometry and on the accuracy of the material parameters. A further missing element in essentially all these prior treatments, however, is a rigorous analysis of the influence of the receiving transducer.

In this chapter, we examine the influence of planar and focused transmitting and receiving transducers in two and three dimensions. The procedure here follows rigorous treatments by Lobkis *et al.* [13] and Zhang and Chimenti [14]. It permits the inclusion of transducer diffraction effects of both the transmitter and receiver in a completely general fashion. Earlier studies found that the coincidence of the RC poles with the observed reflection minima can depend on the frequency and experimental geometry. In some cases, this dependence is particularly strong. On the other hand, under special conditions in identical two-transducer measurements, the beam shape, its sidelobes, and the near- or far-field conditions can be suppressed to the point where they have a relatively minor influence on the receiver voltage. Typically, the beam size, the lateral transducer positioning with respect to each other, and their distance to the sample are major geometrical factors determining the location, width, and depth of minima in the receiver voltage signal.

8.2 Scattering Coefficient Integral

The geometry for the analysis presented here is shown in Fig. 8.1. The system consists of two identical planar transducers of radius a (transmitter and receiver) immersed in liquid and oriented at equal and opposite angles α from the normal. A solid plate of thickness d is located a distance z_0 from the center of the transducer faces. The distance between points of intersection of the transducer acoustical axes on the upper plate surface is denoted by x_i. The origin of coordinates lies on the upper plate surface, as seen in Fig. 8.1.

The problem, as posed in Fig. 8.1, is essentially 2-D, but with a diffraction contribution from the y dependence of the beam function. For isotropic media (or anisotropic media where the (x, z) plane contains a symmetry axis), this diffraction contribution is the only significant effect of the 3-D nature of the beam. According to the Rayleigh integral formula for the field of a planar transducer [15], the particle velocity potential of incident field on the plate surface can be expressed as

$$\Psi_i(x, y) = \frac{1}{2\pi} \int\int_{\sigma_T} V_T(f, \sigma_T) \frac{\exp(ik_f r)}{r} \, d\sigma_T, \tag{8.1}$$

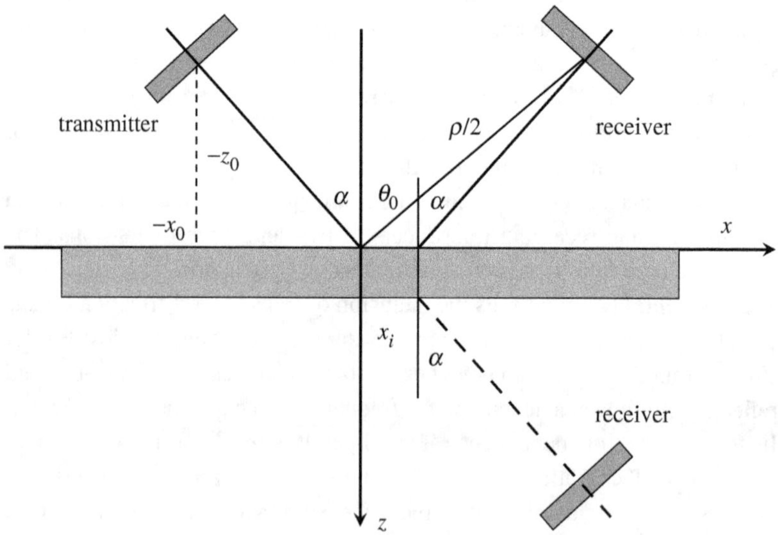

Figure 8.1. Schematic of experimental geometry for analysis of beam reflection from fluid-loaded plate of thickness d; transducer orientation angle α is measured from the plate normal, θ is the polar phase match angle, ρ is the total water path, and x_i is the beam shift parameter, the distance along the interface between the beams' two acoustical axes. The plane wave component that propagates toward the origin has phase-match angle θ_0. The plate is infinite in the y direction.

where $\mathcal{V}_T(f, \sigma_T)$ is the frequency- and position-dependent particle velocity on the transmitter surface σ_T, f is the frequency, k_f is the wavenumber in the fluid, r is the distance between two points (x_T, y_T, z_T) on the transmitter and $(x, y, 0)$ on the plate, so that $r = \{z_T^2 + (x_T - x)^2 + (y_T - y)^2\}^{1/2}$, and subscript T refers to the transmitter. Using Brekhovskikh's plane wave spectral decomposition of a spherical wave $\exp(ik_f r)/r$ in the Rayleigh integral formula [16] permits us to express the incident transducer field Ψ_i in the form of a set of plane waves as

$$\Psi_i(x, y) = \frac{ik_f}{4\pi^2} \int_0^{2\pi} \int_0^{\pi/2 - i\infty} D_T(\theta, \bar{\phi}) e^{-ik_f \sin\theta \{x \cos\bar{\phi} + y \sin\bar{\phi}\}} \sin\theta d\theta d\bar{\phi}, \tag{8.2}$$

where θ and $\bar{\phi}$ are polar and azimuthal integration angles for each plane wave direction of propagation, and $D_T(\theta, \bar{\phi})$ is the transmitter directivity function which depends on its position and orientation

$$D_T(\theta, \bar{\phi}) = \iint_{\sigma_T} \mathcal{V}_T(f, \sigma_T) e^{ik_f z_T \cos\theta + ik_f \sin\theta \{x_T \cos\bar{\phi} + y_T \sin\bar{\phi}\}} d\sigma_T. \tag{8.3}$$

The distinction between $\bar{\phi}$ and ϕ is significant; the former is the out-of-incident plane integration angle (essential only for the 3-D voltage calculation or the diffraction-corrected 2-D voltage). The latter is the fiber orientation angle in the plane of the laminate. When the fiber angle ϕ is different from 0° or 90° in a uniaxial laminate, or when modeling a multi-axial laminate at high frequency, then the 3-D voltage model given below requires an integration over $\bar{\phi}$. The angle θ may also be identified as the phase-match angle of a particular plane wave component of the acoustic beam.

For the scattered field, each spectral component of the incident beam will be weighted by the appropriate value of plane wave scattering coefficient $S(\theta, \bar{\phi}, f; C_{IJ}^*, C_f)$, which in the general case of an anisotropic plate is a function of both angles θ and $\bar{\phi}$, the frequency f, the plate thickness d and its complex elastic stiffnesses C_{IJ}^*, and the elastic property C_f of the ideal fluid surrounding the plate, among other parameters. In what follows, we shall temporarily ignore the dependence of S on the out-of-incident plane angle in the (x, y) plane $(\bar{\phi})$. Then, the scattered field Ψ_r at an arbitrary point (x_R, y_R, z_R) on the receiver surface is given by the appropriate form of Eq. (8.2) with its integrand multiplied by the scattering coefficient $S(\theta, f)$ and an additional phase term $\exp(ik_f z_R \, cos\theta)$. This last contribution arises because the wave propagates over a finite path from the plate to the receiver.

The scattered field is

$$\Psi_r(x,y,z)=\frac{ik_f}{4\pi^2}\int_0^{2\pi}\int_0^{\pi/2-i\infty} S(\theta,f)D_T(\theta,\bar{\phi})e^{ik_f(z_R\cos\theta-\sin\theta\{x_R\cos\bar{\phi}+y_R\sin\bar{\phi}\})} \times\sin\theta d\theta d\bar{\phi}. \tag{8.4}$$

According to the Auld and Kino reciprocity formulas [17, 18], the output signal of the receiving transducer can be represented as an integral of the product of the receiver spatial excitation and the reflected field, carried out in receiver coordinates (x_R, y_R, z_R) over the face of the receiving transducer,

$$V_R = \iint_{\sigma_T} \mathcal{V}_R(f, \sigma_R)\Psi_r(x_R, y_R, z_R)\, d\sigma_R, \tag{8.5}$$

where $\mathcal{V}_R(f, \sigma_R)$ is the frequency- and position-dependent particle velocity on the receiver surface σ_R when it operates as a transmitter. Combining Eqs. (8.5) and (8.4) yields

$$V(x_i, f) = \frac{ik_f}{4\pi^2}\int_0^{2\pi} d\bar{\phi}\int_0^{\pi/2-i\infty} S(\theta, f)D_T(\theta, \bar{\phi})D_R(\theta, \bar{\phi})\sin\theta d\theta, \tag{8.6}$$

where $D_R(\theta, \bar{\phi})$ is the receiver directivity function, in analogy to $D_T(\theta, \bar{\phi})$.

The problem now is to obtain accurate estimates of the transducer directivity functions $D_T(\theta, \bar{\phi})$ and $D_R(\theta, \bar{\phi})$. There are several ways to proceed. As most liquid-coupled or contact ultrasonic inspections employ mechanically damped piezoelectric transducers for essentially all measurements, perhaps the most accurate approximation would be the familiar field expression for a baffled circular piston radiator having a position-independent particle velocity on its radiating surface, *i.e.*, $\mathcal{V}_{T,R}(f, \sigma_{T,R}) = \mathcal{V}_{T,R}(f)$. For this case, the integral of Eq. (8.3) evaluates to an exact result for the directivity function at any transducer orientation angle α

$$D_{T,R}(\theta, \bar{\phi}) = 2\pi a^2 \mathcal{V}_{T,R}(f)\left[\frac{J_1(k_f a\sin\delta)}{k_f a\sin\delta}\right] e^{ik_f(z_0\cos\theta+x_0\sin\theta\cos\bar{\phi})}, \tag{8.7}$$

where $J_1(k_f a\sin\delta)$ is the first-order Bessel function of the first kind, $(-x_0, 0, -z_0)$ are the coordinates of the transmitter center, and $(x_0, 0, \mp z_0)$ are the receiver center coordinates, as seen in Fig. 8.1, and δ is the angle between the wavevector of the spectral component at $(\theta, \bar{\phi})$ and the transducer acoustical axis (or central ray). The angle δ is related to the phase-match, transducer orientation, and azimuthal angles in the problem through the cosine law of spherical trigonometry, $\cos\delta = \cos\theta\cos\alpha + \sin\theta\sin\alpha\cos\bar{\phi}$.

By reciprocity, $D_T(\theta, \bar{\phi})$ and $D_R(\theta, \bar{\phi})$ are equal for nominally identical probes, and we may write the receiver voltage as a function of transducer position and spatial and frequency response as

$$V(x_i, f) = 2ik_f V_T(f) V_R(f) a^4 \int_0^{\pi/2 - i\infty} S(\theta, f) e^{ik_f \rho \cos(\theta - \theta_0)} \sin\theta \, d\theta$$
$$\times \int_0^{\pi} \left[\frac{J_1(k_f a \sin\delta)}{k_f a \sin\delta} \right]^2 e^{-ik_f \rho \sin\theta_0 \sin\theta (1 - \cos\bar{\phi})} d\bar{\phi}. \tag{8.8}$$

This weighted integral is proportional to the output signal of the receiver; therefore, for brevity, we choose to call it simply the *receiver voltage*. In Eq. (8.8), ρ is the distance between the receiver center in reflection and an image source point at $(x_0, 0, z_0)$, equivalent to the separation of the transducer centers measured through the origin of coordinates, as shown in Fig. 8.1. Then, θ_0 is the particular phase-match angle formed between the ray from the transducer center to the origin and the z axis, implying that $\tan\theta_0 = \tan\alpha + x_i/z_0$, as illustrated in Fig. 8.1.

While the Bessel function expression of Eq. (8.8) is a reasonably accurate representation of the directivity function for a damped commercial piezoelectric ultrasonic probe, it is only one possible choice to approximate the experimental beam. Another often used representation is the Gaussian beam [7,8,19], which turns out to be an acceptably good model only for the central lobe of a piston radiator. A third possibility is simply to measure the beam directivity experimentally and use this tabulated function in place of Eq. (8.7). We will see below that the theoretically calculated receiver voltage is almost completely insensitive to the substitution of Gaussian beams in Eq. (8.6) for piston radiator beams of Eq. (8.7), so long as the experiment consists of identical planar transmitter and receiver transducers [20].

Reflected field calculations in the form of Eq. (8.4) are the typical representations seen in most treatments of leaky plate wave scattering. Approximation of the experimental transducer directivity function by a Gaussian beam [7, 8, 21, 22] has often been motivated by convenience and ease of calculation, rather than accuracy of representation. In our case, however, where piston transducers are used for generation *and* detection, the spatially convolved Gaussian beams become an excellent approximation for the combined pistons, even in the nearfield, as we show below. The amplitude of the first sidelobe of Eq. (8.7) is 0.13 of the function maximum at $\delta = 0$, whereas this number is only 0.017 for the combined beams on axis. When the receiver is scanned along the leaky wave propagation direction, its acoustical axis is only displaced, not rotated. For planar transducers, the beam sidelobes interact only with each other, not with the main lobe, whose direction remains unchanged. These considerations suggest the replacement of the product $D_T D_R$ with

$$\left| D_T^G(\theta, \bar{\phi}) D_R^G(\theta, \bar{\phi}) \right| = \pi^2 a^4 V_T(f) V_R(f) e^{-(\bar{b} k_f a \sin\delta)^2}, \tag{8.9}$$

where the dimensionless parameter $\bar{b}$ describes the normalized waist dimension of the Gaussian beam, related to the Fresnel length b by $\bar{b} = b/(2k_f a^2)$. This number is nearly constant over limited range distances. As a weak function of transducer–transducer separation, it can be approximated by $\bar{b} \approx 0.522$ when the separation distance is about one Rayleigh distance $\mathcal{R}(\simeq a^2/\lambda_f)$, where λ_f is the wavelength in the fluid. When Gaussian directivity functions are substituted for piston beams, the voltage expression becomes

$$V(x_i, f) = \frac{ik_f a^4}{2} \mathcal{V}_T(f)\mathcal{V}_R(f) \int_0^{\pi/2 - i\infty} S(\theta, f) e^{ik_f \rho \cos(\theta - \theta_0)} \sin\theta d\theta \times \int_0^{\pi} e^{-(\bar{b}k_f a \sin\delta)^2} e^{-ik_f \rho \sin\theta_0 \sin\theta(1-\cos\bar{\phi})} d\bar{\phi}. \tag{8.10}$$

The analysis leading to Eq. (8.8) and the substitution of piston with Gaussian beams in Eq. (8.10) is, however, valid for any value of $k_f a$, from highly collimated to highly diffracting beams. In fact, as $k_f a \to 0$ both representations, $D_T D_R$ and $D_T^G D_R^G$, converge to point sources [16].

It must be emphasized that the substitution of Gaussian beams in Eq. (8.10) is justified only because we are modeling the combined response of the real transducer beams. The replacement of the piston radiator incident beam by the Gaussian beam in the case of the scattered *field* [7, 8] is not at all accurate. By way of illustration let us examine a worst-case example. In Fig. 8.2, we plot the beam-shift or x_i-dependence (defined in Fig. 8.1) of two incident fields $\Psi_i(x_i, z = \mathcal{R}/2)$: the solid curve corresponds to the field of a piston radiator sampled by a point receiver, and the dotted curve is the same function for a Gaussian beam. Figure 8.2 shows the field of a circular piston beam of radius a (solid curve) calculated for a propagation distance, from source to observation point, equal to one half the Rayleigh distance $\mathcal{R}$. This is the exact downrange distance at which a piston beam has its last on-axis pressure null. A best-fit Gaussian beam (dotted curve) is also shown for comparison. Clearly, the piston nearfield at this range distance is poorly approximated by a Gaussian beam.

In contrast, a similar calculation is shown in Fig. 8.3. Here, at the same downrange distance between source and observation point, we have modeled the x_i-dependence of the receiver voltage for two cases: a piston beam sampled by an identical piston receiver and a Gaussian beam sampled by an identical Gaussian receiver. After the spatial averaging performed by the receiving transducer with an identical beam function, the voltage in Fig. 8.3 shows a remarkable insensitivity to choice of beam model, even at this downrange distance chosen for its pathological behavior. The voltage calculated from Eq. (8.8) for pistons and the equivalent expression for two Gaussian beams lie in almost perfect coincidence, demonstrating the

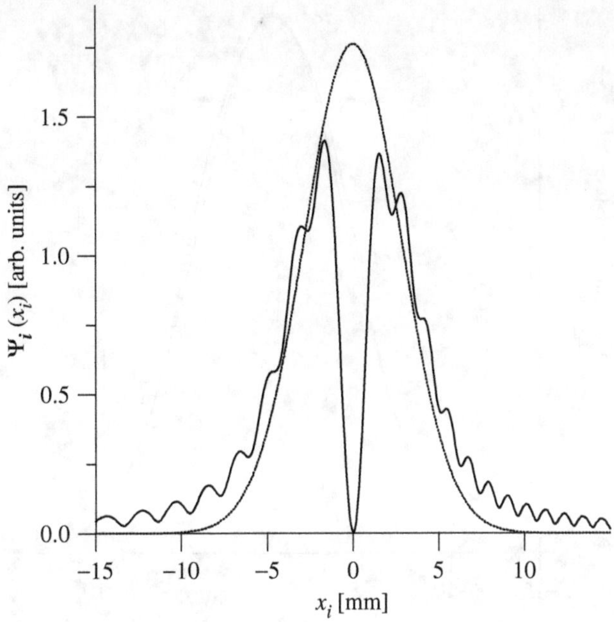

Figure 8.2. Incident field of a piston radiator (solid curve) calculated as a function of slant-range shift distance x_i for a water path equal to half the Rayleigh distance $\mathcal{R} \simeq a^2/\lambda_f$, compared to a Gaussian beam (dotted curve) at the same downrange distance $z = \mathcal{R}/2$ (after Lobkis *et al.* [13]).

assertion made above. Note also that the voltage measurement modeled here is performed in the piston radiator nearfield and yet very accurately modeled by the Gaussian beam substitution, and this in a region where the incident field of the piston transducer has very strong pressure variations, as seen in Fig. 8.2. The clear implication of these results is that the piston beam function of Eq. (8.7) and the Gaussian beam of Eq. (8.9) yield nearly the same signal, when the received voltage of one of a pair of identical transducers is measured in the geometry of Fig. 8.1, even in the nearfield of the transducers. A more complete exposition of this entire phenomenon, together with analytical calculations and many numerical examples can be found in Lobkis and Chimenti [20]. Casually, we can think of a receiver as a low-pass filter of spatial frequency components of the incident beam field; therefore, the rapid field variations on the scale of the sidelobes are less important.

8.2.1 Uniform asymptotics

The numerical evaluation of the integral in Eq. (8.10) is deceptively tedious, owing to the highly oscillatory nature of the integrand. Unless the integration

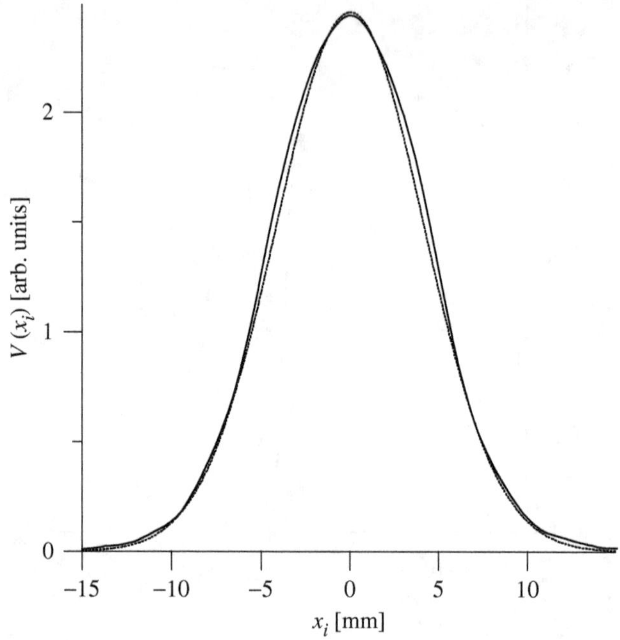

Figure 8.3. Receiver voltage versus beam shift x_i at a range of $z = \mathcal{R}/2$ for combined piston beams (solid curve) and combined Gaussian beams (dotted curve) with beam waist parameter $b = 0.545$ for the conditions of Fig. 8.2 (after Lobkis *et al.* [13]).

path is displaced into the complex plane or some other device employed, numerical evaluation may not be straightforward. Moreover, numerical methods are not necessary, because integrals of this type can be evaluated analytically, with some approximations. The integral in Eq. (8.10) can be performed asymptotically by a steepest descent evaluation along a deformed contour in the complex θ plane. The procedure is uncomplicated, except when the saddle point closely approaches or coincides with a pole of the reflection coefficient, unfortunately the most interesting case. To accommodate this complication, a procedure termed *uniform asymptotics* has been developed [23] and applied to many related acoustic reflection problems [19,24,25]. The method lends itself well to a geometric argument. In Fig. 8.1 the distance from the transmitter center to the origin (labeled $\rho/2$), and its counterpart returning to the receiver center, is the specular ray and can be associated with the saddle point of the θ integrand in Eq. (8.10). Likewise, assume that some spatial frequency in the incident beam in Fig. 8.1 equals the phase-match angle to excite a leaky wave. Then, this wave component corresponds to a pole of the SC and can be associated with that respective contribution to Eq. (8.10). An exception occurs when the saddle point and pole exactly, or nearly, coincide, requiring the method of uniform asymptotics.

If the incident plane contains a symmetry axis, the integral over $\bar{\phi}$ is nearly independent of the reflection coefficient, and it can be calculated at once by the saddle point method in the limit $k_f \rho \gg 1$, which is essentially always satisfied. There are two saddle points, $\bar{\phi} = 0, \pi$ of the $\bar{\phi}$ integrand in Eq. (8.10). This result implies that the really significant contributions to the voltage in Eq. (8.10) come from wave components lying in and near the incident (or (x, z)) plane. The point $\bar{\phi} = \pi$ (backward wave) can be safely ignored for a piston source that is in some sense collimated ($k_f a \sin\alpha > 1$). After the $\bar{\phi}$ evaluation, Eq. (8.10) can be written

$$V(x_i, f) = \frac{ia^4}{2} \mathcal{V}_T(f)\mathcal{V}_R(f) \sqrt{\frac{2\pi k_f}{i\rho \sin\theta_0 + 2\bar{b}^2 k_f a^2 \sin\alpha}}$$
$$\times \int_0^{\pi/2 - i\infty} S(\theta, f) e^{ik_f \rho \cos(\theta - \theta_0) - (\bar{b}k_f a \sin(\theta - \alpha))^2} \sqrt{\sin\theta}\, d\theta. \tag{8.11}$$

Integral expressions such as Eq. (8.11) have been analyzed by a number of workers [16, 23, 24]. Brekhovskikh [16] developed an expression for spherical wave reflection analogous to Eq. (8.11), and it can be adapted to help evaluate the current integral. The point source in Brekhovskikh's analysis must be replaced by Eq. (8.9), the directivity function of the Gaussian beam for which the maximum amplitude occurs at $\theta = \alpha$. Using the steepest descent path suggested by Brekhovskikh [16] for the phase function $k_f \rho \cos(\theta - \theta_0)$, the saddle point θ_s in our case is given by the solution to

$$\sin(\theta_s - \theta_0) - i\gamma \sin 2(\theta_s - \alpha) = 0, \tag{8.12}$$

where the parameter $\gamma = \bar{b}^2 k_f a^2/\rho \approx 0.856(2\mathcal{R}/\rho)$ is related to the ratio of the Rayleigh distance $\mathcal{R}$ to the actual distance $\rho/2$ from transducer to plate.

The exact solution of Eq. (8.12) for the saddle point is a complicated expression and not necessary in most practical cases. Rather, it is sufficient to investigate the solution of Eq. (8.12) for $x_i \ll \rho$, *i.e.*, for a small shift of the transducer axes compared with the distance between the transducers and the plate (see Fig. 8.1). In that case, the asymptotic saddle point solution is

$$\theta_s \approx \alpha + \frac{x_i \cos\alpha}{\rho(1 + 4\gamma^2)}(1 + 2i\gamma). \tag{8.13}$$

This expression includes all geometrical parameters of the system. Thus, the effect of changing the transducer parameters x_i, α, ρ, a, f, or ($\gamma \sim k_f a^2/\rho$) is mathematically equivalent to shifting the saddle point θ_s along both the real and imaginary axes in the complex θ plane. This result may be compared with the point source, spherical wave solution [16].

In the leaky wave region ($x_i > 0$) the imaginary part of θ_s moves toward the positive imaginary direction from the real axis in the complex θ plane and moves closer to the SC poles, denoted θ_p. As θ_s approaches a pole, the effect of a plate mode on the reflected field increases. For $x_i < 0$ the distance between θ_s and the poles increases, and the specular component dominates the receiver voltage. The saddle point is entirely real only when $x_i = 0$, or when the amplitude maximum and stationary phase points coincide, that is when $\theta_0 = \alpha$. This condition, however, does not imply optimal coupling to guided wave modes, which occurs instead when $|\theta_p - \theta_s|$ is minimum.

With the calculation of the saddle point position, it is possible to evaluate Eq. (8.11) by representing this integral as the sum of saddle point and SC pole contributions in the form of a uniform asymptotic solution, such as suggested by Felsen and Marcuvitz [23] and Zeroug [24]. Rewriting their expression in the current terminology,

$$V(x_i,f) = \frac{a^4}{4}\mathcal{V}_T(f)\mathcal{V}_R(f)\sqrt{\frac{2\pi k_f}{\rho(i\sin\theta_0 + 2\gamma\sin\alpha)}}e^{ik_f\rho P(\theta_s)}\left\{\sqrt{\frac{2\pi i\sin\theta_s}{k_f\rho P^{\{2\}}(\theta_s)}}S(\theta_s,f)\right.$$
$$\left. + \sum_{j=1}^{n}\sqrt{\sin\theta_{pj}}\mathrm{Res}[S(\theta_{pj},f)]\left[\sqrt{\pi/(k_f\rho)}\Delta_{pj}^{-1} + i\pi w\left(\sqrt{k_f\rho}\Delta_{pj}\right)\right]\right\}, \tag{8.14}$$

where the saddle point phase function $P(\theta_s)$ and its second derivative $P^{\{2\}}(\theta_s)$ are given by

$$\begin{aligned} P(\theta_s) &= \cos(\theta_s - \theta_0) + i\gamma\sin^2(\theta_s - \alpha) \\ P^{\{2\}}(\theta_s) &= -\cos(\theta_s - \theta_0) + 2i\gamma\cos(2\theta_s - 2\alpha). \end{aligned} \tag{8.15}$$

The so-called "numerical distance" [16, 23] Δ_{pj} between the jth complex pole (of a total of n poles) $\theta_{pj}(= \theta'_{pj} + i\theta''_{pj})$, and the complex saddle point θ_s, is expressed as $\Delta_{pj}^2 = i(P(\theta_s) - P(\theta_{pj}))$. The term $\mathrm{Res}[S(\theta_{pj},f)]$ is the residue of the SC $S(\theta_{pj},f)$ at the jth pole, and the function w in Eq. (8.14) is related to the complementary error function (erfc) and is given in [26] by

$$w\left(\sqrt{k_f\rho}\Delta_{pj}\right) = \exp\left[-k_f\rho\Delta_{pj}^2\right]\mathrm{erfc}\left(-i\sqrt{k_f\rho}\Delta_{pj}\right). \tag{8.16}$$

As the pole approaches the saddle point, equivalent to incidence at a phase-matching leaky guided wave angle, we can approximate the phase function and its derivatives and the "numerical distance" Δ_{pj}. Following Bertoni and others [7, 8, 19] and particularizing to the RC, contributions to Eq. (8.14) come essentially only from the singularities or poles. Therefore, $R(\theta, f)$ (the reflection coefficient in this case) can be approximated by

$$R(\theta, f) \simeq \frac{\Pi_{j=1}^{n}(\theta - \theta_{zj})}{\Pi_{j=1}^{n}(\theta - \theta_{pj})} \tag{8.17}$$

where θ_{pj} is the jth complex pole location, and θ_{zj} is its corresponding zero, related to the pole by $\theta_{zj} \approx \theta'_{pj}$, or the real part of the pole is approximately equal to the corresponding zero. Using this approximate form of the reflection coefficient and taking into account the fact that

$$R(\theta_s, f) + \sum_{j=1}^{n}\{\text{Res}[R(\theta_{pj}, f)]/(\theta_{pj} - \theta_s)\} = 1 \quad \text{and that}$$

$$\text{Res}[R(\theta_{pj}, f)] = i\theta''_{pj}, \tag{8.18}$$

the voltage function Eq. (8.14) can be written as

$$V(x_i, f) = \frac{\pi a^4 V_T(f) V_R(f)}{2\rho(i+2\gamma)} \exp\{i\Phi - (x_i/x_\ell)^2\}$$
$$\times \left\{1 - \sqrt{\pi k_f \rho(i/2+\gamma)} \sum_{j=1}^{n}\left[\theta''_{pj} w\left(\sqrt{k_f \rho(i/2+\gamma)}(\theta_{pj} - \theta_s)\right)\right]\right\}, \tag{8.19}$$

where the phase function $\Phi = k_f \rho - (2k_f \gamma x_i^2 \cos^2\alpha)/(\rho(1 + 4\gamma^2))$, and x_ℓ is a localization parameter given by $x_\ell = a\bar{b}\sqrt{1 + 4\gamma^2}/(\gamma \cos\alpha)$. The structure of the expression in Eq. (8.19) is similar to that of both Bertoni and Tamir [7] for the half-space and Pitts *et al.* for the plate [8]. The essential difference is that Eq. (8.19) models the observed *voltage* (and not simply the field) for an identical transmitter and receiver.

For an elastic half-space [7], the first term in Eq. (8.19) (the pre-factor on the top line multiplied by the first term in braces on the second) represents the contribution of the specularly reflected wave. This signal is located in the region $x_i \approx x_\ell$ and depends on the transducer position and frequency. In the nearfield $\gamma > 1$, we have that $x_\ell \approx a/\cos\alpha$; in the farfield $\gamma < 1$, and the localization is $x_\ell \approx 2\rho/(k_f a \cos\alpha)$, *i.e.*, proportional to the incident beam projection onto the solid surface. The second term in Eq. (8.19) (pre-factor times the second term in braces on the bottom line) represents the contribution from the leaky wave. For the plate, this decomposition into specular and leaky wave components is not so obvious, because as others have noted [8], the reflection coefficient zero lies on the angle axis. The essential difficulty arises from the fact that at a phase-match angle θ_z for leaky wave excitation, the plane wave reflection coefficient vanishes $R(\theta = \theta_z, f) = 0$. In other words, the incident wave component responsible for the leaky wave makes no contribution to the specular reflection (the first term).

8.3 Computational Results

According to the reciprocity principle of Auld and Kino [17,18], it is possible to consider the output signal dependence as an interaction between two acoustical fields on the plate surface: one from the reflected field created by transmitter, and the other the (incident) field created by the receiver when it functions as a transmitter. Expressed by Eq. (8.6) with either piston or Gaussian directivity functions, these fields are actually infinite in physical extent, but we will represent them schematically as elliptical beam footprints; here, the field amplitudes are largest. If we wanted to investigate the properties of only the reflected acoustical field, but not the two-transducer voltage $V(x_i, f)$, it would suffice to replace the receiver beam by a point source. In that case, our results would be identical with a direct evaluation of Eq. (8.6) (putting $D_R = 1$) or with those of [8].

In Figs. 8.4–8.8, the calculated receiver voltage $V(x_i, f)$ is presented at a fixed incident, or orientation, angle α as a coherent sum (vector sum) of specular (dashed) and leaky wave (dotted) components and labeled 1 and 2, respectively. The sum is denoted by the solid curves and labeled 3. The vertical dashed lines labeled with a plate mode indicate the position of SC zeroes, in this case the reflection coefficient. We denote a positive beam shift x_i when the transducers' acoustical axes intersect at a point located below the upper plate surface, and negative otherwise.

The receiver voltage $V(x_i, f)$ dependence calculated from Eq. (8.10) for a negative beam shift of $x_i = -10$ mm is plotted in Fig. 8.4, where the sample is assumed to be a 1.51-mm thick steel plate. The transducer axes

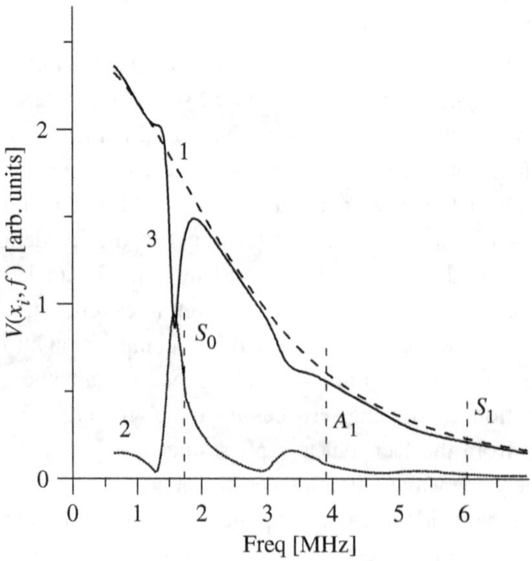

Figure 8.4. Calculated receiver voltage versus frequency for beam shift $x_i = -10$ mm. Dashed curve (1) is specular component, dotted curve (2) is leaky wave, and solid curve (3) is total voltage. Zeroes of the plane wave reflection coefficient are denoted by vertical dashed lines and labeled with mode designation (after Lobkis *et al.* [13]).

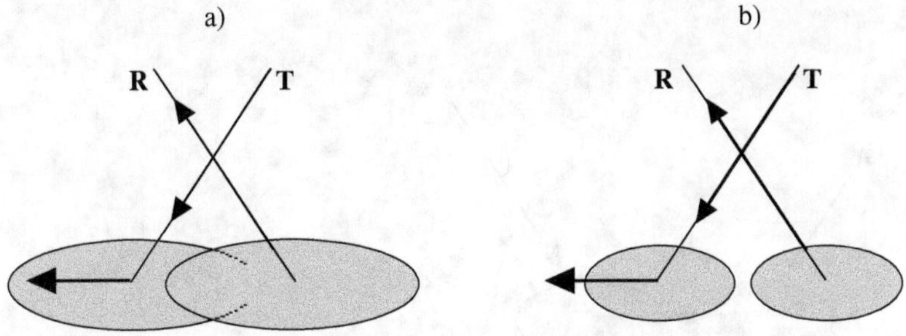

Figure 8.5. Schematic illustration of diffraction effects and beam overlap for transmitter (T) and receiver (R) beams at low (a) and high (b) frequency for $x_i < 0$, the conditions of Fig. 8.4 (after Lobkis *et al.* [13]).

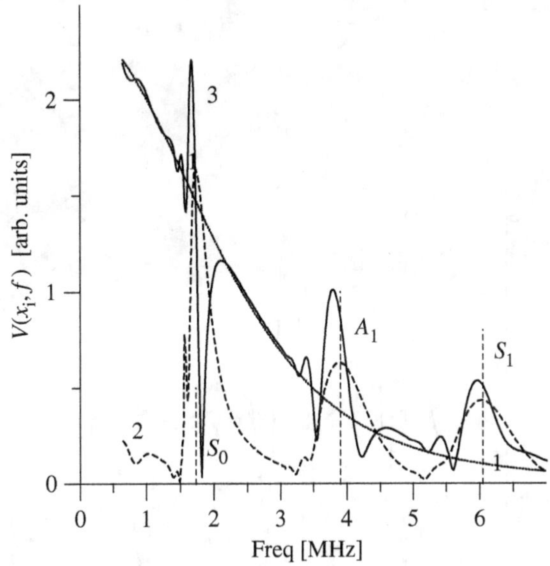

Figure 8.6. Calculated receiver voltage versus frequency for beam shift $x_i = +12$ mm, showing the effects of diffraction on voltage minima. Dotted curve (1) is specular component, dashed curve (2) is leaky wave, and solid curve (3) is total voltage. Plane wave reflection coefficient zeroes are denoted by vertical dashed lines and labeled with their mode designations (after Lobkis *et al.* [13]).

are in the (x, z) plane and oriented to orientation angles of α=25°. For this transducer position only the influence of the S_0 mode on the received voltage is pronounced. The two higher-order modes produce a much smaller effect. The reason for this behavior is illustrated in Fig. 8.5b, where the beams are represented schematically. With a negative beam shift, the effect of two additional modes, A_1 and S_1, on the signal is very weak because they occur at higher frequencies, where the diffraction-limited beam footprint leaves

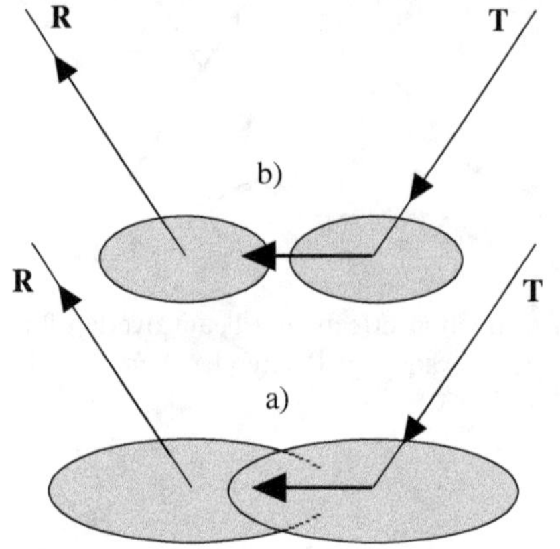

Figure 8.7. Schematic illustration of diffraction effects and beam overlap for transmitter (T) and receiver (R) beams at low (a) and high (b) frequency for $x_i > 0$, the conditions of Fig. 8.6 (after Lobkis *et al.* [13]).

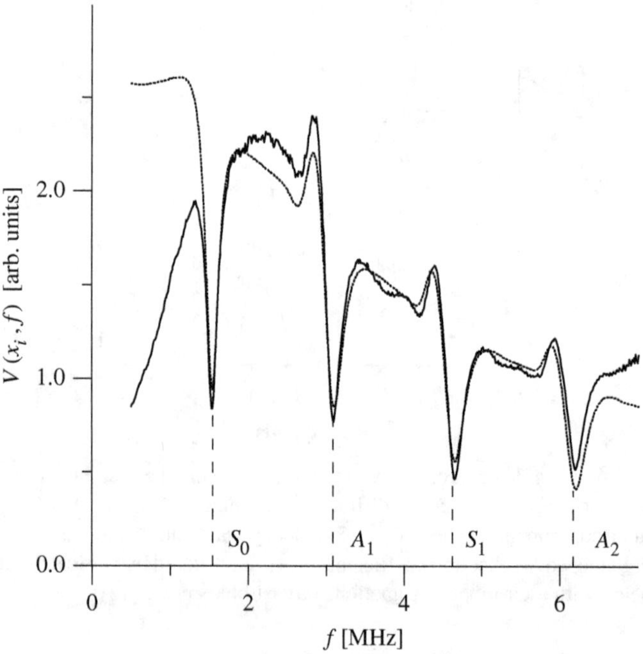

Figure 8.8. Experimental (solid curve) and theoretical (dotted curve) frequency scan of receiver voltage for a pitch–catch geometry with 9.5 mm diameter transducers at $\alpha = 20°$ and $x_i = +5$ mm. RC zeroes are denoted by vertical lines and labeled with their mode designation (after Lobkis *et al.* [13]).

almost no overlap, as in Fig. 8.5b. For low frequencies, the sizes of the two footprints on the plate are larger because the beams spread as they propagate from the transducers. There is consequently a significant region of intersection between them, shown in Fig. 8.5a, and S_0 mode in this region has a strong influence on the received voltage.

More typically, the offset distance x_i is positive. An argument similar to the one above based on geometrical diffraction can be made to explain the frequency dependence of the voltage signal for a positive offset of $x_i = +12$ mm, whose calculated results are presented in Fig. 8.6. For the S_0 mode, the coherent sum at low frequency yields a minimum in $V(x_i, f)$ very near the corresponding RC zero. At higher frequencies, however, the specular part is much smaller, owing to the decrease in beam diffraction, and the leaky wave component dominates the total voltage signal for the A_1 and S_1 mode contributions. Combined, these effects produce a *maximum* in $V(x_i, f)$, instead of a minimum, at the positions of the A_1 and S_1 modes for beam shifts larger than the transducer diameter $2a$. The beam geometry, illustrated in Fig. 8.7a and b, shows this dependence on beam footprint schematically. As shown in [13], there exists an offset distance x_i for which the voltage spectrum (measured or calculated) is nearly identical to the occurrence of the plane wave RC minima and can, therefore, be used effectively to infer the scattering coefficient from a voltage measurement. This offset is nearly the transducer radius a. When $x_i = a$, the voltage minima correspond closely over a wide frequency range to the RC minima, and therefore the plate wave modes. This circumstance is illustrated in Fig. 8.8, where the dotted curve is a calculation according to Eq. (8.6) (using Gaussian beam directivity functions), and the solid curve is the result of a measurement. For the measurement, the results have been normalized to suppress the transducer frequency response and are valid from about 1.25 MHz to about 6.5 MHz. Notice how the calculation follows all the detailed amplitude variations in the measured signal. The depths of the successive voltage minima demonstrate that the plate wave resonances are nearly perfectly predicted by the theory. Moreover, in this case the minima lie almost directly over the RC zeroes.

8.4 Complex Transducer Points

In the previous section, we saw that the application of Gaussian beams to the analysis of scattering measurements produces accurate predictions so long as the experiment is performed with identical reciprocal transducers. Moreover, the receiver voltage is relatively insensitive to the exact form of the directivity functions of receiver and transmitter, as long as they are identical. This fact permits the successful application of Gaussian beams for the more complicated, but perhaps more accurate, piston radiator beams. Once we select Gaussian beams as acceptable substitutes, a mathematically elegant alternative beam representation is available. It is possible to simplify the construction of Gaussian acoustic beams by exploiting the advantages of the so-called "complex source point" [27], or more generally considering both source and receiver, the complex transducer point (CTP) [22–29]. In this formalism, a point source at real coordinates is displaced into the complex plane and produces a Gaussian beam to excellent approximation. As is essentially a point source, the CTP can be used to provide a much simpler, but still quite accurate, model for the voltage calculation in a pitch–catch experimental setup with piston probes.

We begin the calculation of the receiver voltage with the Green's function for an unbounded medium. Let us assume a point source is located at $\mathbf{r}' = (x', y', z')$ and a point receiver located at $\mathbf{r} = (x, y, z)$. Then, the free-space Green's function is

$$G(\mathbf{r}) = \frac{e^{ik_f|\mathbf{r}-\mathbf{r}'|}}{4\pi|\mathbf{r}-\mathbf{r}'|}, \tag{8.20}$$

and it satisfies the inhomogeneous Helmholtz equation

$$(\nabla^2 + k_f^2)G(\mathbf{r}) = -\delta(\mathbf{r}-\mathbf{r}'), \tag{8.21}$$

where $k_f = \omega/V_f$ is the wavenumber and $V_f\left(=\sqrt{C_f/\rho_f}\right)$ is the acoustic wavespeed in the surrounding fluid medium, C_f is the fluid bulk modulus, and ρ_f here is again the mass density in the fluid. The term $\delta(\mathbf{r}-\mathbf{r}')$ is the Dirac delta function whose argument is the vector difference between the source and receiver positions. We attach a subscript f to k, C, and V to indicate that k_f, C_f, and V_f are the wavenumber, elastic property, and sound velocity in the fluid.

We model the combined transmitting and receiving piston transducer fields as Gaussian beams, now conveniently represented by displacing a real point receiver and a real point source into the complex plane [27],

$$\mathbf{r} \to \tilde{\mathbf{r}} = \mathbf{r} + i\mathbf{b} \tag{8.22}$$

$$\mathbf{r}' \to \tilde{\mathbf{r}}' = \mathbf{r}' + i\mathbf{b}' \tag{8.23}$$

where the vectors $\mathbf{b}$ and $\mathbf{b}'$ specify the transducer beam directions and their absolute Fresnel lengths for receiver and transmitter (primed quantities). The magnitude $|b|$ is related to the width parameter W of the beam at its waist (*i.e.*, the $1/e$ radius) through $W = \sqrt{2b/k_f}$, and similarly for b'. (In this calculation, the Fresnel lengths b and b' have dimensions of length). The voltage $V_{\mathcal{R}}(\tilde{\mathbf{r}}; \tilde{\mathbf{r}}')$ can be expressed as [29],

$$V_{\mathcal{R}}(\tilde{\mathbf{r}}; \tilde{\mathbf{r}}') = i\omega\rho_f\gamma(\omega)\frac{e^{ik_f|\tilde{\mathbf{r}}-\tilde{\mathbf{r}}'|}}{4\pi|\tilde{\mathbf{r}}-\tilde{\mathbf{r}}'|}, \tag{8.24}$$

where $\gamma(\omega)$ contains the signal strength of the transmitter and receiver and the temporal frequency spectrum of the transducers and associated electronics, ω is the angular frequency. As shown by Zeroug *et al.* [29], this representation approaches a Gaussian beam in both magnitude and wavefront curvature for paraxial rays. In fact, the CTP remains quite a good approximation for rays somewhat distant from paraxial. A comparison of the two beams near the source is shown in Fig. 8.9. The calculations are performed for a 0.5-inch (12-mm) diameter beam at 5 MHz propagating in water. The Gaussian beam is the continuous curve and the CTP is the dashed curve. Even at values of beam amplitude near zero, the two beams are essentially identical. At no point does the Gaussian beam differ from the CTP by more than 1 part in 10^4. These fractional differences are well below other possible sources of uncertainty in the modeling of piston transducer beams.

The interaction of the acoustic beam with the plate, shown in Fig. 8.10 but with planar transducers, can be synthesized from the spectral decomposition of the complex Green's function in Eq. (8.24), weighted by the plane wave scattering coefficient $S(k_x, k_y)$ [25]

$$V_{\mathcal{R}}(\alpha, x) = -\frac{\omega\rho_f}{8\pi^2}\gamma(\omega)\int\int_{-\infty}^{\infty}\frac{S(k_x, k_y)}{k_z}e^{ik_x(\tilde{x}-\tilde{x}')+ik_y(\tilde{y}-\tilde{y}')+ik_z(\tilde{z}\mp\tilde{z}')}dk_x dk_y, \tag{8.25}$$

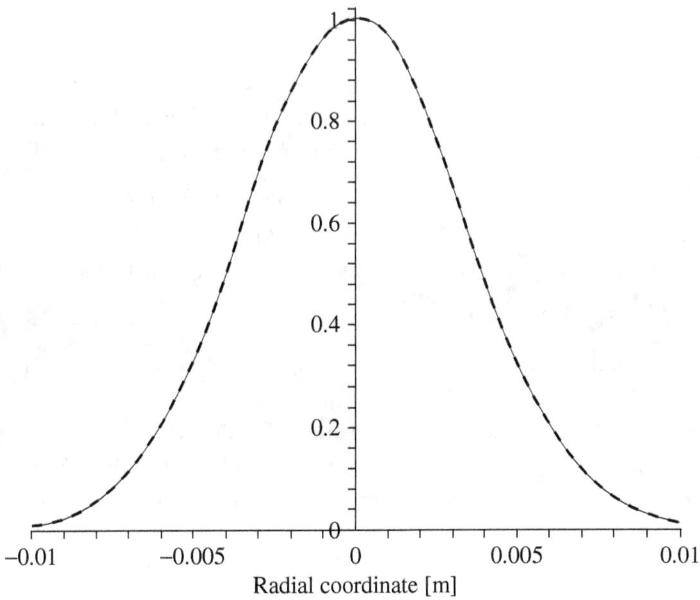

Figure 8.9. Comparison of Gaussian beam (thin solid curve) to complex transducer point (CTP) approximation (heavy dashed curve) for a 1/2-inch diameter beam at 5 MHz in water. The two curves are virtually identical.

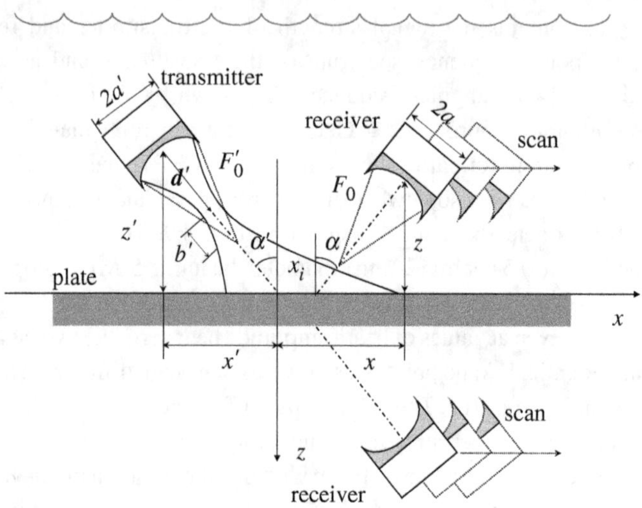

Figure 8.10. Schematic of experimental geometry for analysis of beam reflection from fluid-loaded structures; transducer orientation angle α is measured from the vertical, b' is the transmitter Fresnel length and F_0 and F_0' are the beam focal lengths, x_i is the beam shift parameter (the distance along the $z = 0$ interface between the beams' two acoustical axes). A plane wave component of the beam that propagates toward the plate makes a phase-match angle of θ with the plate normal. The plate is assumed infinite in the y direction (after Chimenti and Fei [31]).

where α is both the transmitter and receiver orientation angle, dk_x and dk_y are the projections of the fluid wavenumber k_f onto the x and y coordinates, respectively, and $k_z = \sqrt{k_f^2 - k_x^2 - k_y^2} = k_f \cos\theta$ is the projection of fluid wavenumber onto the z axis. Here, the integration variable k_y plays a similar role to $\bar{\phi}$ in the angular version of the spectral integral. In the geometry shown in Fig. 8.10,

$$\mathbf{b} = (-b\sin\alpha, 0, \pm b\cos\alpha), \qquad \mathbf{b}' = (b'\sin\alpha', 0, b'\cos\alpha'). \tag{8.26}$$

Figure 8.10 illustrates a reflection or transmission measurement, and the upper sign in the $\pm$ indicates reflection. After substituting Eq. (8.26) into Eq. (8.23) we have for the complex receiver coordinates,

$$\tilde{x} = x - ib\sin\alpha, \quad \tilde{y} = y, \quad \tilde{z} = z \pm ib\cos\alpha, \tag{8.27}$$

and for the complex transmitter coordinates,

$$\tilde{x}' = x' + ib\sin\alpha, \quad \tilde{y}' = y', \quad \tilde{z}' = z' + ib\cos\alpha. \tag{8.28}$$

To analyze the data in terms of incident-plane wavenumber ξ or phase-match angle θ, we treat the data by performing a spatial Fourier transform on the measured (or calculated) voltage $V_{\mathcal{R}}(x, \omega, \alpha,)$

$$\hat{V}_{\mathcal{R}}(\xi, \omega, \alpha) = \int_{-\infty}^{\infty} V_{\mathcal{R}}(x, \omega, \alpha) e^{-i\xi x} dx. \tag{8.29}$$

For the experimental two-transducer configuration described in this chapter, the voltages predicted by 2-D and 3-D models differ in relatively minor ways, so long as the plane of the incident beam contains a symmetry direction in the sample material [30]. When this is not the case, serious differences between the 2-D and 3-D predictions can arise, and for fiber orientation angles ϕ other than 0° or 90° in transversely isotropic or orthotropic laminates, the more general 3-D analysis is essential. One way to understand these differences is to consider that for propagation in off-symmetry directions, leaky plate wave energy deviates out of the incident plane, as shown in Chapter 5. The 2-D approximation could still be applicable for off-symmetry-axis measurements if the double through-transmission or double reflection experimental configuration is used, as described in Chapter 3 and used for some simulations in Chapters 6 and 7. We will briefly sketch out some of these complications at the end of this section. Directly below, however, we present the simple analysis for the 2-D case; the more general and more complicated 3-D analysis, sometimes necessary for composite laminates, can be found in detail in Han and Chimenti [14] for the CTP approach and in Lobkis and Chimenti [30] for the more conventional Gaussian directivity function method. The 2-D analysis is still quite useful in a large variety of situations. Although the 2-D result may not be rigorously valid for quasi-isotropic, multilayer composites at high frequency, it will be sufficiently accurate to permit inference of composite elastic properties if the laminate can be modeled elastically as an orthotropic continuum and the incident plane contains a material symmetry axis, as will be shown next. Otherwise, the more general 3-D analysis must be used [14, 30], possibly combined with the multilayer stiffness-method scattering coefficient derived in Chapter 6.

8.4.1 Two-dimensional voltage calculation

Limiting the model here to the 2-D case, the voltage calculation shown in Eq. (8.25) loses its k_y dependence and becomes simply

$$V_{\mathcal{R}}(x, \omega, \alpha) = -\frac{\omega \rho_f}{8\pi^2} \gamma(\omega) \int_{-\infty}^{\infty} \frac{S(k_x, 0, \omega)}{k_z} e^{ik_x(\tilde{x}-\tilde{x}')+ik_z(\tilde{z}\mp\tilde{z}')} dk_x, \tag{8.30}$$

where $\omega = 2\pi f$, and $k_z = \sqrt{k_f^2 - k_x^2}$ when the k_y dependence vanishes. The assumption here is that the transmitter and receiver produce and detect so-called *sheet beams*, essentially line sources with no directional dependence out of the incident plane. Laboratory transducers in a rectangular shape with aspect ratios of three or four are quite good approximations of a sheet beam. Even circular piston radiators, under some conditions, can be successfully modeled by this approximation. Substituting the voltage expression in Eq. (8.30) into Eq. (8.29) and performing the dx integration yields a Dirac delta function $2\pi\,\delta(k_x - \xi)$ over the incident-plane wavevector. The presence of the delta function makes the dk_x integral simple to evaluate,

$$\hat{V}_{\mathcal{R}}(\xi, \omega, \alpha) = -\frac{\omega\rho_f}{4\pi}\gamma(\omega)\frac{S(\xi, 0, \omega)}{k_z}e^{i\xi(-ib\sin\alpha - \tilde{x}') + ik_z(\tilde{z}\mp\tilde{z}')}. \tag{8.31}$$

The negative sign above in the $\mp$ is for reflection and the positive for transmission. A more detailed version of this calculation, along with an expansion of the piston radiator beam in terms of a weighted sum of Gaussians (expressed mathematically as CTPs), can be found in the paper by Chimenti and Fei [31]. Including the complex coordinates from Eqs. (8.27) and (8.28) and expressing the wavevector as a phase-match angle $\theta(=\arcsin(\xi/k_f))$, or the incident angle of a particular plane wave component of the beam, the magnitude of the voltage in Eq. (8.31) becomes

$$\begin{aligned} |\hat{V}_{\mathcal{R}}(\xi, f, \alpha)| &= \frac{1}{2} f\rho_f\gamma(f)\frac{|S(\xi, 0, f)|}{k_f\cos\theta}e^{k_f(b\cos(\theta-\alpha) + b'\cos(\theta-\alpha'))}, \\ &\approx \frac{1}{2} f\rho_f\gamma(f)\frac{|S(\xi, 0, f)|}{k_f\cos\theta}e^{k_f(b+b')}e^{-k_f b(\theta-\alpha)^2/2}e^{-k_f b'(\theta-\alpha')^2/2}, \end{aligned} \tag{8.32}$$

where the frequency $f = \omega/(2\pi)$ has replaced ω.

Equation (8.31) demonstrates the effect of the extrinsic experimental parameters on the output voltage. The vertical positions of the transmitter and receiver, for example, influence only the phase, not the magnitude of the output voltage. Moreover, the properties of the transmitter and receiver acoustic beams change the angular range of the voltage. Both beams have a Gaussian distribution profile centered on a beam axis whose orientation is determined by the transducer orientation angle α. The angular half-spread of the beam $\Delta\theta$, given by $\Delta\theta = 2[k_f W]^{-1}$, is determined by the transducer Fresnel length b (or the beam waist width W) for a given frequency and coupling fluid properties. The smaller the Fresnel length, the larger the angular spread of the transducer beam. The output voltage is the combined contributions from transmitter and receiver beams. In the technologically important case of an identical transmitter and receiver at the same orientation angle α, we will have $b' = b$ and $\alpha' = \alpha$. When these conditions hold, the result is a particularly simple expression for the 2-D voltage magnitude,

$$|\hat{V}_{\mathcal{R}}(\xi, f, \alpha)| = \frac{1}{2} f \rho_f \gamma(f) |S(\xi, 0, f)| \frac{e^{2k_f b} e^{-k_f b(\theta - \alpha)^2}}{k_f \cos\theta}, \tag{8.33}$$

where the quantities k_f (the fluid wavenumber), ξ and θ are related according to Snell's law $\xi = k_f \sin\theta$. The transducer orientation angle is α, θ is the phase-match angle, b is the transducer Fresnel length, $|S(\xi, 0, f)|$ is the scattering coefficient magnitude, f is the frequency, ρ_f is the fluid density, and $\gamma(f)$ is a frequency-dependent term accounting for the sensitivities of the transducers and associated electronics. The expression in Eq. (8.33) is basically a windowed version of the angular (or ξ) dependence of the scattering coefficient magnitude $|S(\xi, 0, f)|$. The surprising simplicity of the 2-D calculation leading to Eq. (8.33) is a direct consequence of the CTP approach. It is also worth noting that when the Fresnel length b vanishes in Eq. (8.33) above, the exponentials go to unity, implying that the beams are no longer collimated. The resulting simple expression is then valid for a point source and point receiver in the fluid. All beam effects are gone.

8.4.2 Synthetic aperture scanning

The dispersion relation of a guided wave, for example, is recorded as a curve in frequency and wavenumber, as we saw in Figs. 5.4 and 5.5 of Chapter 5. When performing experiments, it is normally straightforward to acquire broadband frequency data, either by collecting the time-domain trace of a scattered impulse excitation, scanning the frequency of a tone-burst signal, or some other means to code a broad frequency bandwidth into a time-domain signal. This approach, however, accounts for only one axis of the dispersion plot—the frequency axis. To obtain a similar broadband signal on the wavenumber axis requires some additional effort. In time and frequency, we have for the experimental frequency domain voltage

$$\tilde{V}_{\mathcal{R}}(x, \alpha, \omega) = \int_{-\infty}^{\infty} V_{\mathcal{R}}(x, \alpha, t) e^{-i\omega t} dt, \tag{8.34}$$

whereas in the wavenumber (or spatial frequency) domain, Eq. (8.29) is the analogous expression. The conjugate Fourier variables in Eq. (8.29) are wavenumber $\xi (= k_x)$ and coordinate x. To construct the experimental signal whose Fourier transform is the broadband *temporal* frequency dependence, as in Eq. (8.34), requires a short impulsive time-domain excitation whose scattered waveform is observed over a long time $(-\infty, +\infty)$, in Eq. (8.34).

By analogy, the broadband wavenumber domain response should employ a *spatially* restricted excitation, observed over a wide *coordinate* range. (Wavenumber ξ and phase-match angle θ are related through $\xi = k_f \sin\theta$.)

As we saw in the previous sub-section, a small beam waist width W does indeed lead to a wide angular range $\Delta\theta$ by diffraction. Later, we shall see that a much more efficient way to accomplish the same result is through focusing. The point is that the wide angular beam spread of an incident beam scattered from a plate must be sampled at many coordinate points over a wide spatial range $-\infty < x_i < +\infty$. The situation is illustrated in Fig. 8.11, showing a transmitter plus one receiver for reflection and one for transmission, shown at various scanning locations. The beam shift distance is shown as x_i, and the transducer orientation angle is α. The x-dependent coordinate scan with a step size of about $\lambda_f/10$, or one-tenth the fluid wavelength, builds the spatial Fourier data that are transformed over $\exp[-i\xi x]$, as shown in Eq. (8.29), to yield the voltage wavenumber spectrum. In practice, it is sufficient to scan (when using 10-mm diameter probes) from a point $x_i \lesssim -10$ mm to $x_i \gtrsim 50$ mm, where the measured signal amplitude is no longer recoverable from the noise. The spatial resolution determines the number of points in the data to be transformed. The finite scan range, or spatial window function, amounts to smoothing in the wavenumber domain, while a finite number of coordinate samples yields a limited interpolation density in the wavenumber domain, as in any discrete Fourier transform.

Although the transducer orientation angle α is fixed during the scan, energy is detected even when the receiver is at the left-most position in the lower left of Fig. 8.11. A chain-dash line from the transmitter center to

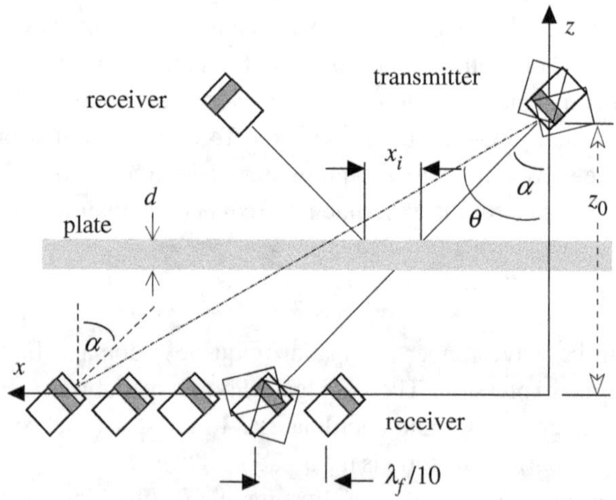

Figure 8.11. Schematic diagram of experimental geometry for synthetic aperture measurements, showing transducer placement and position scanning operation in through-transmission and in reflection. Depending on angular beam spread $\Delta\theta$, transducer orientation angle α may need resetting to cover entire dispersion spectrum. Chain-dash line from transmitter to far left receiver defines the variable phase-match angle θ (after Lobkis and Chimenti [30]).

the receiver in the farthest left position defines the variable phase-match angle θ, as shown. Transmitter beam plane wave components make an angle θ between the z axis and the receiver at any position. When focusing is employed, the range of θ over which detectable energy is observed will be large. For a diffracting planar transducer, $\Delta\theta$ is typically smaller, but never zero. A closer look at Eq. (8.33) shows that the synthetic aperture scan yields a windowed version of the plane wave scattering coefficient. The effect of the transducer directivity functions $D_T(\theta, \bar{\phi})$ and $D_R(\theta, \bar{\phi})$ on the voltage in Eq. (8.33) is negligible, as we saw in Figs. 8.2 and 8.3, another consequence of the synthetic aperture scan. More details on this technique can be found in [13, 14], [30–32].

8.4.3 Focused beams

The CTP representation of a Gaussian beam leads to an especially simple extension, allowing us to express a focused Gaussian beam mathematically. The real point source and point receiver are again displaced into the complex plane, but with the following alteration,

$$\begin{aligned} \mathbf{r} &\to \tilde{\mathbf{r}} = \mathbf{r} + \mathbf{d} + i\mathbf{b} \\ \mathbf{r}' &\to \tilde{\mathbf{r}}' = \mathbf{r}' + \mathbf{d}' + i\mathbf{b}', \end{aligned} \tag{8.35}$$

where the vector $\tilde{\mathbf{r}} = (\tilde{x}, \tilde{y}, \tilde{z})$ is again a complex vector that specifies the location of the CTP in the complex plane and the vector $\mathbf{b}$ specifies the transducer beam direction and its Fresnel length, but now the vector $\mathbf{d}$ is the location of the Gaussian beam waist of the transducer relative to the transducer aperture center, as shown in Fig. 8.10. A similar relationship holds for the focused transmitter point $\mathbf{r}'$ and its constituent parts. The beam width at the waist location W and the distance between the waist location and the transducer aperture center d (which is also the magnitude of the vector $\mathbf{d}$) are given by [33],

$$W = W_0 \frac{\beta}{\sqrt{1+\beta^2}}, \qquad d = \frac{F_0}{1+\beta^2}, \tag{8.36}$$

where $\beta = 2F_0/(k_f W_0^2)$, F_0 is the geometrical transducer focal length, and W_0 is the beam width at the transducer aperture. For most applications, W_0 can be estimated as $W_0 = 0.752a$ [13, 33] for measurements near or larger than the Rayleigh distance $\mathcal{R}$. For focused beams, the voltage in the 2-D approximation is just that given in Eq. (8.30), but with the substitution of parameters shown above. The beam width $\Delta\theta$ is given by $2/(k_f W)$, as mentioned above. The smaller the Fresnel length b (or the beam waist width W), the larger the angular spread of the transducer beam. Practically speaking, it is not possible for the Fresnel length of a real transducer to be identically zero, but it is possible for a realistic Fresnel length to be very small, either by decreasing the transducer aperture size a while maintaining the ratio of a/F_0 constant, or by decreasing the focal length F_0 while maintaining the aperture size a constant. The former relies on transducer diffraction, and the latter relies on the beam-focusing effect to achieve the same result—a wide angular beam aperture. Focusing is clearly the best way to accomplish this, and by using focused transducers with a high ratio of a/F_0, an angular beam spread of tens of degrees can be obtained easily. For planar transducers we simply allow F_0 to be infinite, and we have $d = F_0/(1+\beta^2) = k_f^2 W_0^4 F_0/(k_f^2 W_0^4 + 4F_0^2) \to 0$, implying that the beam waist is at the aperture center, and thus $W = W_0$.

8.4.4 Three-dimensional effects on the receiver voltage

The purpose of this brief treatment is to alert the reader to some of the complications arising in reflection or transmission voltage measurements with collimated beams when the incident plane of ultrasound on an anisotropic plate, such as a composite laminate, does not contain a symmetry axis. There is no need to restate the analytical expressions here because either Eq. (8.10) or Eq. (8.25) represents the completely general 3-D transducer voltage, allowing the scattering coefficient S to be a function of phase-match angle, azimuthal angle, frequency, complex solid viscoelastic stiffness matrix (itself a function of the fiber orientation angle), and fluid properties, $S = S(\theta, \bar{\phi}, f, C^*_{I,J}(\phi), \rho, C_f, \rho_f)$. In the full 3-D analysis, no algebraic simplifications occur, and the two angle integrals over $d\bar{\phi}d\theta$ in Eq. (8.10), or wavenumber integrals over $dk_x dk_y$ in Eq. (8.25), must be carried out numerically. The situation is illustrated schematically in Fig. 8.12. This schematic shows the incident (x, z) plane, containing the transducer central ray, and the beam spread angles $\pm\theta$ within the incident plane and $\pm\bar{\phi}$ out of the (x, z) plane.

In Fig. 8.13, the difference is shown between calculations of transmission functions (air is the immersion medium) of a 3.6-mm thick uniaxial composite laminate oriented with the incident plane normal to the fibers (frame (a)) and in the fiber direction (frame (b)). The solid curves contain the 3-D dependence and the dotted curves do not. In these two cases, because the laminate in either frame (a) or (b) contains a symmetry axis, the difference in the two sets of curves arises from the 3-D diffraction correction to the simple 2-D calculation. The 2-D calculation assumes a "sheet beam" source and receiver, infinite in extent outside of the incident plane. The 3-D calculation assumes finite-dimension probes, but the variation in the scattering coefficient with $\bar{\phi}$ is symmetric about the incident plane. If the incident plane contains a symmetry axis or if the plane wave scattering coefficient $S(\theta, \bar{\phi})$ has $\bar{\phi}$ variations that are small within the integration interval $\Delta\bar{\phi}$, the saddle-point method, presented in Section 8.2.1 earlier,

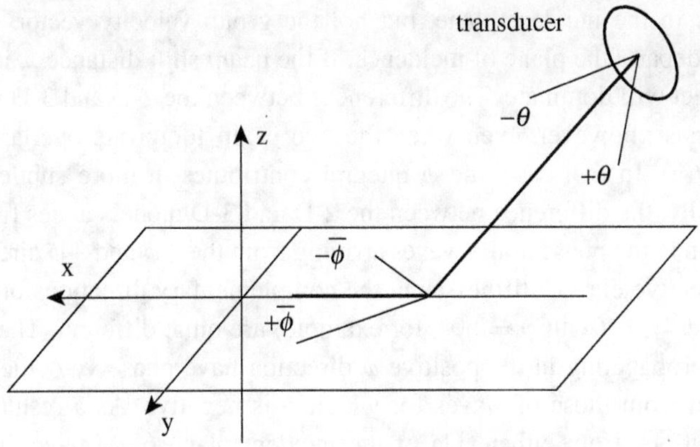

Figure 8.12. Schematic representation of beam spread angles in ($\pm\theta$), and out of ($\pm\bar{\phi}$), the incident (x, z) plane, which contains the transducer central ray (after Lobkis and Chimenti [30]).

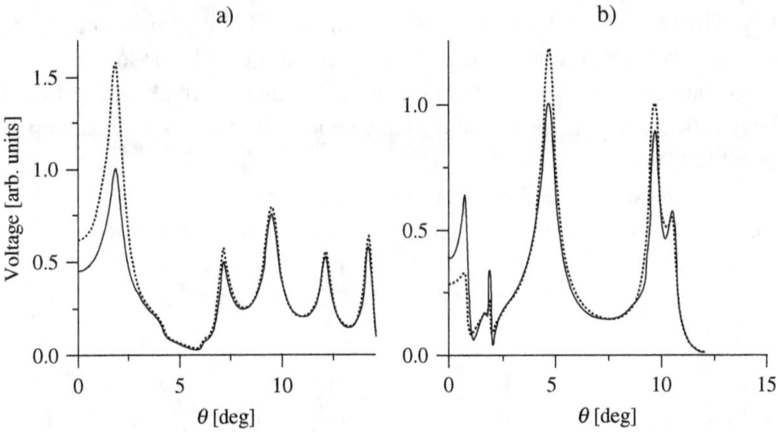

Figure 8.13. Calculated angular transmitted voltage functions for a 3.6-mm uniaxial carbon-epoxy laminate plate versus phase-match angle θ at a frequency of 900 kHz. Both plots show some 3-D effects. The 3-D calculation (solid curves) and 2-D (dotted curves) differ only in minor ways. The fiber orientation angle ϕ in (a) is 90°, and in (b) $\phi = 0°$ (after Lobkis and Chimenti [30]).

may also be used to estimate the out-of-incident plane integral over the $\bar{\phi}$ variable in Eq. (8.10) [30].

A more serious deviation occurs when the fiber orientation angle ϕ differs from either 0° or 90°, particularly at near-normal phase-match angles θ. As mentioned earlier in this chapter and shown in Section 2.6 of Chapter 2, when the wavevector is not aligned with a symmetry axis, the phase and group velocities will have different directions. The phase velocity vector remains in the incident plane, but not the group velocity vector, and so energy departs the plane of incidence. If the beam shift distance x_i is large, this effect will dominate. The differences between the 2-D and 3-D models still persist, however, even when the two beam footprints overlap, as in Fig. 8.7(a). In that case, the $\bar{\phi}$ integral contributes in more subtle ways. Physically, the difference between the 2-D and 3-D models arises from the variation in the phase of the waves arriving from the $+\bar{\phi}$ and $-\bar{\phi}$ directions. The effective elastic stiffnesses in the complementary directions on either side of $\bar{\phi} = 0°$ (with $\phi = 60°$, for example) are quite different. Therefore, waves propagating in the positive $\bar{\phi}$ direction have phase velocities quite different from those of waves for which $\bar{\phi}$ is negative. As a result, wave contributions from either side of the incident plane ($\pm\bar{\phi}$) have different values of phase at the receiver surface. This phenomenon is also the origin of the energy flux deviation detailed in Section 2.6. Coherent summation of these waves results in a voltage signal different from that of the incident-plane calculation, causing the peaks and their amplitudes to shift. If the $\bar{\phi}$ integration is ignored, this effect can easily lead to an incorrect estimate of C^*_{IJ}, when $\phi \neq 0°, 90°$. As we shall see, these 3-D effects surrounding the $\bar{\phi}$ integration are compounded in the case of air-coupled ultrasound, where most measurements are made close to the acoustic normal, owing to the large refraction from air into almost any solid. These topics are taken up in detail in Chapter 9.

Two examples of calculated and measured voltages are shown in Fig. 8.14 for the reflected voltage function in water immersion, where the plate is a

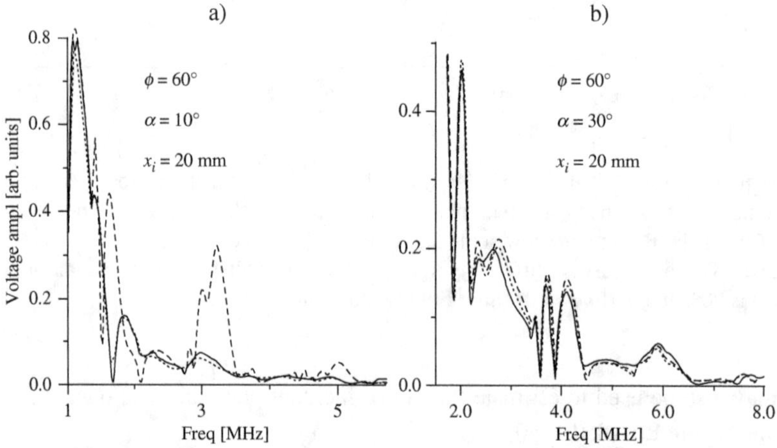

Figure 8.14. Experimental (solid curve) compared to calculated voltage function with 2-D integration (dashed curve) and with 3-D integration (dotted curve) for a 0.92-mm uniaxial carbon-epoxy laminated plate in the $\phi = 60°$ direction at an offset distance of $x_i = 20$ mm and an incident angle of $\alpha = 10°$ (frame (a)), or an incident angle of $\alpha = 30°$ (frame (b)) (after Lobkis and Chimenti [30]).

0.92-mm thick uniaxial laminate oriented at an azimuthal fiber angle of $\phi = 60°$. The incident angle in frame (a) is $\alpha = 10°$ and in frame (b) $\alpha = 30°$. The offset distance (see Fig. 8.1) in both cases is $x_i = 20$ mm. These data and calculations are not synthetic aperture scans, where the receiver undergoes stepwise translation, but rather simple stepped frequency curves with the transducers in a fixed relative position. The solid experimental curves in Fig. 8.14(a) and (b) are both well modeled by the 3-D calculation (dotted curves). The 2-D calculation is less accurate, especially in frame (a), although the disparities are much smaller in frame (b). The reasons for the closer agreement between 2-D and 3-D models at higher incident angles are twofold: first, at the larger transducer orientation angle α in frame (b) the beam's effective lateral (out-of-incident plane) diffraction is narrower ($\Delta\bar{\phi} \propto \alpha^{-1}$). Second, the anisotropy of the quasilongitudinal and quasishear partial waves induced at a general fiber orientation angle ϕ is much stronger for lower incident angles [30].

8.5 Experimental Results

The value of beam focusing in elastic property estimation is to achieve a broad spatial frequency bandwidth, to accompany the broad temporal frequency bandwidth, normally achieved through impulse, chirp, or other types of broadband rf excitation. Beam diffraction alone is insufficient to produce a wide spatial frequency spectrum, unless very small and therefore inefficiently low power beams are used. It is much more efficient to use focusing to broaden the spatial frequency range. By using focused transducers with a high ratio of a/F_0, an angular beam spread of tens of degrees can be obtained easily. The point is to use highly focused transducers to make rapid measurements of the scattering coefficients over a large angular range. Raw ultrasonic transducer data for time and receiver position are shown in Fig. 8.15 for a scan with typical focused transducers having a center frequency of about 5 MHz and a focal length of about 30 mm on an aluminum plate. Taking fast Fourier transforms (FFT) of the data in Fig. 8.15 in both time and space transforms the data to the frequency–wavenumber domain, where the minima in the voltage function of Eq. (8.33) reveal the occurrence of guided wave modes in the reflected (or transmitted) spectrum.

This double-transformed version is shown in Fig. 8.16. The frequency and wavenumber coordinates are labeled and indexed; the numbers around the top and right sides of the figure refer to the angle θ (in degrees) of the plane wave components, or phase-match angles, of the received beam. The lighter portion of the plot (high reflection coefficient magnitude) shows the experimentally usable region covered by the angular spread of

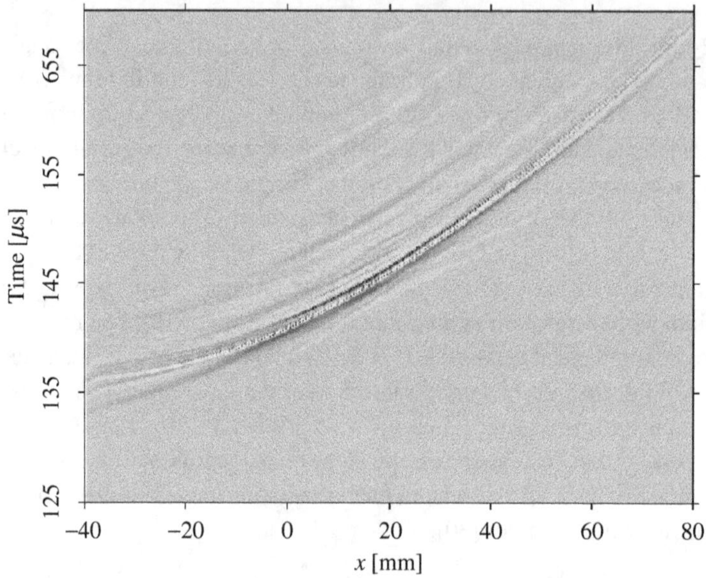

Figure 8.15. Superimposed raw data in the x, t domain for many stepped receiver positions, according to the geometry illustrated in Fig. 8.10 (after Chimenti and Fei).

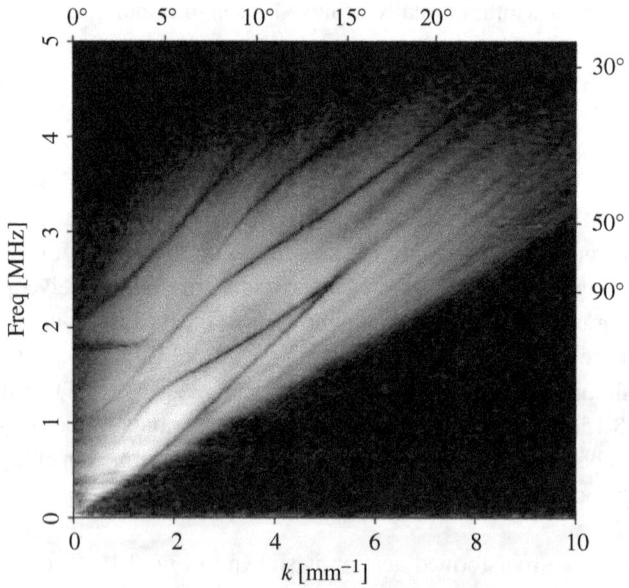

Figure 8.16. Fast Fourier transformed time-position data displayed in the k vs f domain for a 19-mm, 2.25-MHz transducer pair on a 1.6-mm aluminum plate at an orientation angle $\alpha = 24°$. The transducers have a focal length of 25 mm (after Chimenti and Fei [31]).

the focused beams, extending in this case from about 10° to about 50°. The zeroes of the reflection coefficient appear as dark curves in the plot, showing the trace of the guided plate wave modes. Were the plate to be replaced by a perfectly reflecting half-space, only the lighter, high amplitude gray-scale portion would remain, and the dark lines showing the modes would vanish.

The experimental data in Fig. 8.16 have been modeled using Eq. (8.33), and the theoretical prediction presented is in Fig. 8.17. The model calculation is a noiseless version of the dispersion curves represented in the measurements, and because of its noise-free aspect, usable results are seen to much lower phase-match angles than can be reconstructed in the data. Furthermore, the data are modeled assuming a single pair of CTPs as a representation of a pair of piston radiators. As we saw earlier in this chapter, Gaussian beams in transmit–receive behave identically to piston beams under similar conditions. This statement is true for planar radiators, but not rigorously valid for focused beams. In the case of focused piston beams, interference of sidelobes and the main lobe does introduce some diffraction effects, as shown in [31]. These effects can also be modeled by applying a weighted sum of Gaussian beams or CTPs. While the multiple-CTP approach does lead to better detailed comparison between theory and experiment, the improvement is cosmetic only. The significant element of data such as in Fig. 8.16, namely the mode predictions, remains unchanged.

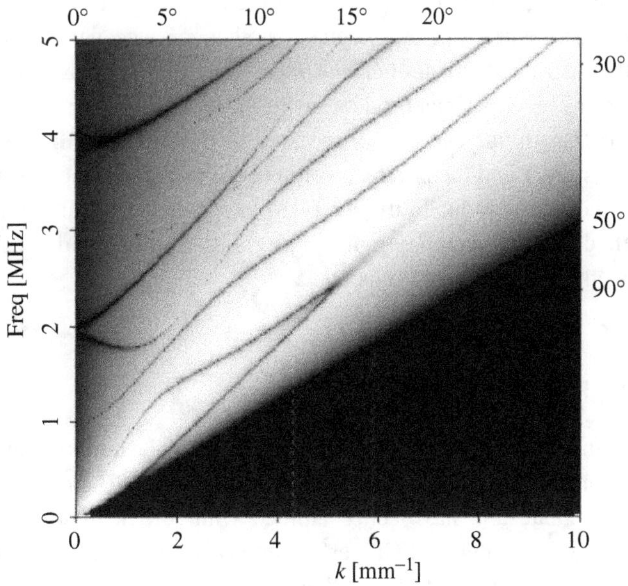

Figure 8.17. Model calculation according to Eq. (8.33) of the data in Fig. 8.16 (after Chimenti and Fei [31]).

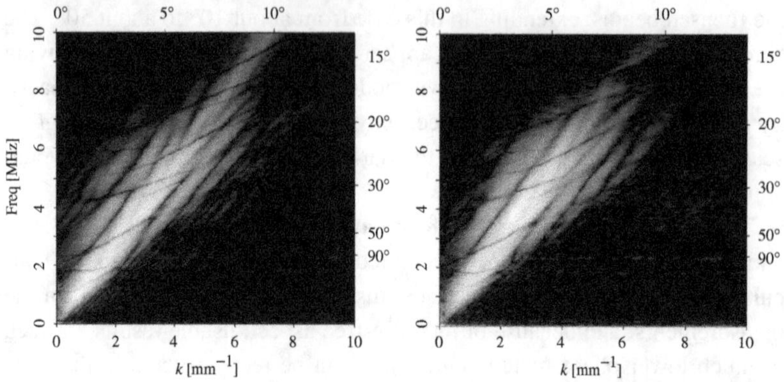

Figure 8.18. Demonstration of electroacoustic reciprocity. In left frame, transmitter is focused and receiver is planar; in right frame, the roles are reversed. Planar probe is 5 MHz, 8.3 mm; focused is 5 MHz, 25 mm, 100 mm focal length. Vertical scales on both plots are frequency in MHz (after Chimenti and Fei [31]).

A striking confirmation of the principle of electroacoustic reciprocity [17] is shown in Fig. 8.18. The mathematical underpinnings of this relation are best seen in Eq. (8.6), where it is clear that the transmitter and receiver directivity functions are simply multiplied by the scattering coefficient to form the integrand of the voltage integral. Due to the simple product, the two beam directivity functions are completely interchangeable under the integral, implying that there is no difference if the transmitter has one set of characteristics and the receiver another, or vice versa. (With the CTPs, this aspect of transmitter and receiver equivalence is not as easy to identify.) The experimental data in Fig. 8.18 have been acquired with a focused transmitter and planar receiver in the left-hand frame, and a planar transmitter and focused receiver in the right-hand frame. To the limit of experimental reproducibility, the two plots in Fig. 8.18 are essentially identical, demonstrating the equivalence of transmitter and receiver in pitch–catch ultrasonic measurements.

8.6 Elastic Stiffness Reconstruction

Applying these techniques to graphite-epoxy composite laminates, we have the k, f domain data of Fig. 8.19. The sample here is a T300/CG914 8-ply uniaxial laminate and has been examined with 5-MHz focused probes. The results of the double FFT performed on the x, t measurements are shown in Fig. 8.19 as image data, where the gray level represents scattered wave amplitude in dB. Several guided wave modes are also labeled on the plot. Discrete points on the curves from these data are extracted and

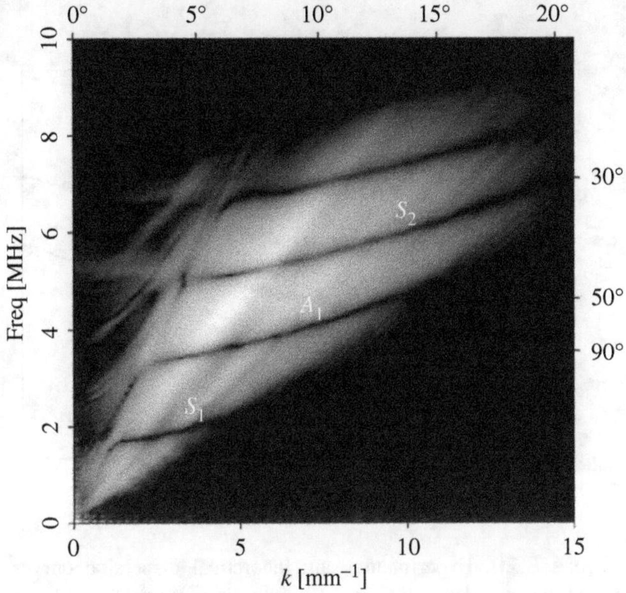

Figure 8.19. Wavenumber–frequency domain experimental dispersion results on 8-ply-thick uniaxial laminate of T300/CG914 composite. Data are acquired in broadband, synthetic aperture scans. Guided wave mode indices are labeled on plot (after Chimenti and Fei [31]).

plotted separately and compared to the results of fluid-loaded plate wave calculations. The data extracted from Fig. 8.19 are plotted as open circles, and the theoretical dispersion curves are shown as dark solid curves for the symmetric modes and medium gray solid curves for the antisymmetric modes. These results are plotted in Fig. 8.20. Similar data derived from measurements on this same plate, but with the incident plane normal to the fibers, yield the results shown in Fig. 8.21. In both Figs. 8.20 and 8.21, the calculated elastic stiffnesses have been adjusted to fit the measurements. Comparing these inferred stiffnesses to independently measured data, we find that the stiffnesses from scattering coefficient measurements differ from contact measurements on average by less than 1%. Similar methods to deduce elastic property data from measurements in the acoustic microscope are demonstrated by Wang and Rokhlin [35].

To simplify the reconstruction of elastic properties of composites from scattering coefficient measurements, a sensitivity analysis [31, 34] performed on the calculated dispersion curves shows where selected stiffnesses are most influential in determining the shape and path of observable guided wave modes. An example of such an analysis is shown in Fig. 8.22. The point of this plot is to illustrate that only a small selection

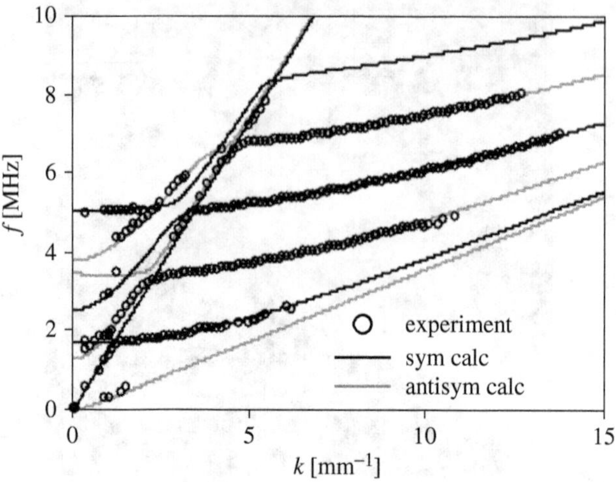

Figure 8.20. Experimental and theoretical dispersion curves based on the data in Fig. 8.19 with the plane of incidence containing the fiber direction. Solid curves are theoretical calculations, and open circles are data derived from Fig. 8.19 (after Chimenti and Fei [31]).

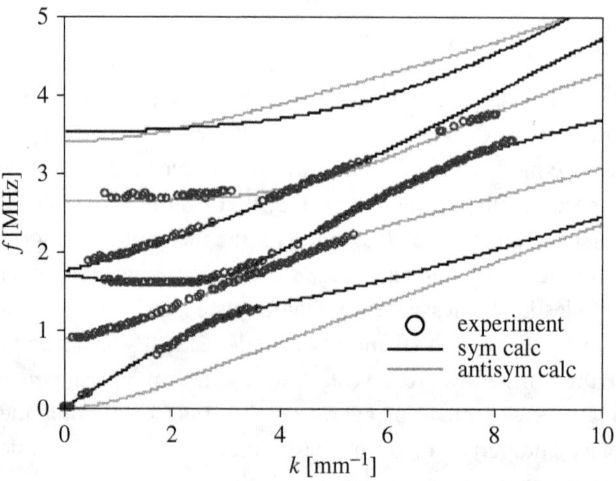

Figure 8.21. Experimental and theoretical dispersion curves based on reflection data with the plane of incidence normal to the fiber direction. Solid curves are theoretical calculations, and open circles are data derived from measurements (after Chimenti and Fei [31]).

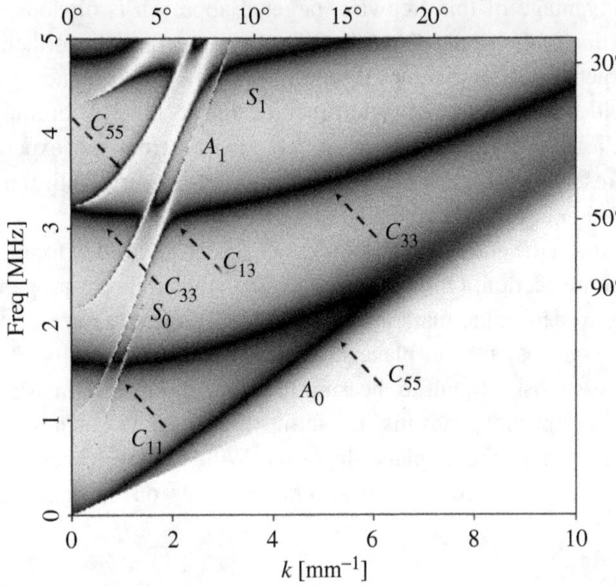

Figure 8.22. Theoretical dispersion curves showing locations of maximum sensitivity of various guided wave modes to several elastic stiffnesses (after Chimenti and Fei [31]).

of guided wave data is essential (or even recommended) in reconstructing the elastic stiffnesses of the underlying plate material. The elastic constant reconstruction procedure can be summarized briefly: 1) estimate the elastic properties using mode cutoff frequencies and other structural characteristics of the dispersion spectrum; 2) isolate the regions that are sensitive to individual elastic stiffnesses based on the sensitivity study; and 3) start with the estimates and reconstruct the elastic constants in order from the selected dispersion data using an optimization algorithm. The choice of the optimization algorithm is not critical here, because we have acceptable initial estimates for the elastic parameters and because the data set is small. In addition, because we reconstruct one parameter at a time, a 1-D optimization algorithm is sufficient and it is also much more reliable than any multi-dimensional routine. In recent work [31], we have used a golden-section algorithm, although any equivalent optimization method would work as well. The following parameter is used as the object function F,

$$F = \sum_{i=1}^{N} (V_i^e - V_i^m)^2, \tag{8.37}$$

where N is the total number of selected data points for reconstruction, and V_i^e and V_i^m are the experimental and model-predicted reflection minima, respectively.

The advantage of this stepwise, targeted approach is obvious. Only a small portion of the whole dispersion data set is ever used; nevertheless, this set of targeted data determines the framework of the dispersion spectrum and provides all the necessary information for accurate determination of all the elastic constants. In fact, as we can see from the experimental results presented earlier, if the targeted data are fit very well, the rest of the dispersion data are fit automatically.

Sometimes, the entire elastic stiffness matrix is not needed for ultrasonic composite inspection. Only a few, or even a single, elastic properties are necessary to determine material quality. Often, the most important of the several stiffnesses is the in-plane stiffness. This one, nominally C_{11}, is the one stiffness most difficult to measure in a conventional normal-incidence reflection or through-transmission set-up because this stiffness governs particle motion in the in-plane direction. With guided waves, however, this property can be mapped at several positions on the laminate [36].

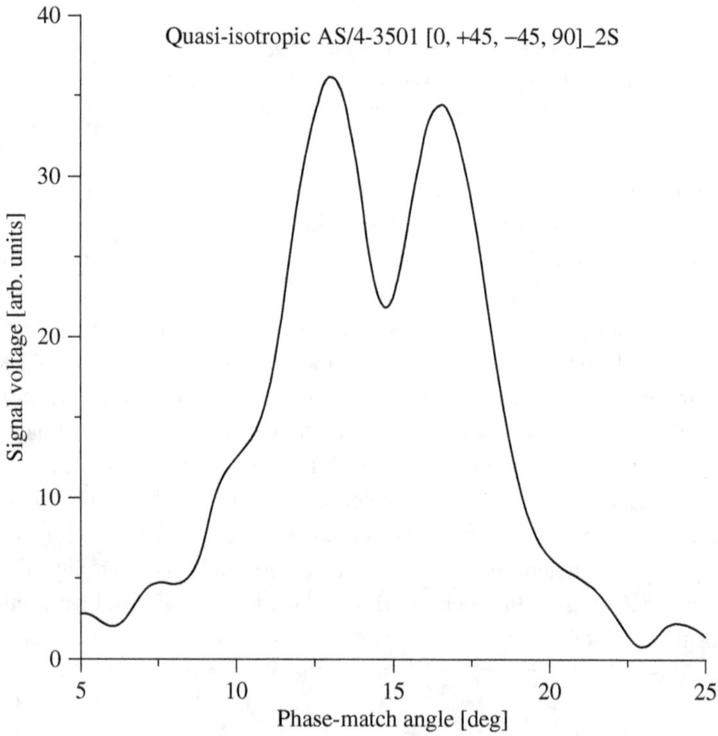

Figure 8.23. Reflection function $S(\theta)$ versus phase-match angle for a quasi-isotropic $[0/+45/-45/90]_{2S}$ graphite/epoxy laminate of AS/4-3501. The minimum in the curve near 14.8 is the result of phase matching of the incident wave to the S_0 guided wave mode. The frequency is 0.5 MHz (after Lobkis *et al.* [36]).

Figure 8.23 shows the result of a synthetic aperture scan at a fixed frequency of 0.5 MHz of a quasi-isotropic graphite-epoxy laminate in a phase-match angle range of 5–25°. The pronounced minimum near 15° indicates the presence of the zeroth-order symmetric Lamb wave. This mode is related to the in-plane stiffness in a very direct way. Repeating this measurement at several points on the laminate gives the NDE engineer a clear picture of the average in-plane stiffness property of the laminate.

Bibliography

1. M. F. M. Osborne and S. D. Hart, "Transmission, reflection, and guiding of an exponential pulse by a steel plate in water. I. Theory," J. Acoust. Soc. Am. **17**, 1–18 (1945).
2. D. C. Worlton, "Experimental confirmation of Lamb waves at megacycle frequencies," J. Appl. Phys. **32**, 967–971 (1961).
3. M. Hattunen and M. Luukkala, "An investigation of generalized Lamb waves using ultrasonic measurements," Appl. Phys. **2**, 257–263 (1973).
4. K. Van Den Abeele and O. Leroy, "On the influence of frequency and width of an ultrasonic bounded beam in the investigation of materials: Study in terms of heterogeneous plane waves," J. Acoust. Soc. Am. **93**, 2688–2699 (1993).
5. L. Cremer, "Theory of sound trapping of thin walls at oblique incidence," (in German), Akust. Z. **7**, 81–104 (1942).
6. A. Schoch, "Sound transmission through plates," (in German), Acoustica **2**, 1–18 (1952).
7. H. L. Bertoni and T. Tamir, "Unified theory of Rayleigh-angle phenomena for acoustic beams at liquid–solid interfaces," Appl. Phys. **2**, 157–172 (1973).
8. L. E. Pitts, T. J. Plona, and W. G. Mayer, "Theory of nonspecular reflection effects for an ultrasonic beam incident on a solid plate in a liquid," IEEE Trans. Sonics Ultrason. **SU-24**, 101–109 (1977).
9. W. G. Neubauer, "Ultrasonic reflection of a bounded beam at Rayleigh and critical angles for a plane liquid–solid interface," J. Appl. Phys. **44**, 48–55 (1973).
10. T. D. K. Ngoc and W. G. Mayer, "Numerical integration method for reflected beam profiles near the Rayleigh angle," J. Acoust. Soc. Am. **67**, 1149–1152 (1980).
11. J. M. Claeys and O. Leroy, "Reflection and transmission of bounded beams on halfspaces and through plates," J. Acoust. Soc. Am. **72**, 585–590 (1982).
12. M. J. S. Lowe and P. Cawley, "Comparison of the modal properties of a stiff layer embedded in a solid medium with the minima of the plane-wave reflection coefficient," J. Acoust. Soc. Am. **97**, 1625–1637 (1995).
13. O. I. Lobkis, A. Safaeinili, and D. E. Chimenti, "Precision ultrasonic reflection studies in fluid-coupled plates," J. Acoust. Soc. Am. **99**, 2727–2736 (1996).
14. H. Zhang and D. E. Chimenti, "Air-coupled transmission coeffcient reconstruction using a 3-D complex-transducer-point voltage model," J. Nondestr. Eval. **22**, 23–36 (2003).
15. A. D. Pierce, *Acoustics: An Introduction to its Physical Principles and Applications* (ASA, New York, 1989), Chap. 5.

16. L. M. Brekhovskikh, *Waves in Layered Media* (Academic, New York, 1975), Chap. 4.
17. B. A. Auld, "General electromechanical reciprocity relations applied to the calculation of elastic wave scattering coefficients," Wave Motion **1**, 3–10 (1979).
18. G. S. Kino, "The application of reciprocity theory to scattering of acoustic waves by flaws," J. Appl. Phys. **49**, 3190–3199 (1978).
19. D. E. Chimenti, J.-G. Zhang, S. Zeroug, and L. B. Felsen, "Interaction of acoustic beams with fluid-loaded elastic structures," J. Acoust. Soc. Am. **95**, 45–59 (1995).
20. O. I. Lobkis and D. E. Chimenti, "Equivalence of Gaussian and piston ultrasonic transducer voltages," J. Acoust. Soc. Am. **114**, 3155–3166 (2003).
21. A. H. Nayfeh and D. E. Chimenti, "Propagation of guided waves in fluid-coupled plates of fiber-reinforced composite," J. Acoust. Soc. Am. **83**, 1736–1743 (1988).
22. S. Zeroug and L. B. Felsen, "Nonspecular reflection of beams from liquid–solid interfaces," J. Nondestr. Eval. **11**, 263–278 (1992).
23. L. B. Felsen and N. Marcuvitz, *Radiation and Scattering of Waves* (Prentice-Hall, Englewood Cliffs, 1973).
24. S. Zeroug, "Complex ray methods for ultrasonic beam propagation and scattering in layered elastic structures," Ph.D. Thesis, Polytechnic Institute of New York, 1993.
25. S. Zeroug and L. B. Felsen, "Nonspecular reflection of two- and three-dimensional acoustic beams from fluid-immersed plane-layered elastic structures," J. Acoust. Soc. Am. **95**, 3075–3098 (1994).
26. *Handbook of Mathematical Functions with Formulas, Graphs, and Tables*, eds. M. Abramowitz and I. A. Stegun, Appl. Math. Ser. **55** (NBS, Washington, 1964).
27. G. A. Deschamps, "Gaussian beam as a bundle of complex rays," Electron. Lett. **7**, 684–685 (1971).
28. S. Zeroug and L. B. Felsen, "Nonspecular reflection of two- and three-dimensional acoustic beams from fluid-immersed cylindrically layered elastic structures," J. Acoust. Soc. Am. **98**, 584–598 (1995).
29. S. Zeroug, F. E. Stanke, and R. Burridge, "A complex-transducer-point model for emitting and receiving ultrasonic transducers," Wave Motion **24**, 21–40 (1996).
30. O. I. Lobkis and D. E. Chimenti, "Three-dimensional transducer voltage in anisotropic materials characterization," J. Acoust. Soc. Am. **106**, 36–45 (1999).
31. D. E. Chimenti and D. Fei, "Scattering coefficient reconstruction in plates using focused acoustic beams," Int. J. Solids Struct. **39**, 5495–5513 (2002).
32. A. Safaeinili, O. I. Lobkis, and D. E. Chimenti, "Air-coupled ultrasonic estimation of viscoelastic stiffnesses in plates," IEEE Trans. Ultrason. Ferroelect. Freq. Contr. (UFFC) **43**, 1171–1180 (1996).
33. R. B. Thompson and E. F. Lopes, "The effects of focussing and refraction on Gaussian ultrasonic beams," J. Nondestr. Eval. **4**, 107–123 (1984).
34. S. I. Rokhlin and D. E. Chimenti, "Reconstruction of elastic constants from ultrasonic reflectivity data in a fluid-coupled plate," in *Review of Progress in Quantitative Nondestructive Evaluation*, eds. D. O. Thompson and D. E. Chimenti, vol. **9** (Plenum Press, NY, 1990), pp. 1411–1418.

35. L. Wang and S. I. Rokhlin, "Time resolved line focus acoustic microscopy of layered anisotropic media: Application to composites," IEEE Trans. Ultras. Ferroelect. Freq. Contr. (UFFC) **49**, 1231–1244 (2002).
36. O. I. Lobkis, D. E. Chimenti, and H. Zhang, "In-plane elastic property characterization in composite plates," J. Acoust. Soc. Am. **107**, 1852–1858 (2000).

9

Air-Coupled Ultrasonics

9.1 Introduction

Ultrasonic material characterization or inspection for defects is conventionally performed using either liquid coupling (water, usually) or some type of gel or oil in contact-mode coupling. Mechanical waves can be transmitted only through some sound-supporting medium from their source (a transducer) to the object under study, and back again. Using distilled, degassed water to couple ultrasound to an object under test works quite well and has many technical advantages, including relatively low signal loss over laboratory or shop dimensions at typical frequencies, almost zero toxicity, and low cost. For many applications, the use of water is acceptable and preferred.

There are, however, certain testing applications for which water can be a disadvantage. These situations include materials that are sensitive to contact with water, such as uncured graphite-epoxy composites or certain electronics. Large objects, whose total immersion is impractical, or objects for which rapid scanning is required might also be unsuitable for water coupling. Recent technological developments are beginning to permit the judicious replacement of water by a far more ubiquitous sound coupling medium—air.

Ultrasonic testing in air has been investigated for more than 30 years [1–6], but recently there has been an upsurge in interest and application because of the availability of much more efficient sound-generating devices designed specifically for operation in air [7–20]. In water- or direct-coupled ultrasonics, one typically employs piezoelectric transducers to generate sound waves because they are well suited to the generation of sound in water or in solids because of their high acoustic impedance. In air, however, we need just the opposite. Air is very compliant, so waves from a high-impedance source couple poorly into air. Much effort has been invested in finding suitable impedance matching materials that will render the familiar piezoelectric probe efficient in air-coupled (A-C) ultrasound. The problem, however, is nearly insurmountable because of the large acoustic impedance difference between air and quartz, for example. Quartz has an acoustic impedance of about 15 MRayl, while air's impedance is about 425 Rayl, a ratio of about 35,000. The challenge is to find a material with an acoustic impedance that nearly equals the geometric average of these two widely disparate values. For a piezoelectric crystal in air it works out to about 80 kRayl. A solid material with that property would be hard to find. Water has an acoustic impedance of 1.5 MRayl, so the ideal material would have to be much more compliant and/or much less dense than water. Natural solid materials come nowhere near this requirement. Only a highly porous solid such as a foam consisting of a highly compliant web with high porosity would have a chance to meet the density and stiffness needs of an ideal matching material. The average cell dimension, however, would also have to be small enough to avoid sound scattering that would dephase the wave. Even if such an ideal matching material were to be found, the composite device would still be efficient only in a narrow frequency band. Another major difficulty of solid piezoelectric materials is that their high stiffness implies a low displacement amplitude, whereas in highly compliant air, a low-stiffness, high-displacement device would be much better to generate the conditions necessary for the inspection of solids. Conventional piezoelectric resonators, therefore, do not make ideal A-C transducers.

9.2 Transduction and Other Challenges

Capacitive film transducers avoid the matching problems inherent with piezoelectric probes [8–10]. A thin polymer film a few microns, or less, in thickness comprises one side of the capacitor, and a fixed back-plate is the other, as shown in Fig. 9.1. In one recent realization of this geometry, the entire capacitive film structure is produced by microelectronic circuit fabrication methods [11]. Across the capacitor are impressed both a constant high-voltage bias and a superimposed radio-frequency ultrasonic signal. The polymer film vibrates with a large displacement from frequencies of about 10 s of kHz up to some MHz, efficiently coupling the electrical impulses to a high-intensity sound wave in the air. Beyond about 2 MHz, sound attenuation in air at 1 atu climbs rapidly, making conventional A-C ultrasonics difficult. Attempts to overcome such limitations have included elevating the operating pressure and positioning the transducers and test piece very close together. Capacitive film transducers also operate as reciprocal devices, detecting the

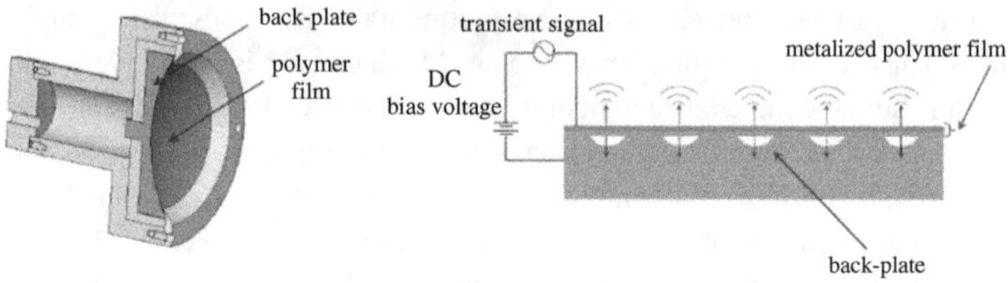

Figure 9.1. Schematic of air-coupled (A-C) transducer construction, showing housing, back-plate, and film (after Song and Chimenti [15]).

waves reflected from, or transmitted through, the object of the test. For this operation, a bias voltage is also necessary. The efficient operation of these devices relies on a thin flexible film, such as metallized polyimide from 1 to 10 μm thick. As the film is so thin, the material and thickness of the metallic coating can be a significant factor in the film's areal density. Dielectric breakdown voltage is also a factor, but using polyimide (Mylar) all but eliminates this concern. Dust and contamination between the film and the backing plate is also a cause of failure, so transducer fabrication under minimal clean-room conditions is an advisable precaution. Despite the success and reasonable durability of capacitive film A-C transducers, other types do exist and are used. Piezocomposites have been tried and do improve impedance match, although only in a rather narrow frequency band [3]. Attempts with sophisticated matching layers have also been investigated [16], but these devices, too, function well only in relatively narrow bandwidths.

As most of the sound amplitude is reflected at the interface between air and a solid, two separate capacitive film transducers (usually on opposite sides of a solid) are generally needed to perform A-C ultrasonics. This phenomenon is explained in more detail below. Recently, a new version of the capacitive transducer has been developed, which permits native focusing of the sound beam to a diffraction-limited circular spot, *i.e.*, focusing without the use of lenses, mirrors, or other extraneous means. This new spherically focused capacitive transducer [14] is ideally suited to ultrasonic inspection technology for defect detection and elastic property characterization in laminated materials.

A-C ultrasonic testing presents subtle challenges to the engineer seeking to employ this technique for defect detection or material characterization. There is much more to the successful application of this technology than simply emptying the immersion tank of water and repeating the test with different transducers. Air is a medium having a sound wavespeed of only 342 m/s, much smaller than that of water at 1485 m/s, and far smaller than the compressional wavespeed of a typical solid, about 2600 m/s for

Plexiglas (a trade name for polymethylmethacrylate, also known as Lucite) or 6300 m/s for aluminum. This large difference in the speeds of sound means that the corresponding sound wavelength in air will be proportionately smaller than the wavelength in the liquid or solid. Moreover, the bending or refraction of the sound as it enters a solid will typically be extreme. The wavevector in the air will be much larger than that in the solid because the wavespeeds are so different. What this means is that all the interesting sound interaction behavior will be finished beyond an incident angle, of about 15° or 20° in most materials. Beyond the shear critical angle, all sound is internally reflected back into the air. This fact implies that focused transducers in air need subtend only relatively small angles to be effective at imaging solids, or in providing a broad angular spectrum of waves, as shown schematically in Fig. 9.2. It means that the incident angles of interest for material characterization will also be similarly limited. The reflection coefficient for an interface of air and aluminum can be seen in Fig. 9.3 and for air and Plexiglas in Fig. 9.4, where the onset of total internal reflection occurs before 20°, even in the extremely soft Plexiglas.

There is another, more insidious effect of the large disparity between the sound wavespeed in air and solids. Coupling of sound waves from one medium to another occurs best when the wavespeeds and mass densities between the two media are about the same. For air and almost any solid,

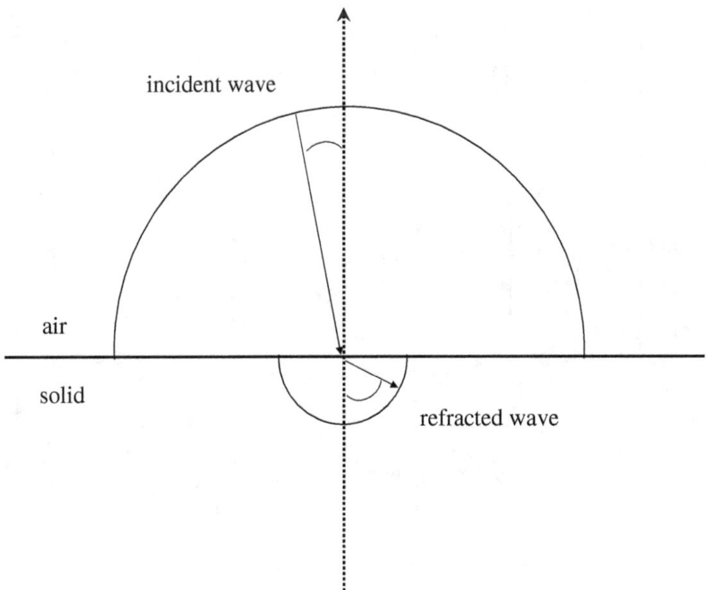

Figure 9.2. Slowness curves for air and a typical solid, showing the relatively small critical angles for sound transmission into the solid.

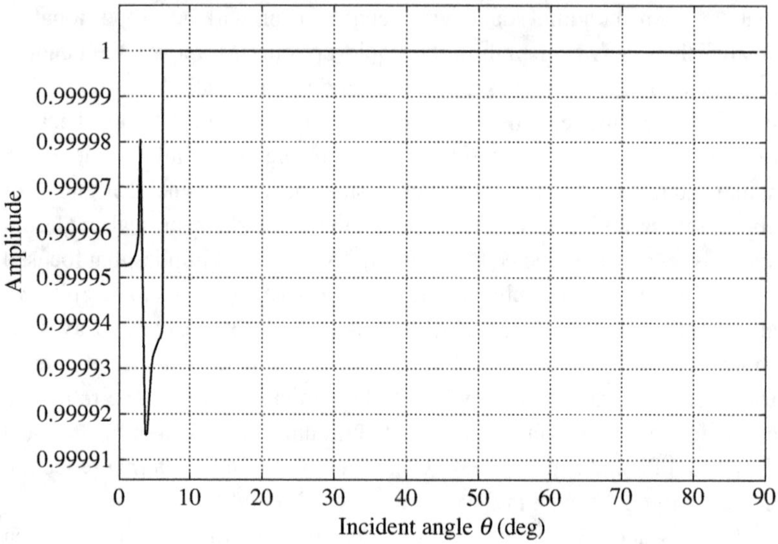

Figure 9.3. Reflection coefficient of an air–aluminum interface, showing very small critical angles.

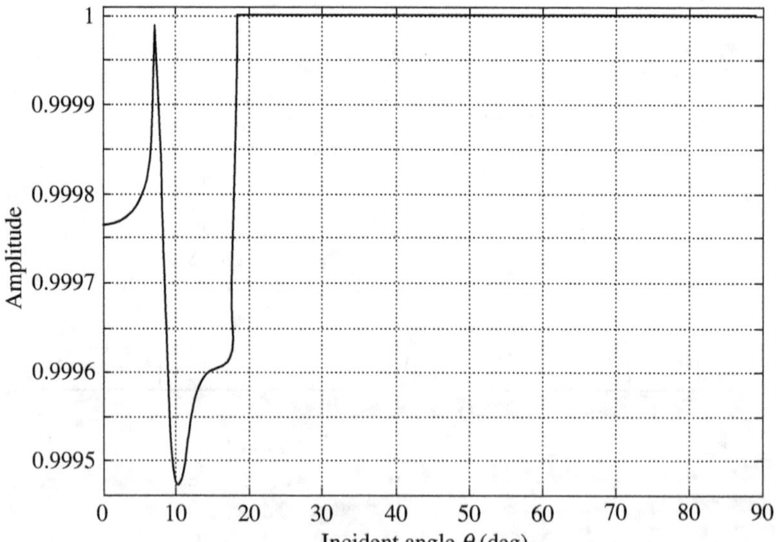

Figure 9.4. Reflection coefficient of an air–Plexiglas interface, showing small critical angles.

however, they could hardly be more different, as mentioned above. The implication is that, in general, at the interface between air and a solid half-space, the insertion losses amount to about −76 dB, and this unfavorable arithmetic applies again as the sound wave returns from the solid back into the air. For defect detection, both the strong interface reflections and the high refraction must be accounted for, already significant technical challenges. When seeking the material elastic properties of a sample from its sound propagation characteristics, however, there are yet additional considerations.

9.3 Material Characterization in Air

To perform material characterization with A-C sound requires a carefully thought out approach. As the transducer is only 10 or 20 sound wavelengths in diameter at typical frequencies, energy from the transducer will experience some diffraction. In fact, in many measurements, the diffraction effects can seriously complicate the interpretation of the experiment, inducing as much structure in the received signal as is caused by the elastic properties of the material being tested. Although the occurrence of maxima in the plate transmission function, related mathematically to the elastic wavespeeds, can be influenced by diffraction, the major effect of the diffracting beam is felt on the imaginary parts of the elastic stiffnesses, *i.e.*, on elastic wave absorption in the material. Measurement of viscoelastic stiffnesses can be quite sensitive to diffraction effects, and ignoring diffraction can lead to incorrect assessment of wave energy losses.

The diffraction of sound from a finite transducer can also be an advantage in A-C ultrasound. Although the reflection of sound is great at solid–air interfaces when the solid is a half-space, the transmission coefficient for plates can be, at certain incident angles, much more favorable. These special angles are exactly the ones at which guided waves are generated in the plate. Figure 9.5 illustrates this behavior between air and a plate of Plexiglas. Note that Plexiglas has properties similar to those of polymer matrix composites in a propagation direction normal to the fibers. The curve labeled 1 is the plane wave transmission coefficient and curve 2 is the finite-beam transmission function for a 2.3-mm thick Plexiglas plate in air. Either curve has far larger maxima, where guided waves exist, than the value of the transmitted sound amplitude for a solid half-space. The two peaks near 15° are the A_0 and S_0 modes. Higher-order modes appear at lower incident angles (higher trace velocity).

The methods described in Chapter 8 have shown considerable success in suppressing the effects of the sound beam characteristics almost completely, leaving the elastic properties of the solid as the only remaining influence

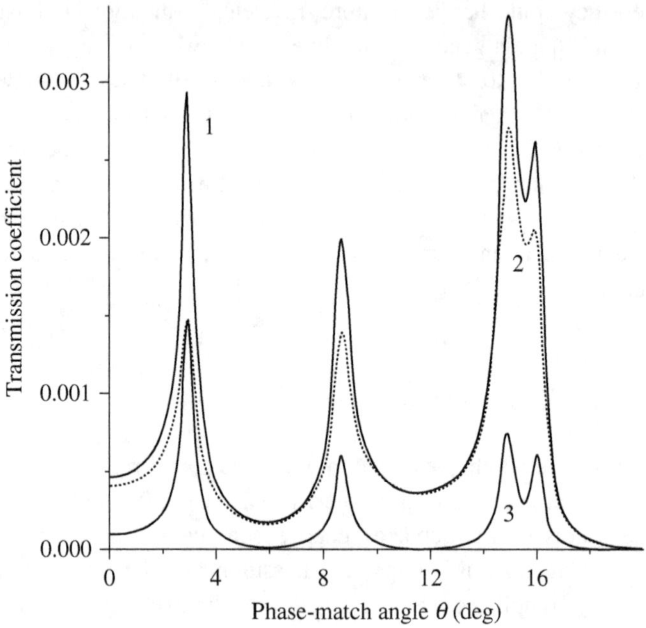

Figure 9.5. Comparison of plane wave displacement transmission coefficient (curve 1) and finite-beam transmission function (curve 2) for air and 2.3-mm thick Plexiglas plate at 700 kHz, showing the relatively small critical angles for sound transmission into the solid. Peaks correspond to guided wave generation in the plate. Curve 3 is the difference between curves 1 and 2 (after Safaeinili *et al.* [6]).

on the processed data (within the angular and frequency windows of the acquired data). The synthetic-aperture method benefits from focused transducers, where the angle data can be collected in a single series of stepped coordinate measurements (related to phase-match angle after a spatial Fourier transform through $\xi = k_f \sin\theta$), as detailed in Section 8.4.2 of Chapter 8. As a focused transducer produces plane wave components at many phase-match angles, only a single measurement is typically necessary. It remains only to scan the coordinate position of one of the two transducers with respect to the other to complete the data acquisition. The result, after processing, is analogous to having made the measurement with an aperture as large as the scan. From such a measurement of the transmission coefficient as a function of frequency, one can infer the elastic properties of solids. An example of a measured transmission function is shown in Fig. 9.6 for a 2.3-mm Plexiglas plate in air at 400 kHz. The solid curve is experimental data, and the dashed curve is a theoretical calculation using optimized viscoelastic stiffnesses. There are three discernible peaks; the first is the S_1 mode, and the larger two are the A_0 and S_0, respectively.

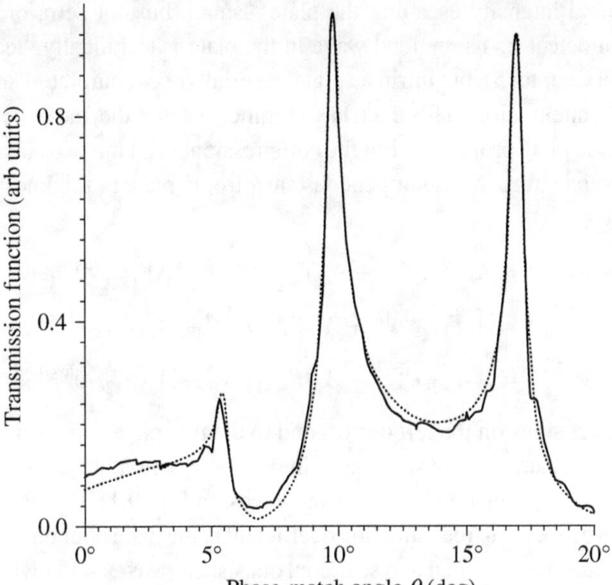

Figure 9.6. Measured transmission function versus phase-match angle θ for a 2.3-mm Plexiglas plate at 400 kHz, showing the experiment (solid curve) and a theoretical estimate (dashed curve) using optimized material parameters (after Safaeinili *et al.* [6]).

This figure demonstrates several of the challenging aspects of A-C ultrasound. There is a very high reflectivity (low transmission) away from the peaks, as expected. The implication for engineers is that A-C ultrasound works best when performed with two transducers deployed on opposite sides of a plate-like structure, to avoid the strong reflection at the front surface, which would overwhelm the receiver circuit electronics. From Fig. 9.6 it is clear that at some incident angles, much higher sound transmission occurs than in general. At these special angles, the plate's ideal, non-lossy, plane wave transmission coefficient experiences total transmission. In this case, even A-C sound becomes reasonably efficient, although the finite-beam, lossy transmission function is still much less than unity (vertical scale is arbitrary). The occurrence of total transmission maxima coincides with the propagation of guided waves in the plate, resulting from the condition of transverse resonance [17]. That is, when partial waves in the transverse plate direction form standing wave resonances, the ideal plate becomes transparent to ultrasound.

The transverse resonance condition, however, is also responsible for guided wave propagation. In A-C ultrasonics, the use of this calculation (where the partial-wave scattering coefficients are computed for solid-vacuum conditions) is actually quite well justified, because of the extremely small sound intensity escaping the plate—small, but not zero, or else we could not detect it. The guided wave in the plate is technically "leaky" (as we saw in Chapter 5), but intrinsic plate material losses completely dominate the wave attenuation. This fact has significance for the measurement of viscoelastic plate properties. For the compressional and the two shear partial waves propagating in a homogeneous anisotropic plate of thickness $2h$, we can write

$$\begin{aligned}
\pm A_\ell e^{-i\alpha_1 h} &= A_\ell \Gamma_{\ell\ell} e^{i\alpha_1 h} + A_{s1} \Gamma_{\ell s1} e^{i\alpha_3 h} + A_{s2} \Gamma_{\ell s2} e^{i\alpha_5 h}, \\
\pm A_{s1} e^{-i\alpha_3 h} &= A_\ell \Gamma_{s1\ell} e^{i\alpha_1 h} + A_{s1} \Gamma_{s1s1} e^{i\alpha_3 h} + A_{s2} \Gamma_{s1s2} e^{i\alpha_5 h}, \\
\pm A_{s2} e^{-i\alpha_5 h} &= A_\ell \Gamma_{s2\ell} e^{i\alpha_1 h} + A_{s1} \Gamma_{s2s1} e^{i\alpha_3 h} + A_{s2} \Gamma_{s2s2} e^{i\alpha_5 h},
\end{aligned} \tag{9.1}$$

where the $\pm$ signs on the left correspond to the two plate wave symmetries. Here, A_ℓ, A_{s1}, and A_{s2} are wave amplitudes for quasilongitudinal (ℓ) waves and the two quasishear (s1, s2) partial waves. In Eq. (9.1), $\Gamma_{\ell s1}$ denotes the plane wave, traction-free reflection coefficient at the plate boundary between incident quasilongitudinal and scattered quasishear waves—likewise for the other coefficients. The exponentials are quasilongitudinal and quasishear partial waves, where the terms α_1, α_3, α_5 are the transverse wavenumbers (across the plate thickness) for quasicompressional and two quasishear partial waves, respectively. The transverse wavenumbers α_k are calculated according to the prescription provided in Section 5.2.2 of Chapter 5.

Equation (9.1) are the relations for transverse resonance in the traction-free anisotropic plate. The reasoning contained in Eq. (9.1) is simply this: a compressional wave at the lower plate surface in the topmost relation is reconstituted from reflected compressional and shear partial waves at the upper surface. Likewise, either of the two shear waves at the lower surface, in the bottom two relations, is composed of reflected compressional and shear wave components existing at the upper surface. In each case, the reconstructed wave amplitudes must be identical to those for that same partial wave at the upper boundary. This argument is the essence of the transverse resonance condition, and it is completely equivalent to the mechanical traction-free boundary conditions on the stresses and displacements.

In air, of course, we do not have exactly traction-free boundary conditions, else there would be no wave amplitude in the air to observe. While this statement is rigorously true, it ignores the fact that the acoustic impedance mismatch at the air–solid boundary is so large, as indicated above, so as to render the plate boundary essentially traction-free. Certainly, the air–solid boundary behaves much closer to a vacuum–solid boundary than a typical liquid–solid boundary does. The example of water, with an acoustic impedance of $1.5 \cdot 10^6$ Rayl, does far more to load the plate mechanically than air at 425 Rayl. Here, too, the technological sword cuts both ways.

On the one hand, the large acoustic impedance mismatch between air and almost any solid implies a weak and noisy acoustic signal. On the other, the fact that air is so tenuous and low in mass density implies that the plate will experience almost zero mechanical loading owing to the presence of air on the plate boundary. In this case, the viscous parts of the viscoelastic stiffnesses can be precisely measured in the near absence of the perturbing influence of the coupling fluid, providing we account rigorously for the beam diffraction of our transducers (including possible 3-D effects, as we saw in Chapter 8). For water loading, the re-radiation of acoustic energy into the coupling liquid can, and does, typically exceed the intrinsic wave losses. In some cases, such as for low-density materials such as composite laminates, the water loading can significantly perturb the apparent real elastic stiffnesses, as demonstrated in Figs. 5.6–5.8.

The prospect for improved viscoelastic stiffness reconstruction with A-C ultrasound is significant. For accurate determination (inversion) of the viscoelastic properties C^*_{IJ} from measurements, of course, we need to ensure that the measured data are actually sensitive to the properties we want to infer. Using the example offered by Plexiglas, the sensitivity of cumulative squared error between predicted and measured curves (such as in Fig. 9.6) to variations in real elastic stiffnesses is shown in Fig. 9.7. The dashed curve shows the error accumulation for fractional changes in the real compressional stiffness, and the solid curve shows that same parameter for the real shear stiffness. Clearly, the error between the two curves is much more sensitive to small variations in the shear stiffness, but even for the compressional stiffness, a variation of 2 or 3 percent is easily discernible in the plotted data. This observation assures stable determination of elastic properties from the experiment. As the critical angle for compressional waves is only about 7° in Plexiglas, most of the sound wave energy transmitted through the plate is carried by shear wave excitation, and this fact helps explain the higher sensitivity of the experiment to shear wave stiffness variations. Analytically, the uncertainty in predicting the real parts of the stiffnesses is no more than 1 percent.

The real advantage of working in air, rather than a liquid such as water, is seen in the reconstruction of the imaginary parts of the viscoelastic stiffnesses, whose sensitivity curves are shown in Fig. 9.8. For the imaginary parts of the viscoelastic stiffnesses, the cumulative squared error for variations in the compressional stiffness is shown as a dashed curve, and that same error for the shear stiffness is shown as a solid curve. As before, the sensitivity of the imaginary shear stiffness is much higher than that for the imaginary compressional stiffness, which has only a very small positive curvature hardly discernible in Fig. 9.8. Nonetheless, although the horizontal scale is coarser for the imaginary stiffnesses, the sensitivity is still

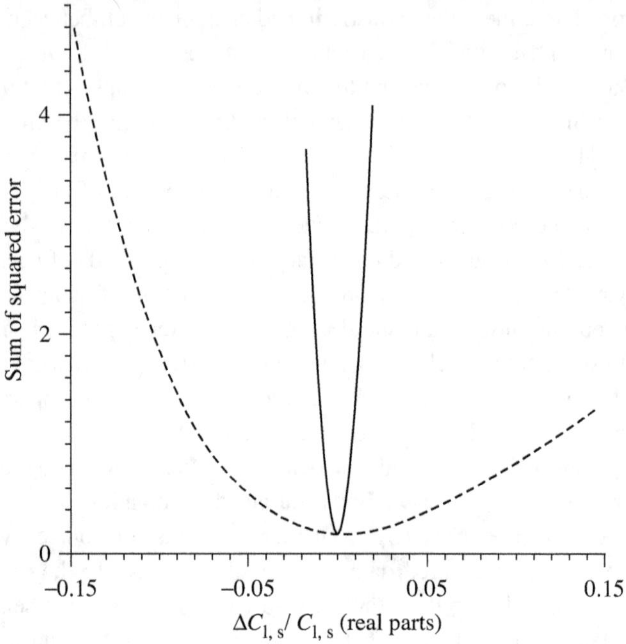

Figure 9.7. Cumulative squared error between predicted and measured transmission function curves for Plexiglas at 400 kHz, as a function of variations in real compressional (dashed curve) and real shear (solid curve) elastic constants (after Safaeinili *et al.* [6]).

excellent, considering other possible sources of systematic error in making such measurements. In this case, we have no loading liquid that introduces a strong wave leakage, and there is no contact transducer bonded to the specimen. It is truly a non-contact measurement, similar to what one would find in the case of laser-generated ultrasound. But for A-C ultrasonics, there is no need to heat the sample locally to high temperatures, which might also introduce errors into the measurement of viscoelastic stiffnesses. What is now clear from the calculations of the last chapter and the results of this one is that without a comprehensive beam model, valid in both near and far fields, and the synthetic aperture scanning technique, a rigorous reconstruction of especially the imaginary viscoelastic stiffnesses would not be possible.

With the information developed above, the attractiveness of using A-C ultrasound for inspection of plate-like structures becomes clear because only plates have transverse resonances, and only double-sided measurements can separate the weak wave transmitted through the plate from the large wave reflected from it. Moreover, the utility of a transducer that contains a range of incident spatial frequency wave components, either a small planar

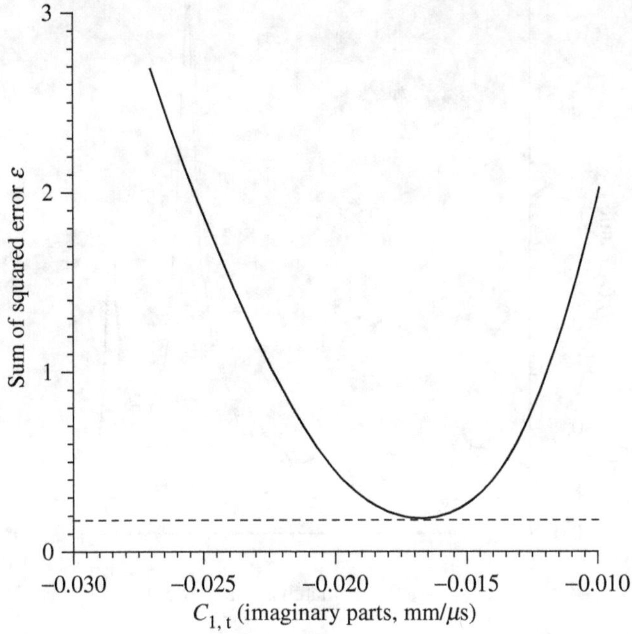

Figure 9.8. Cumulative squared error between predicted and measured transmission function curves for Plexiglas at 400 kHz, as a function of variations in imaginary compressional (dashed curve) and imaginary shear (solid curve) viscoelastic constants (after Safaeinili *et al.* [6]).

probe or a focused transducer, is now also clear. Diffraction or focusing produces a divergent or convergent beam that includes a range of plane wave components. Due to the large acoustic impedance mismatch at any gas–solid interface, the critical transmission properties of plates are crucial for efficient ultrasound inspection in air. For arbitrary materials of arbitrary thickness, the precise incident angles of high transmission cannot be known in advance. Instead, the correct approach is to assume the existence of guided wave modes within a few degrees of normal incidence, and then devise a sound beam with an angular spread which covers that range. This way, there will always be a ray satisfying conditions of high transmission. A further example of a transmission function for a periodically layered composite plate of aramid-epoxy and aluminum is shown in Fig. 9.9. The unit cell of the material consists of about 0.3 mm of aramid-epoxy followed by 0.2 mm of sheet aluminum. The cell is repeated four times, with a final aluminum layer closing the structure. This way, both exposed surfaces are aluminum. The data show rather good agreement with the calculated signal, where the viscoelastic stiffnesses have been adjusted to accommodate the fit.

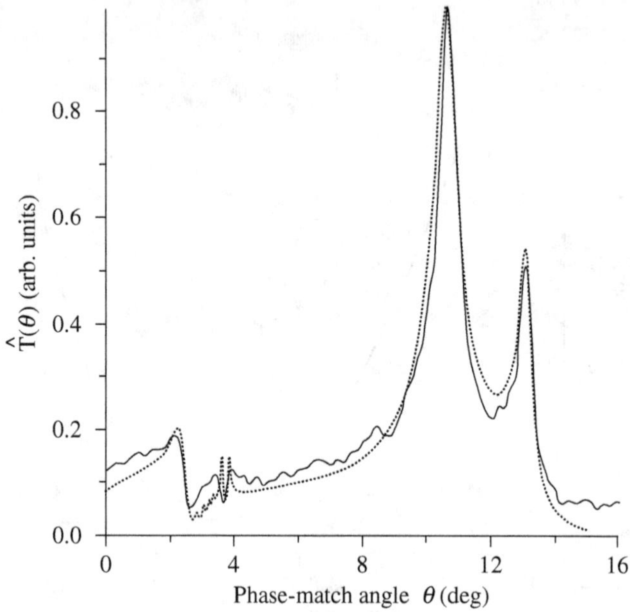

Figure 9.9. Transmission function $\widehat{T}(\theta)$ versus phase-match angle of a periodically layered plate consisting of nine layers of aramid-epoxy and aluminum in a 0.5-mm unit cell size, measured at 700 kHz. Accurate estimates of viscoelastic parameters account for the excellent fit between the measured and calculated signal (after Safaeinili *et al.* [6]).

Nonetheless, there are only four or six stiffnesses to adjust, and all the data must fit with a single choice. The modeling of the peak and trough amplitudes is quite good, even for the more weakly excited modes seen near 2 and 4°.

Composite materials are ideally suited to A-C ultrasonics because of their low density. Coupling sound to these materials can be relatively efficient when incident wave energy arrives at a phase-match angle for one of the guided waves having substantial normal displacements in that wavevector–frequency region of its dispersion. Some examples are shown below. In Fig. 9.10, the measured transmission function for a 1-mm thick uniaxial graphite-epoxy plate scanned normal to the fiber direction at 1.1 MHz is shown. The inferred viscoelastic properties are $C_{22} = 16.25 - 0.305\,i$ GPa and $C_{23} = 7.64 - 0.53i$ GPa. The real parts of these numbers are consistent with water-coupled measurements on this plate. In this figure, the plane wave transmission coefficient and the transmission function are nearly identical. For data scanned parallel to the fiber direction, also at 1.1 MHz, see Fig. 9.11. The data is the solid curve and the calculation is the broken curve. Here, the

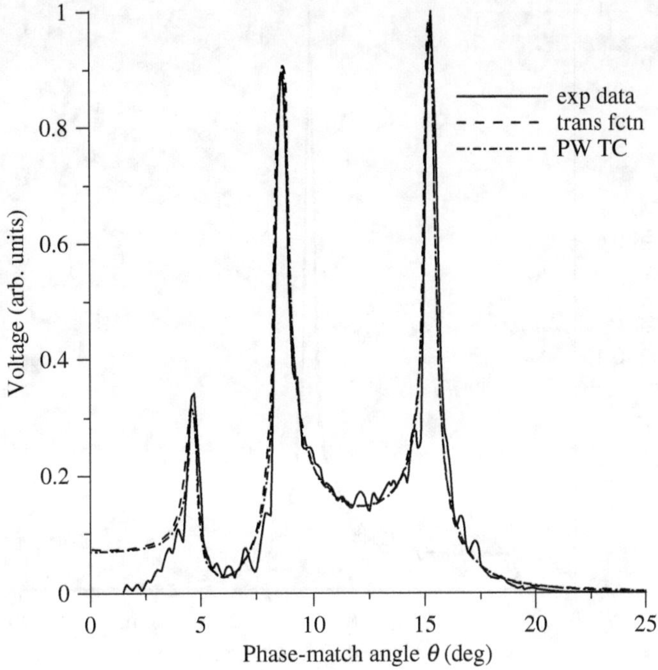

Figure 9.10. Transmission function at 1.1 MHz of uniaxial graphite-epoxy composite plate 1 mm thick with the scan normal to the fibers (after Zhang and Chimenti [22]).

shear elastic constant C_{55} can be inferred. Its value is $8.4 - 0.25i$ GPa. The weakly excited higher-order mode near 2° is barely visible in the data because the finite spatial scan limits the resolution in the transform, or phase-match angle, domain.

Similar experiments have also been performed on biaxial graphite-epoxy composites, where the layup sequence is $[0°/90°]_{3S}$, in the class of balanced orthotropic laminates. As the number of plies in the two directions is equal, lamination theory implies that $C_{11} = C_{22}$, $C_{13} = C_{23}$, and $C_{44} = C_{55}$. Our interrogation frequencies here are low enough that we may expect static lamination theory to apply. In this laminate, there remain six independent stiffnesses: C_{11}, C_{12}, C_{13}, C_{33}, C_{44}, and C_{66}. Figure 9.12 shows the data and transmission function reconstruction for the biaxial composite plate described above measured at 0.41 MHz. At this frequency, the transmission function is dominated by the real part of C_{11} and by C_{44}. The predicted mode near 3° remains unresolved in the measured data because of the finite scan length and the sharpness of the resonance in air. When observed in water coupling, this mode broadens considerably owing to reradiation losses and it becomes much more easily visible. In air at a low frequency, this

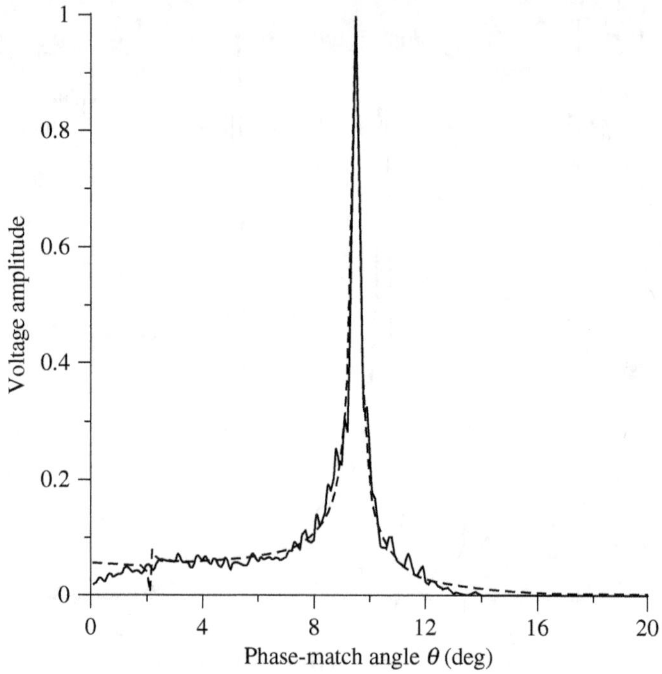

Figure 9.11. Transmission function at 1.1 MHz of uniaxial graphite-epoxy composite plate 1 mm thick with scan direction parallel to fibers (after Zhang and Chimenti [22]).

S_0 mode is not resolvable in conventional scanning with reasonable scan lengths. This mode also contains almost all the dependence on C_{11}, the highest stiffness value. So, to move forward, we will use water-coupled measurements of this mode to infer the real part of C_{11} (71.3 GPa) and then deduce C_{44} from the air-coupled data. The inferred value is $C_{44} = 5.24 - 0.43\,i$ GPa. Using these results, we can proceed to the measurements at a higher frequency of 1.1 MHz, where different portions of the guided wave spectrum are excited. Figure 9.13 shows the result for the same biaxial composite laminate excited at a frequency of 1.1 MHz. With a knowledge of C_{11} and C_{44}, we may now proceed to infer the remaining two constants from the data of Fig. 9.13. The best fit to the data gives $C_{13} = 3.45 - 0.085\,i$ GPa and $C_{33} = 15.83 - 0.49\,i$. The values of the real parts of these measured stiffnesses are fully consistent with predictions of lamination theory for composites of this fiber–resin system and approximate fiber volume fraction. The new element here is the ability to measure accurately the imaginary part of the stiffness, without the disturbing effect of the transducer field. Similar measurements have also been made of more complicated layup sequences, such as quasi-isotropic [22].

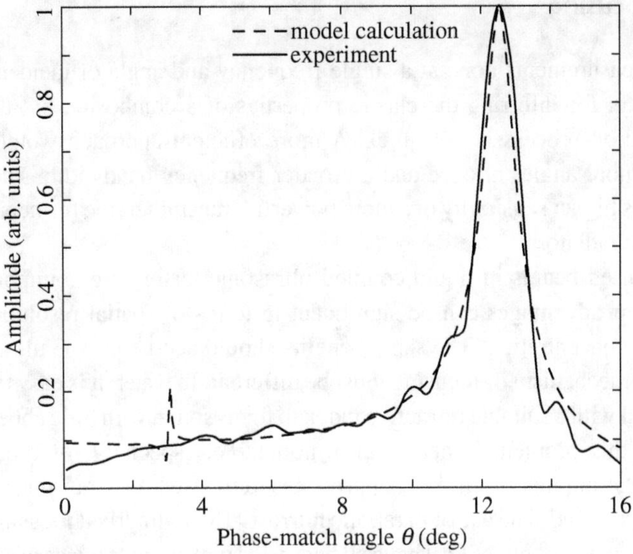

Figure 9.12. Transmission function at 0.41 MHz of a biaxial graphite-epoxy composite plate ($[0°/90°]_{3S}$) The mode near 3° is not resolved in the data because of finite scan length (after Zhang and Chimenti [22]).

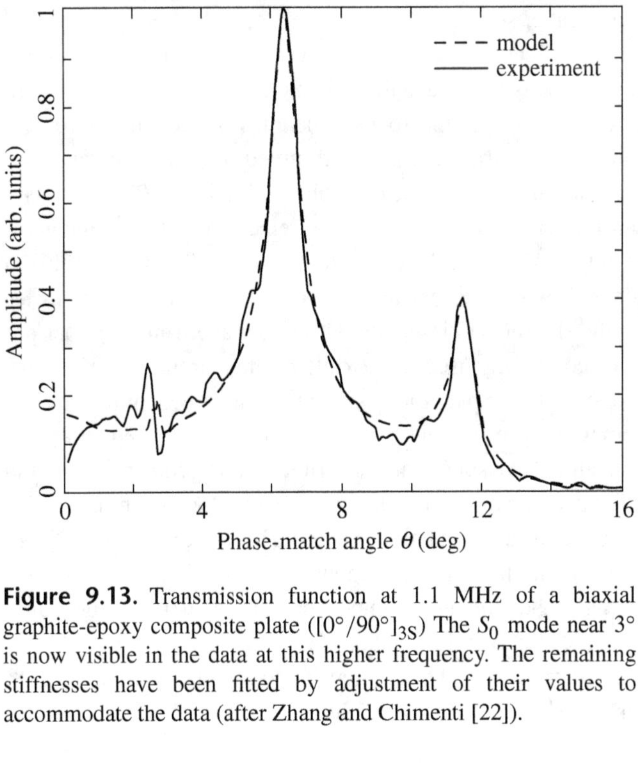

Figure 9.13. Transmission function at 1.1 MHz of a biaxial graphite-epoxy composite plate ($[0°/90°]_{3S}$) The S_0 mode near 3° is now visible in the data at this higher frequency. The remaining stiffnesses have been fitted by adjustment of their values to accommodate the data (after Zhang and Chimenti [22]).

9.4 Focusing

A-C measurements done at a single frequency and angle of incidence can be useful for inferring the elastic properties of a composite, but the data acquisition process is not quick. A more efficient approach would be to cover more angles at once and a broader frequency bandwidth. The latter requires higher sensitivity or a more powerful transmitter; the former implies focused radiators.

Focused beams in liquid-coupled ultrasonic testing are commonplace, and their advantages can be significant in terms of spatial resolution and energy concentration. The same benefits should accrue in A-C ultrasound, but the mechanism of focusing must be different. In water, it is easy to find a material with a suitable refractive index to focus sound with little abberation. In air, this problem is nearly insurmountable, especially for broadband signals. A number of different approaches to focused A-C transducers have been attempted. The use of parabolic mirrors [18], cylindrical focusing [19], and Fresnel zone plates [20] have all been tried with various levels of success. Mirrors produce imperfectly focused beams because they are applied to divergent sound beams in air. Cylindrical focusing works quite well, as long as focusing in one dimension is acceptable. Fresnel zone plates leave much to be desired, because although they produce spherically focused beams, the zone plate blocks a major portion of the radiated energy, leading effectively to a much weaker source.

The only efficient way to focus a beam from an A-C transducer is to perform the focusing in a native fashion, *i.e.*, to conform the radiating surface to the shape of the focused beam. For the capacitive transducer, this amounts to deforming both the dimpled backing plate and the 2 to 10 micron Mylar film to the shape of a spherical radiator. Planar backing plates had been fabricated from silicon using electronic etching techniques [10] or from micro-machined layered electronic media [11]. Neither of these approaches allows for deforming the backing plate. Instead, Song *et al.* [14] replaced the silicon backing plate with one made from a copper/polyimide composite, commonly used for flexible printed circuit material. The copper conductor can also be patterned using much more economical methods than any alternative. The flexible backing plate is conformed to an aluminum fixture machined to a spherical shape of the same geometric curvature as the desired focal length of the finished transducer. Then, the film is deformed by pressing it against a warm ball bearing of similar dimension to the radius of curvature, leaving the film permanently deformed in the shape of a partial sphere. When the film is mounted and biased, it fits snugly against the backing plate, as shown in Fig. 9.14 [15].

The performance of the smaller 10-mm probe excited by a 800-kHz tone burst is shown in Fig. 9.15 in the form of a waterfall plot, clearly

Figure 9.14. Spherically focused A-C capacitive transducers, showing snug fit between Mylar film and flexible backing plate. Frame (a) is a 10-mm diameter device, and frame (b) shows a 50-mm device (after Song and Chimenti [15]).

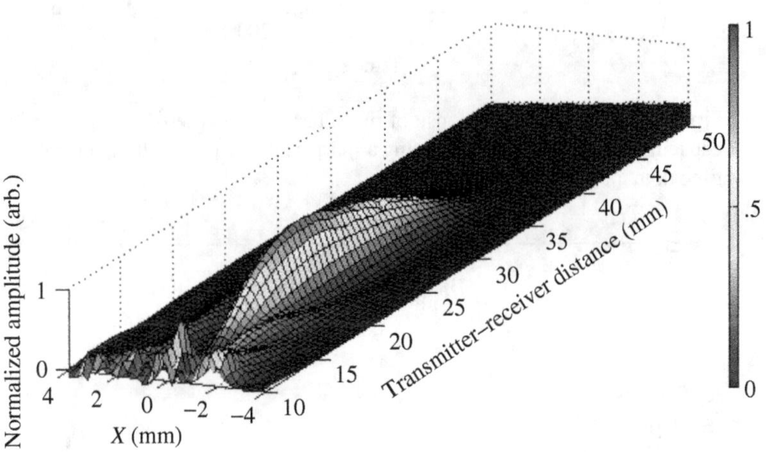

Figure 9.15. Performance of spherically focused 10-mm A-C capacitive transducer at 800 kHz, showing focal region with a peak near 25 mm (after Song and Chimenti [15]).

indicating the focal region, with a maximum near the focal length of 25 mm. Tight focusing together with broadband operation gives these probes wide applicability in material characterization and defect detection. The impulse response of the larger 50-mm diameter probes is shown in Fig. 9.16. The frequency response of the larger 50-mm diameter probes is shown in Fig. 9.17, although these curves are typical of both devices. The very broad bandwidth of the focused transducer is a major plus for the tasks of material characterization and defect detection. As we can see from Fig. 9.17, the 10-dB bandwidth typical of either device extends from 250 kHz to about 1.4 MHz. This zone corresponds to greater than a 100% bandwidth for this transducer, very broad by standards for damped

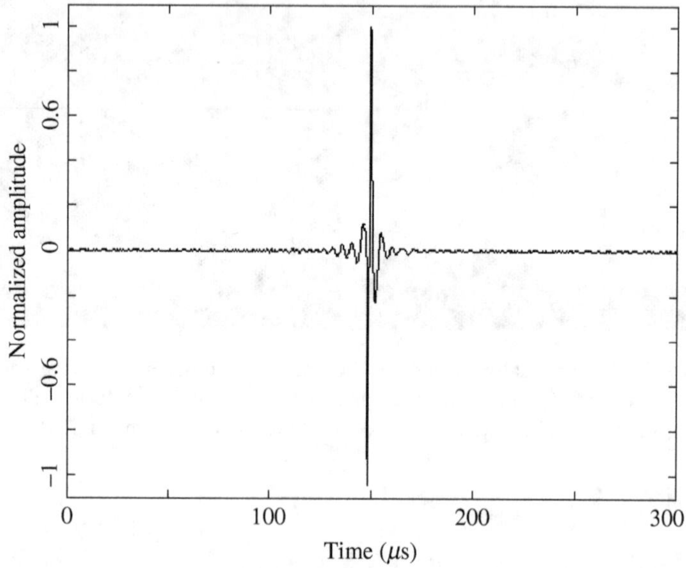

Figure 9.16. Impulse response of the spherically focused 50-mm A-C capacitive transducer, demonstrating a sharp peak and rapid settling in the time domain (after Song *et al.* [14]).

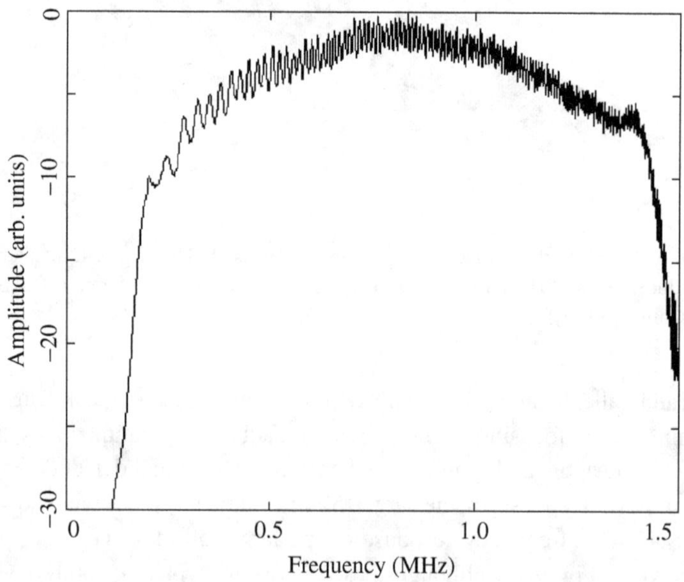

Figure 9.17. Frequency spectrum of the spherically focused 50-mm A-C capacitive transducer, showing a broad, low peak near 800 kHz with a 10-dB bandwidth extending from about 250 kHz to 1.4 MHz. The small, rapid variations in the spectrum are explained in the text (after Song *et al.* [14]).

piezoelectric devices. Superimposed over the average value of the spectral curve are small, rapid amplitude variations. These are a signal artifact caused by an unavoidable design element in the capacitive film point-receiver transducer used in the characterization measurements. The final and perhaps most important performance element of a broadband focused transducer is its diffraction field. Almost no conventional piezoelectric device focused with a plastic lens demonstrates focusing performance limited only by the aperture and the sound wavelength. Reasons for this have to do with abberation caused by a spherical lens surface, imperfect sound axis alignment, inhomogeneous lens medium, or an inhomogeneous sound field from the transducer. Mathematical analysis of the problem of a focused radiator based on the Rayleigh–Sommerfeld calculation assumes a perfectly uniform radiating element, focused by a perfectly refracting lens interface. When these conditions hold, the result for the sound pressure p in the transducer focal plane is the well-known Bessel function expression,

$$p(r, \omega) = -i\rho_g \omega V_0 a^2 \frac{e^{ik_g rd}}{rd} \frac{J_1(k_g a/d)}{k_g a/d}, \tag{9.2}$$

where ρ_g is the gas medium density, V_0 is the uniform piston velocity, a is the piston radius, $k_g(= \omega/V_f)$ is the gas medium wavenumber, r is the radial distance from the central transducer axis, $d = \sqrt{1 + F_0^2/r^2}$, F_0 is the focal length, and J_1 is the first-order Bessel function. The performance of the 10-mm diameter focused capacitive probe is shown in Fig. 9.18. The dash curve in Fig. 9.18 is the theoretical calculation, whereas the curve with open circle data points is the measured data for the 10-mm probe. The close agreement between the two curves indicates the high precision of the focusing for these capacitive transducers. The reason for the excellent agreement in the curves of Fig. 9.18 is related to the manner in which the focusing occurs. As the device's radiating surface has a nearly spherical shape, the sound waves travel almost exactly the same distance to the focal point, suppressing most sources of focusing imperfections. The implication of this result is that critical focusing can be achieved in air, and together with the low sound velocity in air and short wavelengths, some truly remarkable spatially resolved images are possible.

9.5 Techniques and Applications

In this section we show some of the results of A-C ultrasound applied to NDE of composite materials and the use of focused ultrasound to achieve resolution and sensitivity in nondestructive inspections that are generally anticipated only in liquid-coupled measurements. We begin with

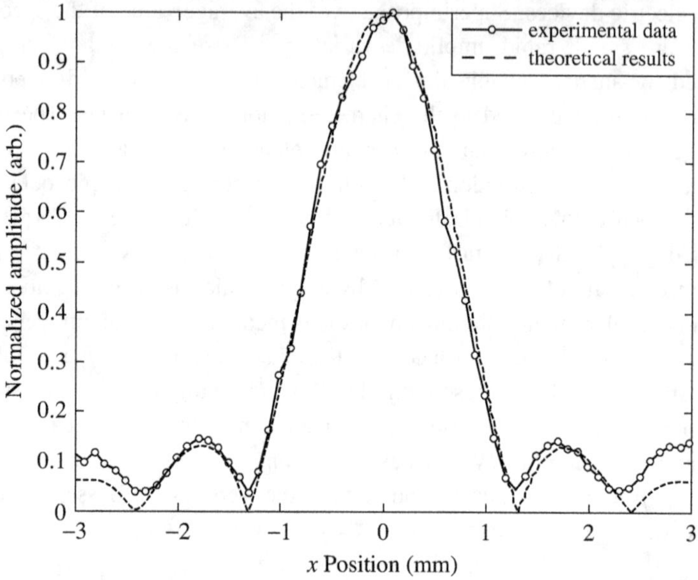

Figure 9.18. Focal-plane pressure amplitude of the 10-mm diameter capacitive probe measured with a quasi-point receiver (open circles) and compared to the corresponding theoretical calculation (dash). The two curves are nearly in perfect agreement, indicating nearly diffraction-limited focusing of the A-C transducer (after Song *et al.* [14]).

a complicated aluminum honeycomb structure terminated by composite facesheets. Internal to the structure are composite or metallic inserts that add hard points for attaching screws or bolts to be used in creating larger, more complex structures. An example of one of these test panels with 1.5-inch composite spool inserts is shown in Fig. 9.19. This C-scan in air has been performed with a broadband excitation of two focused capacitive A-C transducers in a confocal geometry (shown in inset), where the transducer diameter is 10 mm and the focal length is 25 mm. The inserted spools in this particular example are also composite. The C-scan in air clearly shows the cellular structure of the honeycomb, and a halo-like region surrounding each insert is a foam adhesive to stabilize the hard point within the honeycomb, as shown in the inset on lower left of Fig. 9.19. The inserts are well bonded in this example, as indicated by the high transmission over their diameters.

A further detailed study of insert bonding can also be seen in Fig. 9.20, where a second panel is examined with the same transducer pair and broadband excitation. Here, the inserts are titanium, also with foam surrounds, as shown schematically in the figure inset. These lead to similar dark bands around each insert. From the data of Fig. 9.20 it can be seen

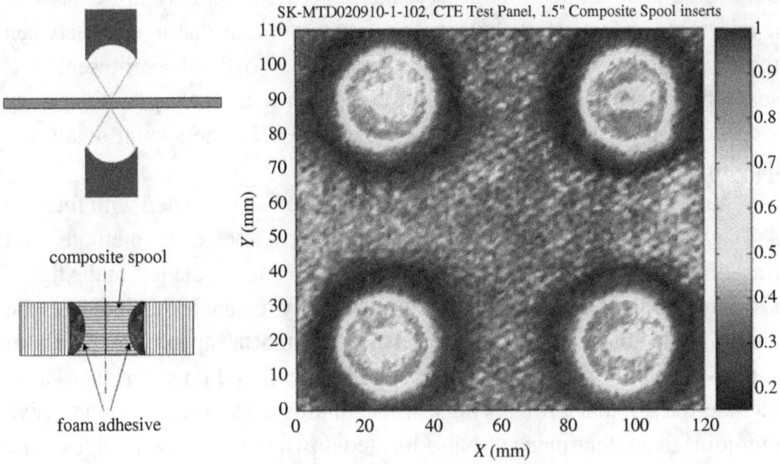

Figure 9.19. Composite panel consisting of aluminum honeycomb with graphite-epoxy facesheets and composite inserts, inspected with 10-mm focused A-C ultrasonic transducers in a confocal geometry using broadband ultrasound. Insets at left show confocal transducer geometry and internal panel construction (after Song and Chimenti [15]).

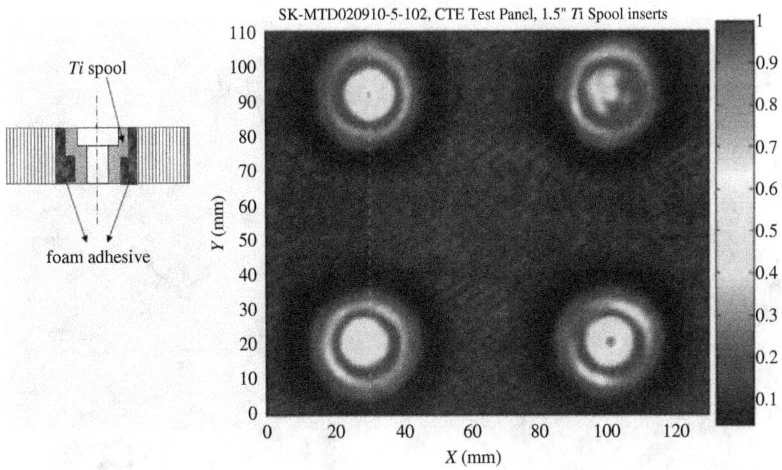

Figure 9.20. Composite panel similar in construction to the one in Fig. 9.19 inspected with 10-mm focused A-C ultrasonic transducers in a confocal geometry using broadband ultrasound. Inset at left shows internal panel construction (after Song and Chimenti [15]).

that three of the inserts are bonded to the facesheets, but the insert at the upper right is only weakly bonded. Considering that the principal structural bond here is at the foam adhesive, facesheet bonding may not constitute a rejectable defect. Nonetheless, the inspection can distinguish between these conditions. Bearing in mind that these two figures represent A-C scans performed in 3/4-inch thick honeycomb with 4-ply graphite-epoxy facesheets at a center frequency of about 750 kHz, the spatial detail and amplitude resolution of the scans is impressive.

Another example of a natural composite material inspected with focused A-C ultrasound is shown in Fig. 9.21, where the same methods and transducers are applied to a specimen of balsa wood. At the left in the figure is a photograph of the sample showing an overlay rectangle that indicates the area of the A-C C-scan. The features in the wood that are apparent in the photo can also be seen in the ultrasonic scan in the frame on the right. These include the circular growth rings, some small inclusions, and an adhesive butt joint to another piece of balsa located down the right-hand edge of the frame near $x = 42$ mm. The large triangular-shaped indication in the A-C scan near $x = 25$, $y = 35$ mm, however, is not at all visible superficially. This defect corresponds to an internal inclusion that must have developed during the growth of the wood. Variations in transmission corresponding to different series of growth rings may give further indication of growth conditions or mineral uptake during those years. The point is that the A-C

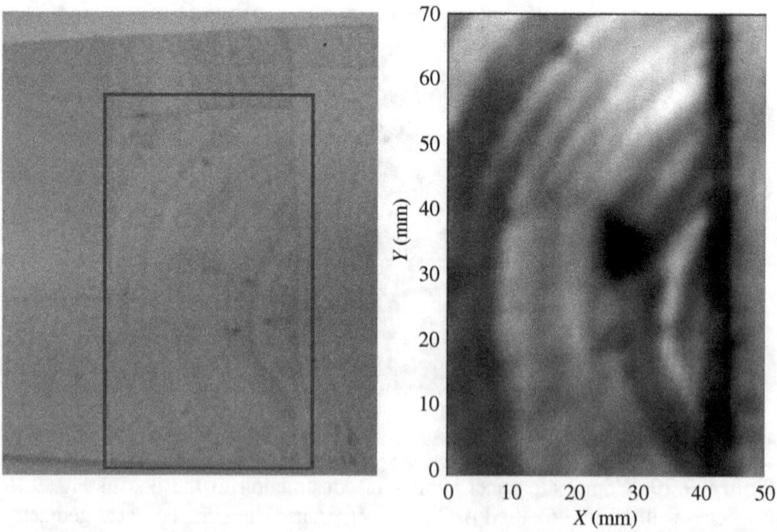

Figure 9.21. One-inch thick balsa wood panel inspected with 10-mm focused A-C ultrasonic transducers in a confocal geometry using broadband ultrasound. At left is a photograph of the sample with the inspected area marked by a rectangle. In the frame at right is the A-C scan (after Song and Chimenti [15]).

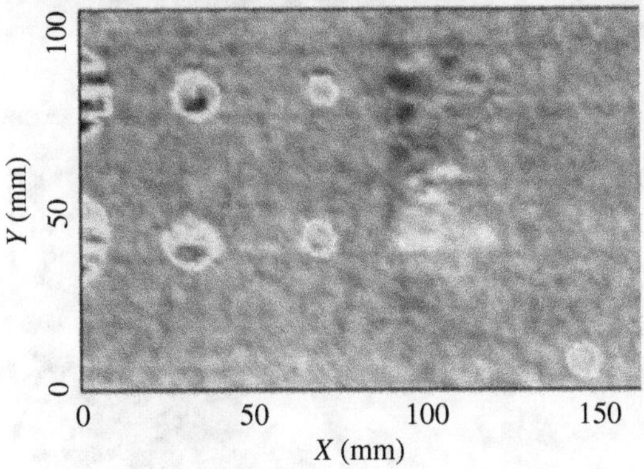

Figure 9.22. Graphite-epoxy, 32-ply uniaxial composite panel inspected with 10-mm focused A-C ultrasonic transducers, as in Fig. 9.21. Artificial delaminations and weak porosity are included in this test panel. Delaminations are at left and lower right, and the porosity is in the center.

scan is replete with detailed amplitude information presented with rather high spatial resolution.

Uniaxial graphite-epoxy in a 32-ply layup is shown in Fig. 9.22 in a A-C scan performed similar to the others in this series. The scan demonstrates that important spatial details of the composite are not lost in A-C ultrasound, especially when critical focusing can be used. Delaminations and low-concentration porosity are artificial defects included in the panel. The delaminations begin at the left-hand edge. One-inch diameter delaminations (partially visible) are followed by two half-inch defects, and then quarter-inch disbonds. The delaminations have been simulated by doubled 1/2-mil teflon inserts. Another quarter-inch disbond is seen at lower right near $x = 150, y = 10$ mm. In the top two rows near $x = 100$ mm are low concentrations of micro-balloons to simulate weak gas porosity from improper cure. These regions are only just discernible. In water-coupled testing with the same sample, no significantly better discrimination for this defect can be produced, unless very high ultrasonic frequency is used.

Additional C-scan ultrasonic data from narrow-band commercial A-C transducers is shown in Fig. 9.23 [21]. The defects here are fabricated using artificially inserted Teflon films. The scan has been made using 400-kHz focused ultrasound. Details down to 0.125 inch (3 mm) are clearly discernible. Further A-C ultrasonic scanning results in an aluminum honeycomb structure (0.375-inch (9-mm) cell size) with four-ply composite facesheets are shown in Fig. 9.24, where the imaging is performed at 120 kHz with 0.75-inch (19-mm) diameter commercial transducers. Here, the sound is transmitted entirely along the cell walls from one facesheet to the other. With the lower frequency, we anticipate a decreased spatial resolution, and this seems borne out by the data. Yet, the image of the damage is sufficiently clear to permit easy identification of the six zones where disbonds have occurred.

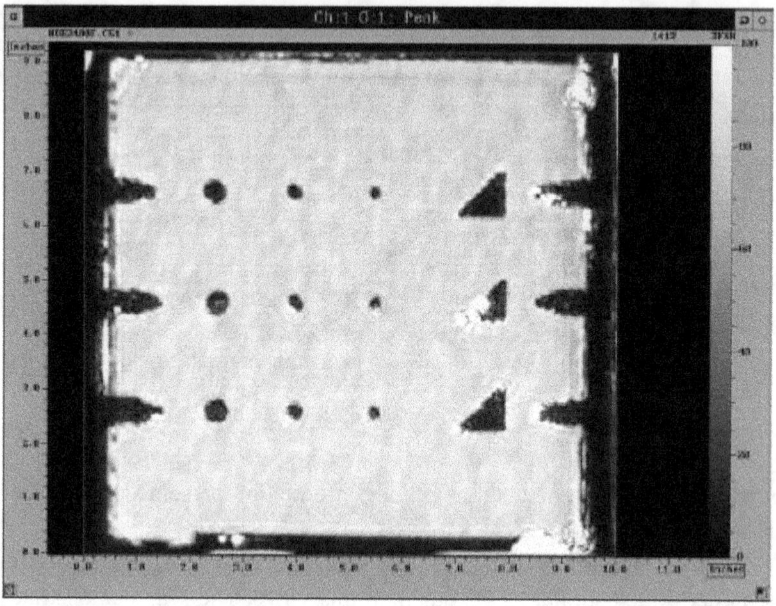

Figure 9.23. An A-C C-scan image of a 10-ply composite laminate embedded with artificial defects made using Teflon film inserts (after Hsu *et al.* [21]).

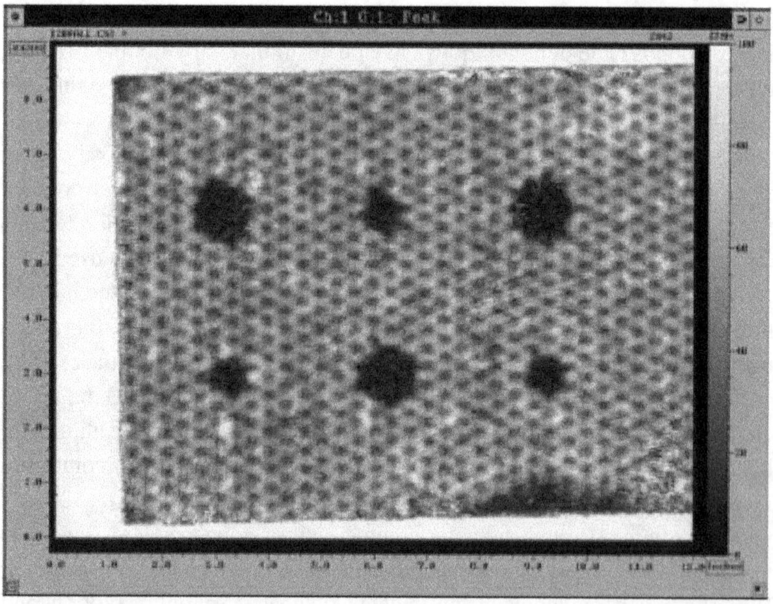

Figure 9.24. An A-C C-scan image of an aluminum honeycomb structure with composite laminate facesheets. This structure has been embedded with artificial disbond defects (after Hsu *et al.* [21]).

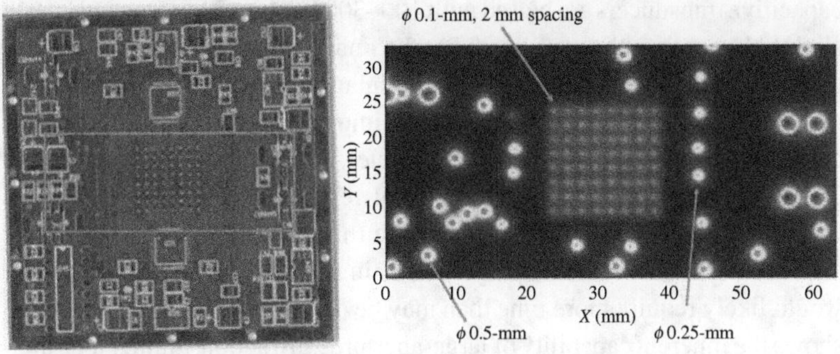

Figure 9.25. Printed circuit board scanned with 10-mm focused A-C ultrasonic transducers in confocal geometry. Feature sizes of solder and contact pads are shown on the figure.

The capabilities of the focused transducers in a confocal geometry are really apparent when the sample features become quite small. Figure 9.25 shows a scan of a section of a 7.5-mm square multilayered printed circuit board with solder-pad features as small as 0.1 mm. At left is a photograph of the board, where the bright rectangle shows approximately the area scanned in the image at right. The image of the grid array at the center of the image is a matrix of contact points for an ultrasonic sensor array. Other solder pads are larger and give rise to stronger signals. Even the 0.1-mm contact pads spaced 2 mm apart are easily imaged and spatially resolved from each other.

A final example of the spatial resolution of focused A-C transducers is seen in Fig. 9.26, where the devices are now the 50-mm diameter transducers. These much larger probes have stronger focusing and higher numerical aperture. The result is an extraordinary resolving power for ultrasonics performed in air at low frequencies. A 250-μm wire is scanned in a confocal geometry. The transmitter is excited with a 1-MHz tone burst, and both

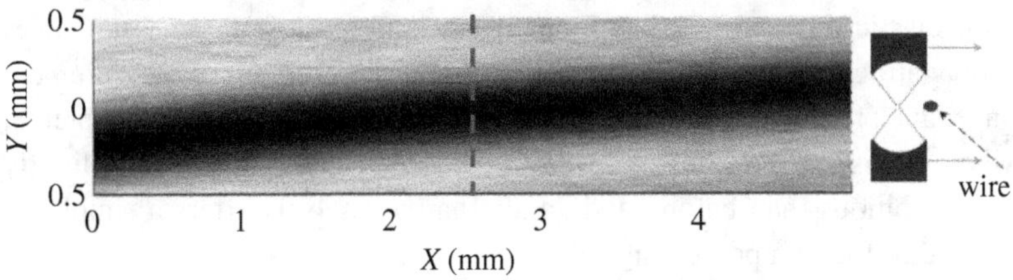

Figure 9.26. Image of a 250-μm diameter wire scanned with 50-mm diameter probes at 1-MHz in a confocal geometry, illustrated in the inset at right. Excellent spatial resolution is seen in the sharp, well defined edges of the wire. Dashed line across the wire was used to analyze the width of the image (after Song and Chimenti [15]).

capacitive transducers are biased with 100–300 V. The region scanned is only slightly larger than the wire itself, but the image clearly shows resolution of the wire edge. The +6 dB points in the image amplitude (measured at the trace of the dashed line from signal minimum at the image center) yields an estimate of 254 μm for the wire diameter, very close to the actual value. These results imply an estimated spatial resolution of about 130 μm in broadband excitation. In order to achieve this resolution, an appropriately small step size is also necessary. While, in practice, such a detailed scan would likely require more time than may be deemed practical, this example shows the inherent capability of large-aperture, diffraction-limited focusing, even in air at 1 MHz.

As we showed in Chapter 8, focusing also plays an important role in materials characterization, where distributed material properties, rather than discrete defects, are the target of the measurements. Just as a broad temporal frequency spectrum can speed the ultrasonic characterization of a plate-like material, so also can a broad spatial frequency spectrum. For materials which are by nature locally planar, such as laminated composites, knowledge of the plate-wave spectral properties can lead indirectly to an inference of the material properties of a composite laminate. Making such measurements using water coupling is both efficient and convenient. As we suggested at the head of this chapter, however, there are some situations where water coupling is not preferred. Inspection of de-bulked, but uncured, composite layups is one example where contact of the composite with water would not be acceptable. In that case, A-C ultrasonic testing is a reasonable alternative.

In the measurements to be presented below, the focusing was accomplished by use of a cylindrical parabolic mirror, which clearly suffers from some abberation, because the mirror attempts to focus a diffracting, rather than parallel, beam from the transducer. Nonetheless, when high spatial resolution is not required, this expedient can be quite serviceable. And in these cases shown below, the focused beam approximates a line source. As our composite samples have high directional anisotropy in their elastic properties, a line focus beam is necessary to generate waves propagating principally in a single direction. The program of measurement and data processing is quite similar to the one described in Chapter 8, where focused beams in water were used. In this case our goals are slightly less ambitious than before, and an alternative analysis offers a simplified algorithm for data processing.

A planar capacitive-film transducer, radiating into a line-focus mirror with the focal line near the surface of the plate, forms both the A-C ultrasonic source and detector. Each transducer is modeled by a line transducer at its focus. The magnitude of the received voltage is proportional to the magnitude of the plane wave transmission coefficient $T(\xi, \omega)$ and the spectral density of the transmitted signal $H(\omega)$

$$|V(\xi, z, \omega)| = \frac{1}{2\pi} \left| T(\xi, \sqrt{k_g^2 - \xi^2}, \omega) H(\omega) \right| . \tag{9.3}$$

This measurement therefore yields directly the transmission coefficient T, subject to the transducer and system frequency bandwidth limitations contained in $H(\omega)$ and subject to the effective spatial frequency window function imposed by the finite angular range of the parabolic mirrors. The expression in Eq. (9.3) may be compared to the voltage in Eq. (8.33) in Chapter 8, which is also a windowed scattering coefficient. Maxima in the transmission coefficient correspond to high-Q transverse resonances within the plate and thereby are very closely related to the dispersion spectrum of the guided modes [6, 22].

The interpretation of the measurement is also subject to the limitations of the above analysis and model. An infinite plate is implicitly assumed, but the real plates measured are finite. More importantly, the angular range of the parabolic focusing mirrors is limited, so the assumption of line-source/line-receiver is not fully justified. The limited angular range could be analyzed with the complex transducer point method [22, 23], as we saw in Chapter 8. Alternatively, an individual measurement is sensitive only to a range in (ω, k_x). If that range is not large enough to subtend all the modes of interest, repeated measurements at multiple incident mirror angles can be used to obtain a more complete image, as was the case for the planar transducer data [6] presented earlier.

To see how these ideas work in practice, let us examine several scans performed in air using line focusing and the methods described above to prepare the data for presentation. Comparisons are made to the guided wave dispersion curves, calculated separately using best estimates of the plate elastic properties, either from tabular data or independent measurements. The first example is a 5.46-mm thick Plexiglas plate in air shown in Fig. 9.27. The measured data values here constitute the magnitude of the complex voltage in frequency–wavenumber space, plotted as a 2-D image, where the bright regions are zones of high transmission, and the dark areas denote low sound transmission. Superimposed over the data are the calculated curves of the guided wave dispersion spectrum. These curves show excellent agreement with the measurements, although not all portions of the calculated spectrum are represented in the measurement. The reasoning here is identical to the case of liquid-coupled measurements. For some branches of some modes, for example, the first-order antisymmetric mode near cutoff at low $k/(2\pi)$, the branch is dark although the calculation predicts a mode there. At higher values of $k/(2\pi)$, between 0.2 and 0.3 MHz, the data show a bright zone lying directly under the prediction. At low wavenumber, in this mode, the out-of-plane particle displacement is near zero, so the

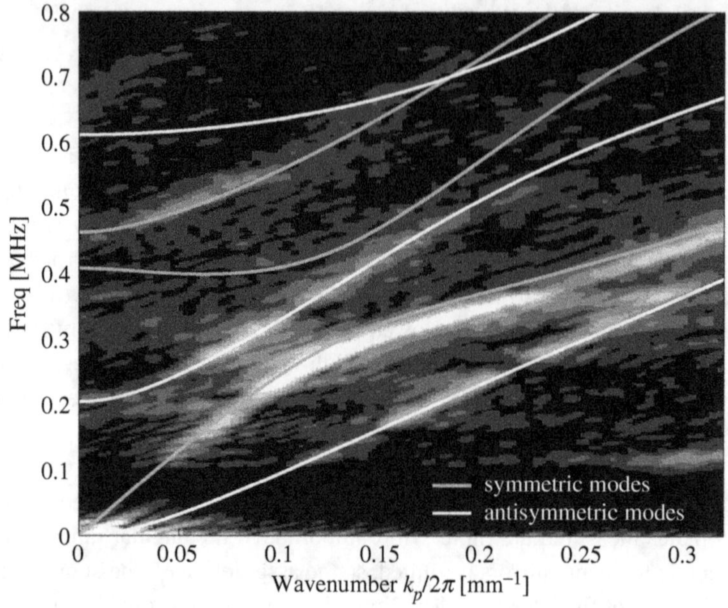

Figure 9.27. Guided wave dispersion spectrum of a 5.46-mm Plexiglas plate measured using line-focused transducers in air. The magnitude of the band-limited complex transmission coefficient data are presented in a 2-D image. The bright areas represent high transmission, and the dark areas are low transmission. Superimposed over the spectrum are the discrete dispersion curves of the symmetric and antisymmetric guided wave modes (after Holland and Chimenti [27]).

guided wave cannot be detected. At higher values of wavenumber, between $k/(2\pi) = 0.08$ and 0.2 1/mm, the mode is detected because in this zone the normal particle displacement is much larger. The same is true for the other regions where the experimental data seems to drop out. A theoretical prediction of the transmission measurement shown in Fig. 9.27 can be seen in Fig. 9.28, again plotted as a gray scale image, where the magnitude of the transmission function, calculated from Eq. (9.3), is recorded as a brightness level. In the prediction these same regions of apparent missing mode branches occur again, demonstrating that the calculation incorporates this frequency-dependent coupling behavior arising from variations in the out-of-plane particle displacement.

Transducer focusing, narrow-band operation, or other means to concentrate acoustic energy are not the only way to elevate the sensitivity of an A-C ultrasonic scan. As in water-coupled ultrasonics, samples can be thought of as "immersed" in the fluid of air, they behave elastically as fluid-coupled plates. Yet, because the immersion medium is so tenuous, internal solid damping mechanisms almost always dominate plate wave energy losses in

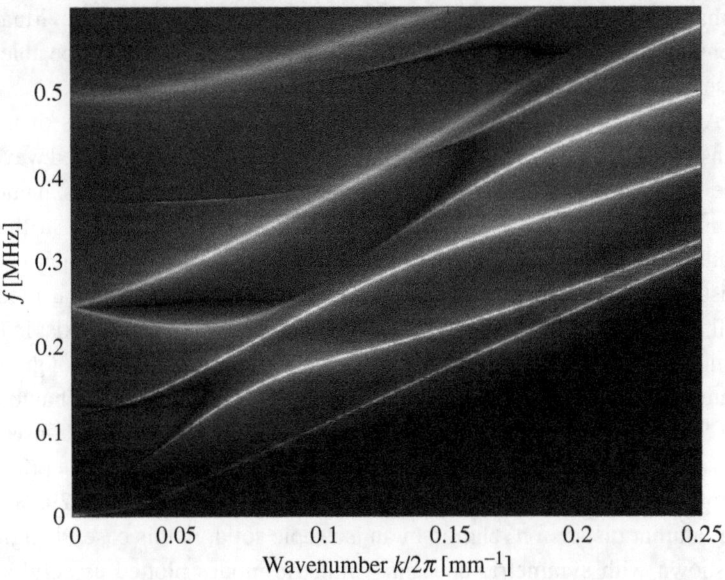

Figure 9.28. Guided wave dispersion spectrum of a 5.46-mm Plexiglas plate calculated from Eq. (9.3). The magnitude of the predicted transmission function is presented as a 2-D gray-scale image. The bright areas represent high transmission, and the dark areas are low transmission. Regions of missing mode branches are the result of small out-of-plane particle displacements and, hence, low transmission (after Holland and Chimenti [27]).

A-C ultrasonics. This is not true when the immersion fluid is water. In that case, the water density and sound speed are close enough to most solids to cause significant plate wave energy loss as radiation into the fluid, the so-called "leaky waves". Once a propagating wave exists in the plate, its wave potential amplitude in the fluid at the plate surface will decrease with coordinate x as $\exp[-k_p''x]$, where the term k_p'' is the imaginary part of the complex plate-wave propagation constant. For a particular plate wave mode and at a constant frequency, the imaginary plate wavenumber will vary approximately as the ratio of the fluid material density ρ_f over the solid density ρ_s, $k_p'' \propto \rho_f/\rho_s$, as shown by Merkulov [24]. This fact suggests that although both water and air may be treated as fluids and that plates immersed in them should display the same wave dynamics, the difference in their densities, a factor of 833, is large enough to imply qualitative differences in behavior.

Wave mode coupling between bulk waves in air and guided waves in the plate exists because of the presence of the air, but its low density also means that the loading or perturbation of the guided wave modes in the air-immersed plate will be completely negligible. The implication here is that wave phenomena in the plate-in-vacuum will also occur, in virtually identical fashion, in the air-immersed plate—except that we will be able to observe and exploit these phenomena through the leaky waves.

As we have seen in this chapter and the previous one, zones of high transmission through a plate occur simultaneously with leaky guided waves. The same wave mechanics that lead to propagating guided waves, namely transverse resonance, also imply high sound beam transmission at those points. One of the properties of guided waves, as shown in Chapter 5, is the existence of mode cutoffs, *i.e.*, frequencies below which particular modes will not propagate. Only the two fundamental modes, A_0 and S_0 exist in the limit of zero frequency. Soon after the complete solution spectrum of guided plate waves was revealed and explained (see *e.g.*, [25]), unusual behavior in the S_1 mode of most materials was observed [26]. This behavior consisted of a pronounced upturn in the mode branch at low wavenumber just prior to cutoff. The situation is illustrated in a sample of Plexiglas Fig. 9.29. Here, the familiar dispersion relation of an isotropic solid, in this case Plexiglas, is shown with symmetric and antisymmetric modes plotted as solid and dashed curves, respectively. In the S_1 mode near zero wavenumber, there is a region where the curve drops and then rises as $k/(2\pi)$ goes to zero. The average slope from the origin to any point on the dispersion curve will be the mode's phase velocity, which is finite throughout the whole region plotted on the graph, except at $k = 0$, where the phase speed is infinite. The *group velocity*, by contrast, is the local slope of the mode. Due to the rise in the curve as $k \to 0$, however, the group velocity of the S_1 mode will pass

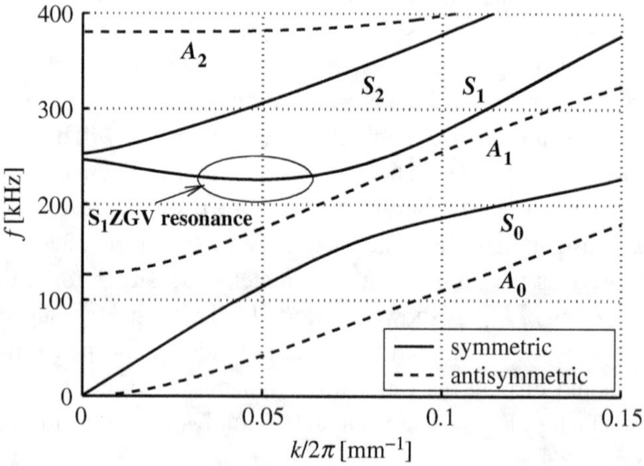

Figure 9.29. Guided wave dispersion spectrum of a 5.46-mm Plexiglas plate calculated from theory. The zero group-velocity (ZGV) zone is denoted by an oval on the S_1 mode (after Holland and Chimenti [27]).

from a positive value to the right of the circled region, through zero in the oval, and on to a negative value to the left of the oval. As the S_1 mode curve reaches its minimum, the group velocity will be zero, the region labeled as S_1ZGV. In this ZGV zone, however, the phase speed is neither zero, nor infinite [28].

While the phase speed governs the advance of a point of constant phase on the wave, the group velocity gives the speed of the advance of a wave packet, or the energy content of the wave. So, for this ZGV region, the wave inside the packet is moving, but the packet—the wave energy—is essentially stationary. This circumstance is to be distinguished from a conventional standing wave resonance, as illustrated using experimental data in Fig. 9.30. The top trace is the transmitted signal when the plate is excited with a broadband waveform (bandwidth 100–1000 kHz) from a capacitive foil transducer focused using parabolic reflectors with the central ray at near-normal incidence. The strong filtering effect of the plate-wave modes is clearly evident. A spectrum of this same signal is shown in the bottom trace, whose resonant frequency is unmistakably separated by more than 20 kHz from the transverse longitudinal resonance caused by the mode cutoff. The standing wave resonances occur at the

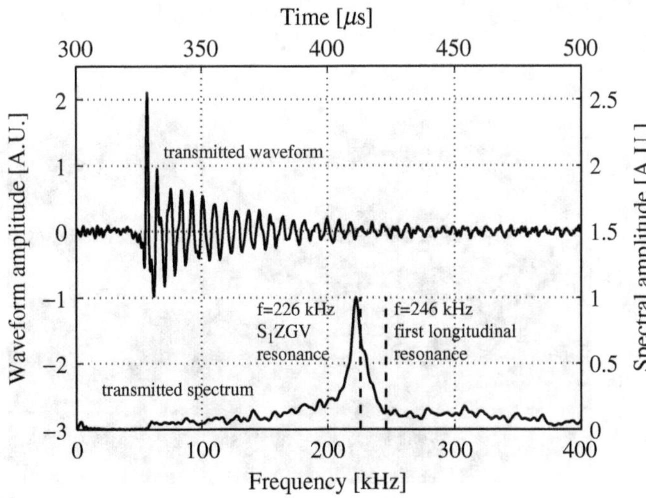

Figure 9.30. Experimental guided wave signal and its dispersion spectrum for the Plexiglas plate of Fig. 9.29. The top trace is the signal transmitted through the plate of a broadband excitation. The lower trace is the spectrum of this transmitted signal, showing the selective filtering of the zero group-velocity (ZGV) mode. The ZGV mode point is clearly distinguished from the standing-wave cutoff resonance in the lower trace and separated from it by 20 kHz (after Holland and Chimenti [27]).

points of the mode cutoffs, and they correspond to fitting integral half-wavelengths of either shear or longitudinal waves across the plate thickness. (For the first-order symmetric mode, this is a single half-wavelength of the through-thickness longitudinal mode.) At the mode cutoffs the phase velocity is infinite, the group velocity is zero, and the wavenumber vanishes, implying there is no propagating guided wave at these points. In the ZGV region, by contrast, the phase velocity is finite, and a propagating guided wave exists. As a way of concentrating energy for the purpose of enhancing defect detection, the ZGV region is much more favorable than a standing-wave plate resonance. Due to the limited frequency range in the ZGV region, there will be a significant range of wavenumbers (or equivalently, incident angles) that will phase match into the S_1 guided wave at those frequencies, especially when using a sound source with a broad spatial frequency bandwidth, such as a focused probe. This effect is illustrated with experimental data in Fig. 9.31. Of all the modes excited by the focused transducer, the S_1ZGV mode is definitely the brightest, meaning that more wave energy is concentrated here than elsewhere. In addition to the data, theoretical predictions of the guided wave modes (calculated as transmission maxima) have been superimposed as chain-dot curves. This energy concentration effect substantially enhances the detectability of defects when the ZGV mode is used to assess some

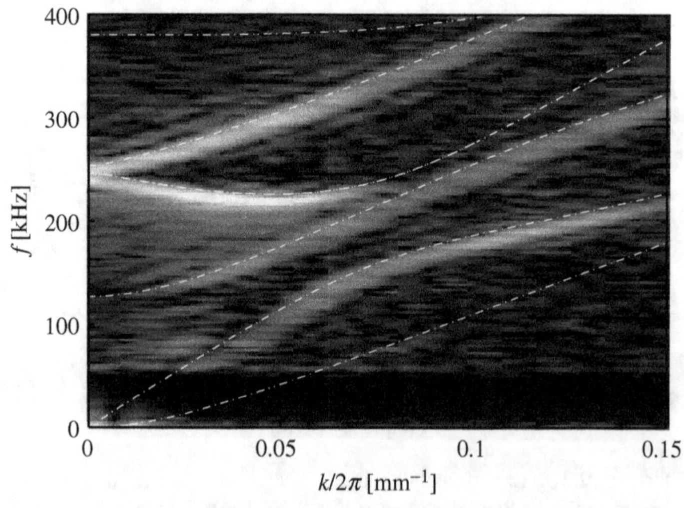

Figure 9.31. Experimental guided wave dispersion spectrum of Plexiglas, showing the large enhancement in the mode amplitude at the ZGV point, denoted by the bright region. Theory curves of the guided wave modes (transmission maxima) are shown in chain-dot. No other region of the curve shows the high wave amplitude seen in at the ZGV mode point (after Holland and Chimenti [27]).

plate characteristic. We can see the result of this enhancement in the following examples.

The simplest kind of defect that falls squarely into the difficult-to-detect category is the marginal change in thickness. This defect type can be simulated by adding to the Plexiglas plate a few narrow strips of cellophane tape. Such a defect is almost impossible to detect at the frequencies conventionally used in A-C ultrasonics, because the additional thickness represented by the tape is a small fraction of a wavelength. To perform such a scan the focused transducers are conveniently oriented to the center of the resonant wavenumber and the frequency is fixed, instead of broadband. In the left-hand frame of Fig. 9.32, a 218-kHz C-scan of a series of strips of Scotch tape was placed on 5.46-mm-thick Plexiglas, for a thickness change of 1% (0.005λ). Strips of width 8.8, 13.1, and 18.4 mm are visible from top to bottom in Fig. 9.32. A further strip 3.7 mm wide was too small to resolve at the top of the frame and does not appear in the image. Measurements performed at frequencies more than 14 kHz away from the S_1 ZGV resonance show none of the strips at all. The right-hand frame of Fig. 9.32 shows a 186-kHz C-scan of an 8.1-mm-thick graphite-epoxy composite plate with buried Teflon inserts that simulate disbond flaws. The inserts, from bottom

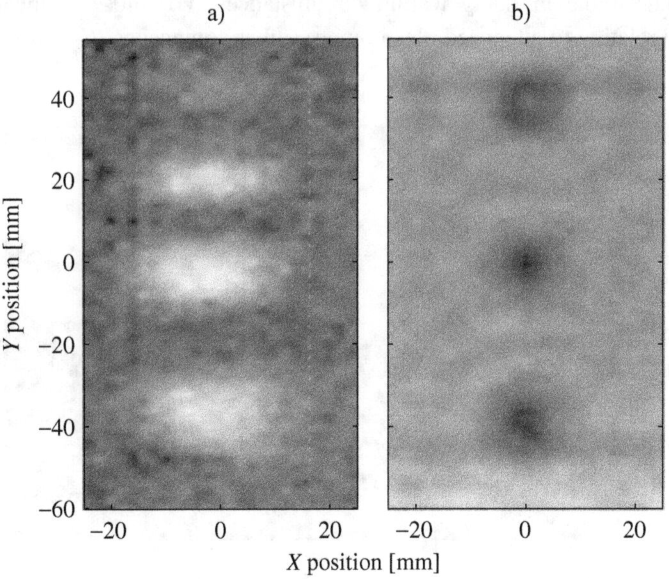

Figure 9.32. Left-hand frame: experimental ZGV C-scan at 218 kHz image of cellophane tape strips on a 5.46-mm-thick plate of Plexiglas. Tape widths are 8.8, 13.2, 18.4 mm from top to bottom; right-hand frame: experimental 186-kHz ZGV C-scan of Teflon inclusions in a carbon fiber-epoxy laminate 8.1 mm thick (after Holland and Chimenti [29]).

to top of the frame, have diameters of 6.4, 4.8, and 3.2 mm. While the 186-kHz scan does not differentiate between sizes of the inserts, it does detect the presence of each simulated flaw, even the smallest. Only at the 186-kHz resonant frequency were all the simulated flaws clearly visible. At other frequencies, only a subset of the flaws were visible. The S_1 ZGV Lamb resonance couples extremely efficiently from air into a solid medium, enhancing defect detection over other modes by at least 10 dB.

A further demonstration of consistency in detection using A-C ultrasonics is seen in Fig. 9.33, where a series of Teflon inserts from 3.2 to 6.4 mm have been embedded at different depths into a cross-ply 8-mm thick graphite-epoxy laminate. Each defect has been detected consistently as a dark zone in an otherwise light background of high transmission. When this same sample is scanned with a commercial A-C transducer at lower frequency, some defects appear dark, and others appear *brighter* than the background. The reason for this difference has to do with diffraction, but having an explanation does not lead to reliable defect detection. Only the S_1 ZGV approach shows all defects darker than the background, consistent with intuition.

A-C ultrasonic testing is a technology ready to assume an important role in the nondestructive inspection of advanced composite materials. Granted, water is an efficient sound coupling medium, but it can also be inconvenient or undesirable in many testing circumstances. For those applications, many relating to advanced plastic-matrix fiber composites, A-C ultrasonic

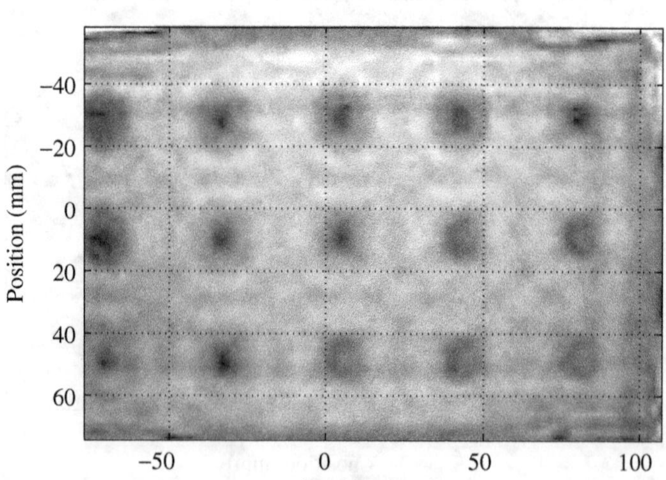

Figure 9.33. Experimental 184-kHz C-scan at the ZGV point of a 8-mm graphite-epoxy laminate with a series of Teflon inserts from 3.2 to 6.4 mm and located at various depths in the sample (after Holland and Chimenti [29]).

testing offers a new and useful alternative, thanks to improved transducer technology, to purpose-built electronics, and to developing methodology designed specifically to exploit this new, but challenging, possibility for nondestructive evaluation.

Bibliography

1. M. Luukkala, P. Heikkila, and J. Surakka, "Plate wave resonance—a contactless test method," Ultrasonics **9**, 201–208 (1971).
2. M. Luukkala, and P. Merilainen, "Metal plate testing using airborne ultrasound," Ultrasonics **11**, 218–221 (1973).
3. D. Reily and G. Hayward, "Through-air transmission for ultrasonic non-destructive testing," in *Proceedings of the IEEE Ultrasonics Symposium*, pp. 763–766 (1991).
4. D. E. Chimenti and C. M. Fortunko, "Characterization of composite prepreg with gas-coupled ultrasonics," Ultrasonics **32**, 261–264 (1994).
5. W. A. Grandia and C. M. Fortunko, "NDE applications of air-coupled ultrasonic transducers," in *1995 IEEE International Ultrasonic Symposium Proceedings*, vol. 1, pp. 697–709 (1996).
6. A. Safaeinili, O. I. Lobkis, and D. E. Chimenti, "Air-coupled ultrasonic estimation of viscoelastic stiffnesses in plates," IEEE Trans. Ultras. Ferroelect. Freq. Contr. (UFFC) **43**, 1171–1180 (1996).
7. T. Yano, M. Tone, and A. Fukumoto, "A 1-MHz ultrasonic transducer operating in air," in *Acoustical Imaging*, eds. A. J. Berkhout, J. Ridder, and L. F. van der Wal, **14** (Plenum, New York, 1985), pp. 575–584.
8. M. Rafiq and C. Wykes, "The performance of capacitive ultrasonic transducers using V-grooved backplate," Meas. Sci. Technol. **2**, 168–174 (1991).
9. S. Schiller, C.-K. Hsieh, C.-H. Chou, and B. T. Khuri-Yakub, "Novel high frequency air transducers," in *Review of Progress in Quantitative NDE* Eds D. O. Thompson and D. E. Chimenti, vol. **9** (Plenum Press, New York, 1990), pp. 795–802.
10. D. W. Schindel, D. A. Hutchins, L. Zou, and M. Sayer, "The design and characterization of micromachined air-coupled transducer capacitance," IEEE Trans. Ultras. Ferroelect. Freq. Contr. (UFFC) **42**, 42–50 (1995).
11. M. I. Haller and B. T. Khuri-Yakub, "A surface micromachined electro-static ultrasonic air transducer," IEEE Trans. Ultras. Ferroelect. Freq. Contr. (UFFC) **43**, 1–6 (1996).
12. K. Suzuki, K. Huguchi, and H. Tanigawa, "A silicon electrostatic ultrasonic transducer," IEEE Trans. Ultras. Ferroelect. Freq. Contr. (UFFC) **36**, 620–627 (1989).
13. H. Carr and C. Wykes, "Diagnostic measurements in capacitive transducers," Ultrasonics **31**, 519–532 (1993).
14. J.-H. Song, D. E. Chimenti, and S. D. Holland, "Spherically focused capacitive-film, air-coupled ultrasonic transducer," J. Acoust. Soc. Am. **119**, EL1–EL6 (2006).
15. J.-H. Song and D. E. Chimenti, "Design, fabrication and characterization of a spherically focused capacitive air-coupled ultrasonic transducer," Int. J. Appl. Sci. Eng. **4**, 1–19 (2006).

16. O. Kraub, R. Gerlach and J. Fricke, "Experimental and theoretical investigations of SiO2-aerogel matched piezo-transducers," Ultrasonics **32**, 217–222 (1994).
17. B. A. Auld, *Acoustic Fields and Waves in Solids*, 2nd ed (Krieger, Malabar, 1992).
18. S. D. Holland, S. V. Teles, and D. E. Chimenti, "Air-coupled, focused ultrasonic dispersion spectrum reconstruction in plates," J. Acoust. Soc. Am. **115**, 2866–2872 (2004).
19. T. J. Robertson, D. A. Hutchins, and D. R. Billson, "Capacitive air-coupled cylindrical transducers for ultrasonic imaging applications," Meas. Sci. Technol. **13**, 758–769 (2002).
20. D. W. Schindel, A. G. Bashford, and D. A. Hutchins, "Focusing of ultrasonic waves in air using a micromachined Fresnel zone-plate," Ultrasonics **35**, 275–285 (1997).
21. V. Kommareddy, J. J. Peters, V. Dayal, D. K. Hsu, "Air-coupled ultrasonic measurements in composites," in *Review of Progress in Quantitative Nondestructive Evaluation*, eds. D. O. Thompson and D. E. Chimenti, vol. **23** (AIP Press, New York, 2004), pp. 859–866.
22. H. Zhang and D. E. Chimenti, "Air-coupled transmission coefficient reconstruction using a 3-D complex-transducer-point voltage model," J. Nondestr. Eval. **22**, 23–37 (2003).
23. S. Zeroug, F. E. Stanke, and R. Burridge, "A complex-transducer-point model for emitting and receiving ultrasonic transducers," Wave Motion **24**, 21–40 (1996).
24. L. G. Merkulov, "Damping of normal modes in a plate immersed in a liquid," Soviet Physics–Acoust. **10**, 169–173 (1964).
25. R. D. Mindlin, "Waves and vibrations in isotropic elastic plates," in *Structural Mechanics*, eds. J. N. Goodier and N. J. Hoff (Pergamon Press, New York, 1960).
26. I. Tolstoy and E. Usdin, "Wave propagation in elastic plates: Low and high mode dispersion," J. Acoust. Soc. Am. **29**, 37–42 (1957).
27. S. D. Holland and D. E. Chimenti, "Air-coupled acoustic imaging with zero-group-velocity Lamb modes," Appl. Phys. Lett. **83**, 2704–2706 (2003).
28. S. I. Rokhlin, "Interaction of Lamb waves with elongated delaminations in thin sheets," Int. Adv. Nondestr. Test. **6**, 263–285 (1979).
29. S. D. Holland and D. E. Chimenti, "High contrast air-coupled acoustic imaging with zero-group-velocity Lamb modes," Ultrasonics **42**, 957–960 (2004).

Index

Note: Page numbers followed by "*f*" and "*t*" denote figures and tables, respectively.

www.ingramcontent.com/pod-product-compliance
Lightning Source LLC
LaVergne TN
LVHW011905200926
848484LV00002B/536
9780195079609